Polar Codes

IEEE Press
445 Hoes Lane
Piscataway, NJ 08854

Polar Codes

From Theory to Practice

Mohammad Rowshan
University of New South Wales, Sydney, New South Wales, Australia

Emanuele Viterbo
Monash University, Melbourne, Victoria, Australia

IEEE SERIES ON DIGITAL & MOBILE COMMUNICATION
John B. Anderson, *Series Editor*

IEEE Press
WILEY

Published by John Wiley & Sons, Inc., Hoboken, New Jersey.

Library of Congress Cataloging-in-Publication Data applied for

Hardback ISBN: 9781119911739

Cover Design: Wiley
Cover Image: © Mohammad Rowshan, Wiley

Set in 9.5/12.5pt STIXTwoText by Straive, Chennai, India
Printed and bound by CPI Group (UK) Ltd, Croydon, CR0 4YY

C9781119911739_261025

To the ones who raised us, taught us, supported us, and challenged us — thank you.

Contents

About the Authors

Mohammad Rowshan is an Engineering ECA Fellow in the School of Electrical Engineering and Telecommunications at the University of New South Wales (UNSW), Sydney, Australia. He received his BEng (Hons) in Electrical and Electronic Engineering from the University of Nottingham in 2015, graduating with First Class Honours and ranked first in his cohort; his MSc in Integrated Circuit Design Engineering from the Hong Kong University of Science and Technology (HKUST) in 2016, where he was awarded the Excellent Student Scholarship and the Arthur and Louise May Scholarship for Young Engineers; and his PhD in Electrical Engineering from Monash University in 2021. His research interests include error-correcting codes, wireless communications, and hardware architecture design.

Emanuele Viterbo is a Professor in the Department of Electrical and Computer Systems Engineering at Monash University, Australia. He served as Head of Department and Associate Dean of Graduate Research in the Faculty of Engineering at Monash University. Prof. Viterbo obtained his degree and a PhD in Electrical Engineering, both from Politecnico di Torino, Turin, Italy. From 1990 to 1992, he worked at the European Patent Office, The Hague, The Netherlands, as a patent examiner in the field of dynamic recording and error-control coding. In 1997–1998, he was a post-doctoral research fellow in the Information Sciences Research Center of AT&T Research, Florham Park, NJ, USA. Prof. Viterbo is a Fellow of the IEEE (2011), ISI Highly Cited Researcher (2009), Member of the Board of Governors of the IEEE Information Theory Society (2011–2016), Conference Committee Chair of the IEEE Information Theory Society (2016–2018). He served as an Associate Editor for IEEE Transactions on Information Theory, European Transactions on Telecommunications, and the Journal of Communications and Networks. He was the General Co-chair of the 2021 IEEE International Symposium on Information Theory and the 2009 IEEE

Information Theory Workshop, and TPC Chair of the 2014 IEEE Information Theory Workshop. His main research interests are in waveform design, polar coding, sphere decoding, lattice codes for Gaussian and fading channels, algebraic coding theory, algebraic space–time coding, digital terrestrial television broadcasting, NAND-flash memory storage, and digital magnetic recording.

Preface

"You are not a drop in the ocean,
You are the entire ocean in a drop."

— Rumi

Polar codes represent a monumental achievement in the history of information and coding theory, hailed as the first provably capacity-achieving error-correcting codes. Introduced by Erdal Arıkan in his seminal 2009 paper, polar codes revolutionized the field by demonstrating how channels could be transformed or "polarized" to achieve reliable communication at rates approaching the channel capacity. This innovation was built upon a foundation of fundamental information-theoretic principles, including channel combining and splitting, and marked a turning point in coding theory with its explicit construction and practical applicability.

What sets polar codes apart is their ability to bridge theoretical elegance and real-world application. By explicitly constructing codes that exploit the polarized nature of channels, Arikan's work laid the groundwork for a coding scheme that is both mathematically rigorous and practically implementable. The introduction of polar codes has inspired a resurgence of interest in coding theory, spurring research into their decoding algorithms, properties, and potential applications.

The practical significance of polar codes has been solidified through their adoption in the 5G New Radio (NR) standard for control channel communications. Their selection in such a critical technological framework underscores their advantages, including high reliability, elegant structure, and scalability. This milestone demonstrates the impact of polar codes beyond academia, establishing them as a cornerstone of modern communication systems and affirming their relevance in industry and next-generation wireless networks.

This book is designed for beginners and aims to provide a clear, comprehensive, and accessible introduction to polar codes. To achieve this, we include sufficient background on communication systems and coding theory, making it easier for readers to understand polar codes in the context of other classical coding families. Throughout the book, we have strived to simplify complex concepts while preserving the depth and rigor necessary for meaningful engagement with the subject. Connections between polar codes and other codes, such as Reed–Muller codes, convolutional codes, and cyclic codes, are highlighted to give readers a broader perspective of the field.

The book is structured to gradually build the reader's understanding of polar codes and their applications:

- Chapter 1 introduces the basics of communication systems and information theory, providing the foundational concepts required for understanding error-correction coding.
- Chapter 2 explores linear codes and major coding families, including Reed–Muller codes, convolutional codes, and cyclic codes, establishing connections between polar codes and their predecessors.
- Chapter 3 covers the fundamentals of soft-decision decoding, an essential component for understanding the decoding processes of polar codes.
- Chapter 4 introduces the fundamentals of polar codes, including the concept of channel polarization and successive cancellation decoding.
- Chapter 5 discusses the key properties of polar codes that influence their performance and utility.
- Chapter 6 explores various construction techniques for polar codes, highlighting methods to optimize their performance in different scenarios.
- Chapter 7 broadens the perspective by discussing decreasing monomial codes as a larger family that includes polar codes and Reed–Muller codes. This chapter also covers permutation groups, ensemble decoding, and weight structures.
- Chapter 8 delves into list decoding of polar codes and its variants, a major advancement in enhancing their error-correcting performance.
- Chapter 9 discusses the special patterns or constituent codes that, by identification of them, can simplify and accelerate the decoding process. These patterns are also called special nodes.
- Chapter 10 focuses on alternative decoding algorithms, including belief propagation decoding, sequential decoding, sphere decoding, ordered statistics decoding (OSD), and guessing random additive noise decoding (GRAND).
- Chapter 11 examines rate-compatible polar codes, a crucial topic for applications requiring flexibility in data rates.
- Chapter 12 discusses polar-coded modulation, including techniques such as bit-interleaved coded modulation and multilevel coding.

- Chapter 13 provides performance comparisons of various polar code constructions and decoding algorithms, alongside their comparisons with LDPC and Turbo codes.
- Chapter 14 introduces readers to MATLAB's 5G Toolbox, showcasing functions for polar encoding, rate matching, and decoding, equipping readers with practical tools for implementation.
- The Appendices include overviews of conceptual channels in 5G, channel coding from 2G to 5G, along with MATLAB and Python scripts for hands-on exploration of polar codes.

How to access the scripts

The scripts provided in Chapter 14 and Appendix C are available on GitHub.

Readers are encouraged to access the repository at github.com/mohammad-rowshan/Book-Polar-Codes-From-Theory-to-Practice to explore and experiment with the code.

How to Use this Book

This book is designed to be flexible for readers with varying backgrounds. The table of contents and index facilitate navigation. Readers may need to consult earlier chapters to build a strong foundation, and backward references are provided within the text to guide their learning. The first three chapters provide foundational background for polar codes, covering fundamental concepts, coding schemes, metrics, and algorithms.

Key Features of this Book

- The mathematical concepts are presented in an accessible manner, emphasizing intuitive understanding over formal proofs. While rigorous proofs are omitted, the underlying rationale and justification for key results are explained.
- To enhance reader comprehension, a number of numerical examples are included throughout the book to illustrate key concepts and facilitate understanding.
- To provide a historical perspective on the concepts and techniques used in polar codes, this book delves into their historical roots and integrates relevant techniques from other coding schemes.
- To illustrate the practical application of polar codes, the adaptation to 5G standards throughout the relevant chapters is explored.
- The use of the MATLAB 5G Toolbox in simulations has been discussed. Additionally, short MATLAB and Python algorithms relevant to different chapters of the book have been provided in the appendix.

Genesis

This book is the culmination of several years of dedicated research and writing. The initial concept was conceived in late 2021 after an invitation from Wiley, and the writing process commenced in early 2022. While the journey has been long and arduous, it has been immensely rewarding. The delay in completion can be attributed to the demanding nature of concurrent academic responsibilities. Despite these challenges, we persisted with unwavering commitment to produce a comprehensive and accessible resource on polar codes.

We hope this book serves as a valuable resource for students, researchers, and engineers, whether they are embarking on their journey into the world of polar codes or seeking a deeper understanding of this field. By providing both theoretical insights and practical tools, this book aims to inspire readers to explore the fascinating world of polar codes and contribute to the ongoing advancements in coding theory and its applications.

September, 2025 Sydney and Melbourne, Australia

Acknowledgment

We extend our sincere gratitude to the anonymous reviewers whose insightful comments and suggestions have significantly improved the quality of this book. Their constructive feedback has helped us to clarify complex concepts, strengthen the overall narrative, and enhance the overall presentation. We also want to thank the reviewers of the original proposal for their encouragement and support. Their positive feedback motivated us to pursue this project with renewed vigor.

We would like to thank Fariba Abbasi Aghdam for her initial interest and willingness to contribute to this project. However, due to unforeseen circumstances, she was unable to participate.

1

Introduction

"Information is the resolution of uncertainty."

— Claude Shannon

Reliable communication over noisy channels is a cornerstone of modern information systems, and channel coding serves as a critical mechanism to achieve this reliability. This chapter provides the foundational knowledge necessary to understand the role of channel coding in communication systems, setting the stage for the rest of the book.

We begin by exploring the need for channel coding in ensuring reliable transmission over noisy channels and by reviewing the structure of a generic communication system. This is followed by a discussion of various channel models, including the binary symmetric channel (BSC), binary erasure channel (BEC), additive white Gaussian noise (AWGN) channel, and fading channel. These models illustrate the diverse challenges posed by different communication environments.

The fundamentals of information theory are introduced next, focusing on key concepts such as the meaning of information, discrete entropy, its properties, and mutual information. The Bhattacharyya parameter, a measure of channel reliability, is also explained to provide deeper insights into channel characteristics.

The chapter culminates with an in-depth discussion of channel capacity and the celebrated Shannon Channel Coding Theorem, which defines the theoretical limits of reliable communication. Additional topics include the block error rate for finite-length codes, the Shannon limit on power efficiency, and the computational complexity associated with coding and decoding processes.

By the end of this chapter, readers will have a comprehensive understanding of the role of channel coding in communication systems, the differences between channel models, the fundamentals of information theory, and the principles underpinning the channel coding theorem. This foundational knowledge will

Polar Codes: From Theory to Practice, First Edition. Mohammad Rowshan and Emanuele Viterbo.

enable readers to appreciate the critical role of coding theory in modern communication systems and serve as a basis for exploring advanced topics in the chapters that follow.

1.1 Reliable Transmission over Noisy Channels

Digital communications and storage are essential parts of our modern lives. We rely on them for everything from sending emails to streaming videos. However, these systems are susceptible to errors that can corrupt the data and make them unusable.

Errors in data transmission and storage systems can be caused by a variety of factors, including:

- Random noise: This is caused by the inherent randomness of the physical world, such as the thermal noise of electronic devices.
- Interference: This is caused by other signals that are present in the same transmission medium, such as radio waves or electromagnetic radiation.
- Channel fading: This is caused by changes in the properties of the transmission medium, such as the attenuation of radio waves due to obstacles or the variation of the refractive index of the atmosphere.
- Physical defects: These are caused by imperfections in the physical components of the data transmission or storage system, such as scratches on a disc or defects in a memory chip.

There are two main strategies for combating errors in digital communications and storage: *automatic repeat request* (ARQ) and *forward error correction* (FEC). ARQ systems detect errors in the received data and request that the data be resent. This is a simple and effective way to ensure reliable data transmission, but it can also lead to increased latency and bandwidth usage. FEC systems, on the other hand, not only detect but also correct errors in the received data. FEC systems can achieve higher reliability than ARQ systems with lower latency, but they are also more computationally expensive. The choice of which error control strategy to use depends on the specific application. For example, FEC is often used in real-time applications where retransmission is not possible, such as live streaming. ARQ is often used in applications where latency is less important than channel bandwidth, such as file transfer over a network. In this book, our focus is on FEC, which is the subject of coding theory.

Coding theory originated in the late 1940s with the work of Claude Shannon, Marcel Golay, and Richard Hamming. Shannon's landmark paper *A Mathematical Theory of Communication* in 1948 laid the foundations for both information theory and coding theory. He introduced the concept of channel capacity, which is the maximum rate at which information can be transmitted over a noisy channel without errors. He also showed that arbitrarily reliable communication is possible at any rate below the channel capacity.

One way to achieve reliable communication is to use error-correcting codes. These codes add redundant bits to the data being transmitted, which can be used to detect and correct errors that occur during transmission. Hamming developed a simple but effective error-correcting code in 1947. His code is now known as the Hamming code. Other examples of communication channels include wireless communication devices and storage systems such as digital video disc (DVD)s and Blu-ray discs. Coding theory is essential for ensuring reliable communication over these channels.

Figure 1.1 illustrates a simple communication channel, ignoring the modulation and demodulation stage for the sake of simplicity. At the source of the communication (transmitter), a message represented as $\mathbf{u} = [u_1 \cdots u_k]$ is intended to be transmitted. However, if this message u is sent directly through the channel without any modifications, the presence of noise could distort u to the extent that its accurate recovery becomes unfeasible. The fundamental concept of coding theory revolves around enhancing the message's resilience by introducing redundancy. This additional information aims to enable the recovery of the original message even in the face of noise-induced corruption during transmission. The process of adding redundancy occurs at the encoder stage, resulting in an augmented message known as a codeword, illustrated as $\mathbf{x} = [x_1 \cdots x_n]$ in the figure. This codeword is then transmitted through the channel, where noise, represented as an error vector $\mathbf{e} = [e_1 \cdots e_n]$, distorts the codeword, leading to the formation of a received vector $\mathbf{y}$.

Subsequently, the received vector undergoes decoding, where the errors are rectified. The surplus redundancy is removed, yielding an estimation $\tilde{\mathbf{u}}$ of the original message. Ideally, $\tilde{\mathbf{u}}$ matches $\mathbf{u}$, ensuring successful recovery. Notably, there exists a direct correspondence between codewords and messages. In many instances, the primary focus shifts from the message $\mathbf{u}$ to the codeword $\mathbf{x}$. From this perspective, the decoder's task involves deriving an estimate $\tilde{\mathbf{y}}$ from $\mathbf{y}$ with the hope that $\tilde{\mathbf{y}}$ aligns with $\mathbf{x}$.

In the case of a DVD or Blu-ray disc, the message source encompasses audio, music, video, or data intended for storage on the disc. The disc itself serves as the communication channel, the DVD or Blu-ray player operates as the decoder, and the recipients consist of listeners or viewers. By refining the communication process through the introduction of redundancy and error correction, coding

Figure 1.1 Error correction coding over a noisy channel.

theory plays a vital role in maintaining the integrity of transmitted information across various contexts and communication channels.

The implementation of error correction comes at a cost. The introduced redundancy functions as an additional layer that consumes transmission resources, such as channel bandwidth or transmission power. Consequently, our objective is to minimize this overhead by minimizing the required redundancy.

To quantitatively assess the amount of redundancy, we introduce the concept of coding rate, denoted as R. This rate is defined as the ratio between the length of the original message and the length of the resulting codeword. For example, if a specific coding scheme produces a codeword of length n from a message of length k, the coding rate can be expressed as:

$$R = k/n$$

In this context, the coding rate reaches its highest value of 1 when no redundancy is incorporated, which basically leaves the message uncoded. The trade-off between coding performance and coding rate is pivotal. With the inclusion of more redundancy, the capability to correct errors becomes more robust, yet this comes at the expense of a reduced coding rate.

An optimal code strives to strike a balance between these factors. It aims to maximize the error correction capability while maintaining the coding rate as close to 1 as possible. This equilibrium ensures that transmission remains efficient in terms of resource utilization while effectively combating the detrimental effects of noise and errors during communication.

So far, we have established some understanding of the advantages offered by channel coding. Next, we discuss some common models for the channels, and in Section 1.4, we delve into the theoretical limit for code rate to have reliable communication.

1.2 Channel Models

There exits various channel models that are used for different applications depending on the specific characteristics of the communication channel and the requirements of the application. In the following, we review the major channel models.

1.2.1 Binary Symmetric Channel (BSC)

The *BSC* is a discrete memoryless channel (DMC) that has two binary symbols as inputs and outputs: 0 and 1. It is symmetric because the probability of receiving a 0 when a 1 is transmitted is the same as the probability of receiving a 1 when a 0 is transmitted. This probability is called the *crossover probability* and is denoted by p.

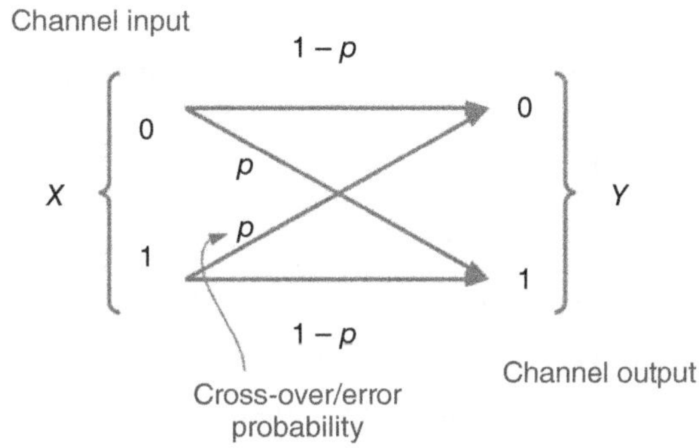

Figure 1.2 Binary symmetric channel.

The probability of no error is $1 - p$. Hence, the channel transition probabilities for the BSC are:

$$\begin{cases} P(Y = 0|X = 0) = 1 - p, \\ P(Y = 0|X = 1) = p, \\ P(Y = 1|X = 0) = p, \\ P(Y = 1|X = 1) = 1 - p, \end{cases} \tag{1.1}$$

Figure 1.2 illustrates the BSC model. The *error probability* of a BSC can be computed as

$$\begin{aligned} P_e &= P(X = 0, Y = 1) + P(X = 1, Y = 0) \\ &= P(Y = 1|X = 0)P(X = 0) + P(Y = 0|X = 1)P(X = 1) = p \end{aligned}$$

since $P(X = 0) + P(X = 1) = 1$.

1.2.2 Binary Erasure Channel (BEC)

Erasure is a special type of error with known location. The BEC is a communication channel that can either transmit a bit correctly or erase it. If the bit is erased, the receiver will not know what the bit was. The BEC output consists of three symbols: 0, 1, and e, where e represents an erased bit. Figure 1.3 illustrates the BEC model.

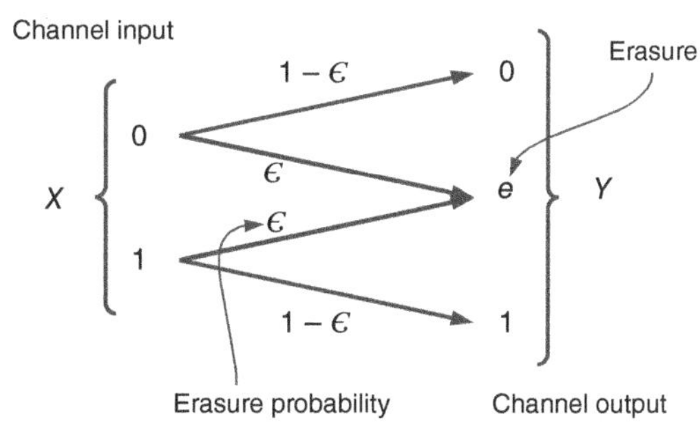

Figure 1.3 Binary erasure channel.

The probability that a bit is erased by the BEC is called the erasure probability and is denoted by ε. The channel transition probabilities for the BEC are:

$$\begin{cases} P(Y=0|X=0) = 1-\varepsilon, \\ P(Y=e|X=0) = \varepsilon, \\ P(Y=1|X=0) = 0 \\ P(Y=0|X=1) = 0 \\ P(Y=e|X=1) = \varepsilon \\ P(Y=1|X=1) = 1-\varepsilon. \end{cases} \tag{1.2}$$

1.2.3 Additive White Gaussian Noise Channel (AWGN)

The AWGN channel model (as shown in Figure 1.4) is used when the inputs and outputs are real- or complex-valued. In an AWGN channel, the signal is corrupted by white noise z that is characterized by a constant spectral density and a Gaussian distribution. The Gaussian distribution is a probability distribution that is often used to model the thermal noise (also known as Johnson–Nyquist noise) introduced by electronic components of the receiver. The probability density function (PDF) of the Gaussian distribution is given by:

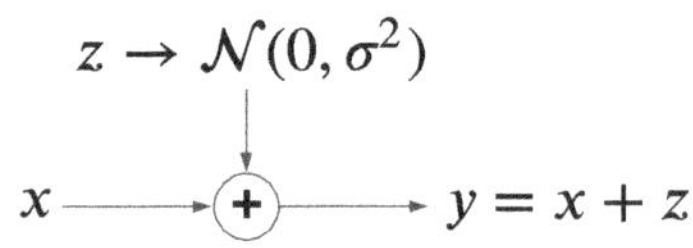

Figure 1.4 Additive white Gaussian noise channel.

$$P(z) = \frac{1}{\sqrt{2\pi\sigma^2}} \exp\left(-\frac{z^2}{2\sigma^2}\right), \tag{1.3}$$

where σ^2 is the variance of a Gaussian random process. As a result, the observed signal at the receiver is given by

$$y = x + z, \tag{1.4}$$

where $z \sim \mathcal{N}(0, \sigma^2)$. Given the normalized Gaussian random variable $n \sim \mathcal{N}(0, 1)$, we can alternatively obtain $z = \sigma \cdot n$ and hence y as

$$y = x + \sigma \cdot n. \tag{1.5}$$

Furthermore, knowing $\sigma^2 = N_0/2$ for the AWGN channel, if we have the signal-to-noise ratio (SNR) $= E_s/N_0$ of the channel, assuming E_s is normalized to $E_s = 1$, we have

$$y = x + \sqrt{\frac{1}{2\frac{E_s}{N_0}}} \cdot n, \tag{1.6}$$

where $\frac{E_s}{N_0}$ is in linear form, not in dB. Since SNR [dB] $= 10\log(\text{SNR})$, to convert the SNR from dB to linear, we can use $\text{SNR} = 10^{\frac{\text{SNR [dB]}}{10}}$. If the given SNR is per

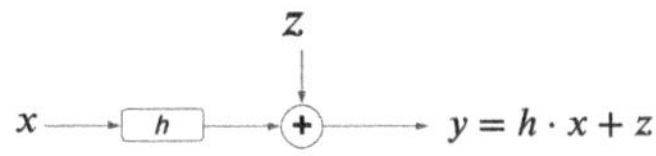

Figure 1.5 Fading channel along with additive Gaussian noise.

information bit that is denoted by E_b/N_0, then we can use the relation $\frac{E_s}{N_0} = R \cdot \frac{E_b}{N_0}$. In the case of higher-order modulation with constellation size $M > 2$, we need to also add the number of bits per symbol, i.e. $m = \log_2 M$. Hence, given SNR per information bit E_b/N_0 and M, then we will have

$$y = x + \sqrt{\frac{1}{2mR\frac{E_b}{N_0}}} \cdot n. \tag{1.7}$$

1.2.4 Fading Channel

The physical wireless channel introduces scaling due to power decay with distance and disturbances via the channel gain coefficient h, as illustrated in Figure 1.5.

As a result, the observed signal at the receiver is given by

$$y = h \cdot x + z. \tag{1.8}$$

We assume that the scaling factor h varies slowly compared to the duration of the waveform. The channel gain coefficient h is often modeled using a Rayleigh or Rician PDF. The most common fading channel models are

- Flat independent fading channel,
- Block fading channel.

In a flat independent fading channel, the attenuation of the signal remains constant over a single symbol duration but changes from symbol to symbol. This means that the channel is memoryless and that the attenuation of any symbol is independent of the attenuation of any other symbol. In a block fading channel, the attenuation of the signal remains constant over a block of symbols, but changes from block to block. This means that the channel has memory and that the attenuation of any symbol is correlated to the attenuation of the previous symbols in the block. It is important to note that fading channels can be further classified based on the characteristics of the fading, such as slow fading (e.g. Rayleigh fading) and fast fading (e.g. frequency-selective fading).

1.3 Fundamentals of Information Theory

This section reviews the basic concepts of information, discrete entropy, mutual information, their properties, and chain rules. We will use these concepts in Chapter 4. This requires knowledge of probability and random variables.

Continuous entropy and continuous mutual information are very closely related to discrete entropy and discrete mutual information, but they are not in the scope of this section. For the sake of brevity, we skip the proofs; however, interested readers can find some references in the bibliographical notes of this chapter.

1.3.1 The Meaning of Information

Information is a phenomenon that has meaning for living creatures. The human brain has the ability to interpret events, concepts, and objects and to evaluate them based on their information content. This means that information is not just a physical phenomenon but also a mental one. It is something that we perceive and understand and that has meaning for us.

The information content of an event is related to how surprised we are when it occurs. We are more surprised when something unexpected happens, or when something with a low probability of occurrence takes place. This suggests that the information content of an event is inversely related to its probability of occurrence. In other words, the more probable an event is, the less information it conveys. Conversely, the less probable an event is, the more information it conveys. This is because a probable event is something we already expect, so it does not tell us anything new. However, an improbable event is something we did not expect, so it tells us a lot about the world.

In human communication, information is conveyed through speech, writing, and pictures. Speech is a sequence of letters, words, and sentences. Some letters and words are more common than others, so they convey less information. For example, the letter "e" is the most common letter in the English language, so it conveys the least information. Conversely, the letter "q" is the least common letter in the English language, so it conveys the most information. The same principle applies to words and sentences.

We can model an information source as a random variable (RV). Random variables fall into two primary categories: discrete and continuous depending on their range set. Discrete random variables exhibit a countable number of values within their range sets. Conversely, continuous random variables present an uncountable number of values within their range sets.

Let X represent a discrete random variable. Its range set is designated as $\mathcal{X} = \{x_i\}$, with individual symbols x_i. For this discrete RV X, the probability mass function $p(x)$ is introduced, defined as

$$p(x) = \text{Prob}(X = x), \tag{1.9}$$

which quantifies the likelihood of the discrete RV producing the symbol x. The probability associated with the symbol x_i is denoted as $p(x_i)$. Considering that information and probability are inversely related, the informational significance of symbol x_i can be quantified through a unitless quantity as

$$I\left(x_i\right) \propto \frac{1}{p\left(x_i\right)}. \tag{1.10}$$

Observe that in (1.10) when $p\left(x_i\right)$ becomes very small, $I\left(x_i\right)$ becomes very large, which suggests the use of a logarithmic scale. We use base 2 since the information content of an event is transmitted using bit sequences. This allows us to assign "bits" as a unit of measure as follows:

$$I\left(x_i\right) = \log_2\left(\frac{1}{p\left(x_i\right)}\right) \text{ bits}. \tag{1.11}$$

The logarithmic scale also gives the natural property that the information of a pair of independent symbols is the sum of the information of each symbol. This can be easily verified observing that $p(x_i, x_j) = p(x_i) \cdot p(x_j)$.

1.3.2 Discrete Entropy

Entropy is defined as the average information content of the discrete random variable X, i.e.

$$H(X) = E[I(X)] = \sum_{x_i \in \mathcal{X}} p\left(x_i\right) I\left(x_i\right) = -\sum_{x_i \in \mathcal{X}} p\left(x_i\right) \log\left(p\left(x_i\right)\right),$$

recalling that the expected value of a function of a random variable is $E[g(X)] = \sum_{x_i} p\left(x_i\right) g\left(x_i\right)$.

In the following, we will omit the range set in the sums to simplify notation.

Entropy represents the mean information contained within an information source, specifically the average information content of a random variable representing the source. Imagine receiving symbols from this information source; a higher entropy indicates a substantial inflow of information from the source. On the contrary, a lower entropy suggests a lesser amount of information being received from the same source. For example, if we have a fair coin, the entropy of the coin toss is 1 bit. This is because there are two possible outcomes (heads or tails), and each outcome is equally likely. We have the maximum amount of uncertainty, or least amount of information, about the outcome of the coin toss. If we have a biased coin, where heads are more likely than tails, the entropy of the coin toss will be less than 1 bit. This is because we have less uncertainty, or more information, about the outcome of the coin toss. If we know that the coin is biased toward heads, we can be more confident that the outcome will be heads.

Let X and Y be two discrete random variables with marginal and joint probability mass functions $p_X(x)$, $p_Y(y)$, and $p_{X,Y}(x, y)$, respectively. The *joint entropy* for the discrete random variables X and Y is defined as

$$H(X,Y) = \sum_{x_i,y_j} p(x_i,y_j)\log\left(\frac{1}{p(x_i,y_j)}\right) = -\sum_{x,y} p(x,y)\log(p(x,y)).$$

The conditional entropy of the discrete random variable X for a given value y_j of another discrete random variable Y is defined as

$$H(X|y_j) = -\sum_{x_i} p(x_i|y_j)\log(p(x_i|y_j)) = -\sum_{x} p(x|y)\log(p(x|y)),$$

which is interpreted as the amount of average information provided by a single symbol of random variable X if a single symbol of random variable Y is known. If X and Y are not independent random variables, it is obvious that $H(X|y) < H(X)$.

The conditional entropy of the discrete random variable X given another discrete random variable Y, i.e. all the values (symbols) of Y are known, is defined as

$$H(X|Y) = \sum_{y_j} p(y_j)H(X|y_j) = \sum_{y} p(y)H(X|y),$$

which can be considered as the total amount of average information per symbol required to know the random variable X assuming that the random variable Y is known.

1.3.3 Properties of the Discrete Entropy

If X and Y are two discrete random variables, and $H(X)$ and $H(Y)$ are the corresponding entropies, then we have the following properties:

1) Discrete entropy is a non-negative quantity, i.e. $H(X) \geq 0$.
2) If the number of elements in the range set of X equals to $|\mathcal{X}|$, the maximum entropy of X equals to

$$H_{\max}(X) = \log(|\mathcal{X}|).$$

3) For the joint entropy $H(X,Y)$, we have the property

$$H(X,Y) \leq H(X) + H(Y),$$

where equality occurs if X and Y are independent random variables, i.e. $p(x,y) = p(x)p(y)$.
4) For N discrete random variables, property 3 can be generalized as

$$H(X_1,X_2,\ldots,X_N) \leq H(X_1) + H(X_2) + \cdots + H(X_N),$$

where equality occurs if $X_1,X_2,\ldots,X_N$ are independent random variables.
5) Conditional entropy satisfies

$$H(X|Y) \leq H(X).$$

Chain rule for entropy. Let $X_1, X_2, \dots, X_n$ be drawn according to a joint pdf $P_{X^n}(x^n) = P_{X_1,\dots,X_n}\left(x_1, \dots, x_n\right)$, where $X^n = \left(X_1, \dots, X_n\right)$ and $x^n = \left(x_1, \dots, x_n\right)$. Then

$$H\left(X_1, X_2, \dots, X_n\right) = \sum_{i=1}^{n} H\left(X_i | X_{i-1}, \dots, X_1\right),$$

where $H\left(X_i | X_{i-1}, \dots, X_1\right) := H\left(X_1\right)$ for $i = 1$. The above chain rule can also be written as:

$$H\left(X^n\right) = \sum_{i=1}^{n} H\left(X_i | X^{i-1}\right),$$

where $X^i = \left(X_1, \dots, X_i\right)$.

Chain rule for conditional entropy.

$$H\left(X_1, X_2, \dots, X_n | Y\right) = \sum_{i=1}^{n} H\left(X_i | X_{i-1}, \dots, X_1, Y\right)$$

1.3.4 Mutual Information

Mutual information can be used to quantify the relationship between two random variables. It is a measure of the reduction in uncertainty about one random variable that we gain by knowing the value of the other random variable. Mutual information can be used to measure the relationship between any two random variables, regardless of their type or distribution. Let us formally define it.

The mutual information between two random variables X and Y is defined as

$$\begin{aligned} I(X;Y) = H(X) - H(X|Y) &= -\sum_{x} p(x) \log(p(x)) \\ &+ \sum_{x,y} p(x,y) \log(p(x|y)) \text{bits/symbol.} \end{aligned}$$

Recall that $\sum_{x,y} p(x,y) \log(p(x)) = \sum_{x} p(x) \log(p(x))$ and $p(x|y) = \frac{p(x,y)}{p(y)}$. Hence, by substitution and rearrangement, we obtain

$$I(X;Y) = H(X) - H(X|Y) = \sum_{x,y} p(x,y) \log\left(\frac{p(x,y)}{p(x)p(y)}\right) \text{ bits/symbol.}$$

The mutual information can equivalently be written as

$$I(X;Y) = E\left[\log\left(\frac{p(X,Y)}{p(X)p(Y)}\right)\right] \text{ bits/symbol.}$$

The *conditional mutual information* between the discrete random variables X and Y given Z is calculated as:

$$\begin{aligned} I(X;Y|Z) &= E\left[\log \frac{p(X,Y|Z)}{(p(X|Z)p(Y|Z)}\right] \\ &= \sum_{x,y,z} p(x,y,z) \log \frac{p(x,y|z)}{p(x|z)p(y|z)}. \end{aligned}$$

Now, let us review the properties of mutual information as follows:

1) Symmetry property states that

$$I(X;Y) = I(Y;X).$$

2) Mutual information is non-negative, i.e.

$$I(X;Y) \geq 0.$$

Alternative expressions for mutual information in terms of entropy and conditional entropy are:

$$I(X;Y) = H(Y) - H(Y|X), \tag{1.12}$$

$$I(X;Y) = H(X) + H(Y) - H(X,Y). \tag{1.13}$$

Chain rule for mutual information.

$$I\left(X_1, X_2, \ldots, X_n; Y\right) = \sum_{i=1}^{n} I\left(X_i; Y|X_{i-1}, \ldots, X_1\right),$$

where $I\left(X_i; Y|X_{i-1}, \ldots, X_1\right) := I\left(X_1; Y\right)$ for $i = 1$.

1.3.5 Bhattacharyya Parameter

The Bhattacharyya parameter, named after Anil Kumar Bhattacharyya, is a measure of the similarity between two statistical distributions. Specifically, it quantifies the overlap between PDFs. In the context of communication channels, the Bhattacharyya parameter is related to the probability of error. It is commonly used to evaluate the error performance of a communication system in the presence of noise or other impairments.

The Bhattacharyya parameter (denoted by B) is calculated using the following formula:

$$B = \frac{1}{4}\sum_{i=1}^{M} \sqrt{P_1(i) \cdot P_2(i)}. \tag{1.14}$$

Here, $P_1(i)$ and $P_2(i)$ are the probabilities of the occurrence of the ith symbol in the two distributions being compared and M is the total number of symbols in the common alphabet. The factor 1/4 is often dropped. When comparing two distributions representing different signal constellations or modulation schemes, a lower Bhattacharyya parameter indicates better distinguishability and, consequently, better error performance. It is inversely related to the probability of error, with a smaller B corresponding to a lower probability of error. So, in essence, the Bhattacharyya parameter is related to the probability of error, but it serves as a measure of the distinguishability between probability distributions rather than a direct measure of the error probability.

In Chapter 4, we use the Bhattacharyya parameter as a measure of the reliability of a channel $W : \mathcal{X} \to \mathcal{Y}$, a generic binary discrete memoryless channel (B-DMC) with input alphabet $\mathcal{X} = \{0,1\}$, arbitrary output alphabet $\mathcal{Y}$, and transition probabilities $W(y|x), x \in \mathcal{X}, y \in \mathcal{Y}$. In this context, the Bhattacharyya parameter as a measure of the reliability of channel W is denoted and defined as

$$Z(W) \triangleq \sum_{y\in\mathcal{Y}} \sqrt{W(y|0)W(y|1)}. \tag{1.15}$$

Depending on the type of channel W, the Bhattacharyya parameter can be obtained differently. For a BEC, the Bhattacharyya parameter is

$$Z(W) = \varepsilon, \tag{1.16}$$

where ε is the erasure probability. For a BSC, we have

$$Z(W) = 2\sqrt{p(1-p)}. \tag{1.17}$$

Finally, for an AWGN channel, given the variance of the channel noise σ^2 and the signaling that modulates $\{0,1\}$ to real values $\{1,-1\}$, the Bhattacharyya parameter is computed as

$$\begin{aligned} Z(W) &= \int_{-\infty}^{\infty} \sqrt{p(y|0)p(y|1)}dy \\ &= \int_{-\infty}^{\infty} \sqrt{\frac{1}{\sqrt{2\pi\sigma^2}} e^{-\frac{(y-1)^2}{2\sigma^2}} \frac{1}{\sqrt{2\pi\sigma^2}} e^{-\frac{(y+1)^2}{2\sigma^2}}} dy \\ &= e^{-\frac{1}{2\sigma^2}}. \end{aligned} \tag{1.18}$$

Similar to Section 1.2.3, knowing $\sigma^2 = N_0/2$ and assuming $E_s = 1$, we can rewrite $Z(W)$ in terms of $\mathrm{SNR} = E_s/N_0$ as $Z(W) = e^{-\frac{1}{2\sigma^2}} = e^{-SNR}$. Note that SNR in this equation is not in dB, but it is in linear form. Since SNR is commonly used in dB, we can use the following equation instead:

$$Z(W) = e^{-\frac{1}{2\sigma^2}} = e^{-\mathrm{SNR}} = \exp\left(-10^{\frac{\mathrm{SNR\,[dB]}}{10}}\right). \tag{1.19}$$

1.4 Channel Capacity and Channel Coding Theorem

Shannon's noisy channel coding theorem states that it is possible to transmit data over a noisy channel at any desired rate, as long as the coding scheme is designed correctly. The theorem is based on the concept of channel capacity, which is the maximum rate at which data can be transmitted over a noisy channel without errors. The channel capacity is determined by the properties of the channel, such as noise level and bandwidth. Shannon defined the discrete channel capacity as the maximum of the average mutual information, which is computed over all possible probability distributions associated with the input symbol alphabet as

$$C = \max_{p(X)} I(X;Y) = \max_{p(X)} [H(X) - H(X|Y)]. \tag{1.20}$$

From (1.20), since $I(X;Y) \geq 0$, we can conclude that $C \geq 0$. Furthermore, since $\max H(X) = \log|\mathcal{X}|$, where $\mathcal{X}$ is the channel input alphabet, then $C \leq \log|\mathcal{X}|$ with equality holding when the channel is error-free.

Formally, the *channel coding theorem* states that for any DMC with input $\mathcal{X}$, output $\mathcal{Y}$ and capacity C, there exists a code with a sufficiently long block length n such that the transmission rate R can be made arbitrarily close to the channel capacity C, and the probability of error ϵ in decoding can be made arbitrarily small. In practical terms, this means that by using longer block codes, the rate can approach channel capacity ($R \to C$), and with sufficiently long codes ($n \to \infty$), the probability of error can be made as small as desired ($\epsilon \to 0$). The channel coding theorem provides a theoretical foundation for designing codes that achieve reliable communication in the presence of noise and disturbances. The goal of coding theory is to find codes that meet the conditions of Shannon's theorem. Many different codes have been proposed, and the best code to use depends on the specific characteristics of the channel.

Now, let us review the capacity of the channels discussed in Section 1.2. The capacity of a BSC with crossover probability p is given by

$$C = \max_{p(X)} I(X;Y) = 1 - H(p) = 1 - p\log p + (1-p)\log(1-p) \text{ bits/channel use,} \tag{1.21}$$

where p indicates the probability of error or the probability of transmitting a symbol and receiving another. The capacity of the BEC (in bits per channel use), in which the parameter e represents the erasure and the probability of erasure $0 \leq \epsilon \leq 1$, is as

$$C = \max_{p(X)} I(X;Y) = 1 - \epsilon \text{ bits/channel use.} \tag{1.22}$$

This means that on average, the $1 - \epsilon$ fraction of symbols can be transmitted successfully without being erased.

The capacity of AWGN channels (in bits per second per Hertz) is given by

$$C = \log_2\left(1 + \frac{P}{N_0}\right), \tag{1.23}$$

where P is the average transmitted power, and N_0 is the power of the AWGN. The term $\frac{P}{N_0}$ is called the signal-to-noise ratio (SNR) of the channel.

Finally, the capacity of a fading channel is more complex than that of other channels due to the time-varying nature of the channel caused by fading.

The capacity of a fading channel is usually considered in terms of ergodic capacity, which represents the average capacity over different fading realizations. The capacity of a fading channel (in bits per second per Hertz) is given by

$$C = \mathbf{E}_h \left[\log_2 \left(1 + \frac{P|h|^2}{N_0} \right) \right], \tag{1.24}$$

where E_h denotes the expectation over all possible channel fading states, P is the average transmitted power, $|h|^2$ is the squared magnitude of the fading coefficient, and N_0 is the power of the AWGN. The term $\frac{P|h|^2}{N_0}$ represents the instantaneous SNR) in the fading channel. The expectation over the fading states captures the average behavior of the channel.

1.5 Block Error Rate for Finite Length Codes

In this section, we discuss a non-asymptotic approximation for the maximal sustainable rate of an AWGN channel as a function of block length N and error probability ϵ, i.e. $R^*(N, \epsilon)$. The discussion will be limited to introducing an approximate expression rather than the full information-theoretic derivation. The interested readers are referred to the suggested references in the bibliographical notes.

As an asymptotic limit, it is known from classic texts that for a general class of channels, the channel capacity C is the highest rate R at which information can be transmitted reliably, given that the block length N is allowed to grow without bound. That is,

$$\lim_{N \to \infty} R^*(N, \epsilon) = C. \tag{1.25}$$

However, at a fixed finite block length, we have to backoff from capacity and choose a rate smaller than capacity $R < C$ to maintain the desired error probability. Figure 1.6 illustrates the achievable rate for a desired block error rate $R^*(N, \epsilon)$ at different finite block lengths N.

The gap to the channel capacity can be characterized by a coefficient called channel dispersion, which measures the stochastic variability of the channel relative to a deterministic channel with the same capacity. The channel dispersion coefficient is defined as

$$V = \lim_{\epsilon \to 0} \lim_{N \to \infty} \sup \frac{(NC - \log M^*(N, \epsilon))^2}{N} \rho. \tag{1.26}$$

where $\rho = \frac{1}{2 \ln \frac{1}{\epsilon}}$ and $M^*(N, \epsilon)$ is the maximum cardinality of a codebook of block length N which can be decoded with block error probability $P_e \leq \epsilon$. As can be seen in (1.26), the channel dispersion is measured in squared information units per

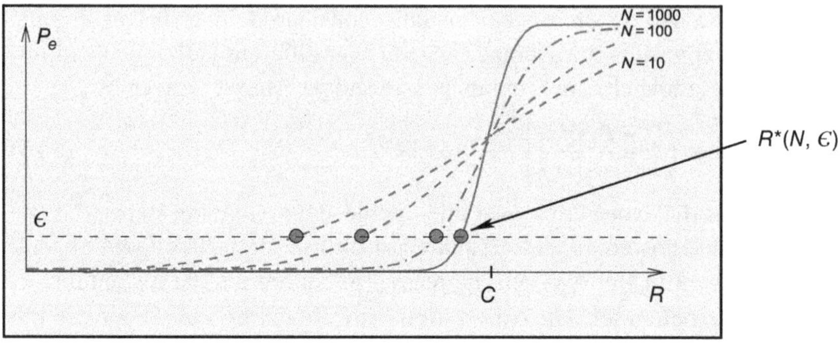

Figure 1.6 Trade-off between the reliability and the code rate: the maximum code rate for a desired ϵ at different blocklength N.

channel use. From the definition of dispersion and its unit, one can picture it as the "variance of the channel capacity." Hence, for finite block-length N, we can consider the maximum code rate to approximately follow a normal distribution of $R \sim \mathcal{N}(C, \frac{V}{N})$ as demonstrated in Figure 1.7 that results in

$$R + \text{constant} \cdot \sqrt{\frac{V}{N}} \lesssim C. \tag{1.27}$$

To be more precise, the maximum code rate via the so-called normal approximation (NA) based on the gap to capacity C is

$$R = C - \sqrt{\frac{V}{N}} \log_2(e)\, Q^{-1}(\epsilon) + \frac{\log_2(N)}{2N}. \tag{1.28}$$

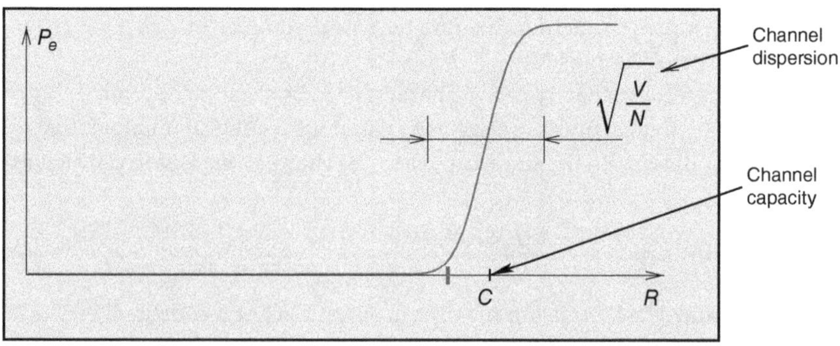

Figure 1.7 A sketch of channel dispersion relative to channel capacity.

This approximation is known as the Polyanskiy–Poor–Verdù (PPV) meta-converse approximation. By rearranging this approximation, we can get a tight approximation for the block error rate ϵ (roughly for block-lengths larger than 100 bits) as

$$\epsilon^*(N,R) = Q\left(\frac{C - R + \frac{\log_2(N)}{2N}}{\sqrt{\frac{V}{N}} \cdot \log_2(e)}\right). \tag{1.29}$$

In the case of the binary-input AWGN channel the capacity and the channel dispersion coefficient can be obtained by using the Laplace transform as

$$C = 1 + \frac{H'(0)}{\ln(2)}, \tag{1.30}$$

and

$$V = H''(0) - (H'(0))^2, \tag{1.31}$$

respectively, and the derivatives of $H(\cdot)$ for every SNR Ω can be numerically evaluated by

$$H^{(\ell)}(s) = \frac{1}{\sqrt{2\pi\Omega}} \int_{-\infty}^{\infty} e^{-\frac{1}{2\Omega}(x-\Omega)^2} e^{-s \cdot h(x)} (-\ln(1 + e^{-2x}))^{\ell}\, dx. \tag{1.32}$$

where ℓ denotes the order of derivative and $h(x) = \ln\left(1 + e^{-2x}\right)$.

The MATLAB script C.1 in Appendix C provides the NA (1.29) for the binary AWGN channel.

1.6 Shannon Limit on Power Efficiency

In this section, we look at the minimum energy per bit required by a transmitter to achieve reliable communication in a band-limited communication system, regardless of modulation or coding scheme. By considering a particular modulation or encoding scheme, the limit is called the constrained Shannon limit.

We are interested in designing a communication system that requires the smallest possible power for reliable transmission of information, that is, with an acceptable bit-error rate (BER). This performance measure is usually illustrated in terms of BER versus E_b/N_0, where E_b/N_0 is the ratio of signal energy per information bit E_b to noise power spectral density per bit N_0. The *power efficiency* η_P refers to the E_b/N_0 that is required at the receiver input to achieve a certain BER.

We learned earlier that the Shannon channel capacity of an AWGN channel with a bandwidth B [Hz], received power P [W], and a constant noise power spectral density $\sigma_n^2 = \frac{N_0}{2}$ [W/Hz] is given by

$$C_{\text{awgn}} = B \log_2\left(1 + \frac{P}{BN_0}\right) \quad \text{bits/s}, \tag{1.33}$$

Here, $P = RE_b$ is the average power, where E_b [J] is the average signal energy per information bit, and R [bits/s] is the data transmission rate. Hence, we have

$$\frac{P}{BN_0} = \frac{RE_b}{BN_0}. \tag{1.34}$$

Keep in mind that this is a simplified relationship and that actual system behavior may vary depending on the modulation scheme, the coding scheme, and the specific characteristics of the channel. Furthermore, the equation assumes that the SNR–block error rate (BLER) relationship is linear over the range of interest, which may not be the case in all situations.

The channel capacity restricts the data rate R to be smaller than the capacity C to achieve reliable transmission. Here, we define the ratio of the data rate R to the bandwidth B as *spectral efficiency*, $\eta_B = \frac{R}{B}$ [bits/s/Hz]. Now, given the requirement $R < C$, by substituting (1.34) into (1.33) and rearranging it, we have the following:

$$\frac{R}{B} < \frac{C}{B} = \log_2\left(1 + \frac{RE_b}{BN_0}\right). \tag{1.35}$$

Then, rewriting (1.35) gives a constraint on the SNR per bit E_b/N_0 for reliable data transmission as

$$\frac{E_b}{N_0} > \frac{2^{\eta_B} - 1}{\eta_B}, \quad \eta_B = \frac{R}{B}. \tag{1.36}$$

Observe that in the case of infinite bandwidth, the spectral efficiency η_B approaches zero. As $\lim\limits_{\eta_B \to 0} \frac{2^{\eta_B}-1}{\eta_B}$ gives the indeterminate form of 0/0, using the L'Hopital rule and simplification, we get the minimum required E_b/N_0 for infinite bandwidth as

$$\frac{E_b}{N_0} > \lim_{\eta_B \to 0} 2^{\eta_B} \ln 2 = \ln 2 = 0.693, \quad \text{or} \quad \frac{E_b}{N_0} > -1.59 \text{ dB}. \tag{1.37}$$

Note that the power efficiency limit is independent of the BER. In fact, this limit determines the minimum possible E_b/N_0 to achieve an arbitrarily small BER.

The MATLAB script in C.2 in Appendix C gives the Shannon limit. Also, for more information on modulation schemes, we invite the readers to see Chapter 12.

Before closing this section, let us find the relationship between SNR or $E_s/N_0 = P/N_0$ and R. E_s/N_0 and E_b/N_0 are both SNR metrics used in digital communications. E_s/N_0 is the ratio of the signal energy per symbol to the noise power spectral density. E_b/N_0 is the ratio of energy per bit to the noise power spectral density.

The relationship between E_s/N_0 and E_b/N_0 depends on the modulation scheme used. For example, for binary phase-shift keying (BPSK), E_s/N_0 is equal to E_b/N_0. For quadrature phase-shift keying (QPSK), E_s/N_0 is equal to $2E_b/N_0$. For the 16-quadrature amplitude modulation (16-QAM), E_s/N_0 is equal to $4E_b/N_0$.

In general, the relationship between E_s/N_0 and E_b/N_0 can be expressed as follows:

$$\frac{E_s}{N_0} = \frac{E_b \cdot \log_2(M) \cdot R}{N_0}, \tag{1.38}$$

where R is the code rate, M is the constellation size, and $m = \log_2 M$ is the transmission rate in bits per symbol (or bits per channel use).

Denote by SNR_1 the initial SNR corresponding to a specific BLER and by R_1 the initial code rate. Now, if you change the code rate to R_2, the relationship between the new SNR (SNR_2) and the initial SNR $\left(\text{SNR}_1\right)$ can be expressed as follows:

$$\text{SNR}_2 = \text{SNR}_1 + 10 \cdot \log_{10}\left(\frac{R_2}{R_1}\right). \tag{1.39}$$

This formula is derived from the fact that increasing the code rate generally requires a higher SNR to maintain the same level of performance (in terms of BLER).

1.7 Time and Space Complexity

Anyone can implement the same algorithm to solve an a encoding or decoding problem in various ways. In addition, there may be multiple algorithms available to solve a single problem. Therefore, we require a metric to compare different algorithms and their implementations. The big O notation denoted as $O(f(n))$ is used to describe how the run-time in terms of the number of operations (steps), or the space/memory requirements of algorithms, represented by the function $f(n)$, grows as the input size n increases. Here, the capital letter O stands for the *order of approximation*.

This notation focuses on the dominant factor that influences an algorithm's complexity while disregarding constant factors and lower-order terms. Hence, the big O notation only describes the asymptotic behavior of complexity, not the exact complexity. This attribute makes it an effective tool for comparing and analyzing algorithms, free from the complexities of implementation details.

Example 1.1 Let us consider $O(n^2)$, pronounced as "Big O squared", which is the complexity of the selection sort algorithm. Selection sort is a sorting method that iterates through the list of n elements, ensuring that each element at a given index i is the ith least/greatest element in the list. This sorting algorithm can be implemented by two nested loops where the outer loop iterates through the list (n iterations in total), while for the ith element in the list, the inner loop finds the smallest/largest element in the remaining part of the list (by $n - i$ iterations). Observe that this actually gives a geometric sum, meaning the inner loop repeats

Table 1.1 Example time complexities.

Complexity	Name	Example
$O(1)$	Constant	Array access
$O(\log n)$	Logarithmic	Binary search
$O(n)$	Linear	Linear search
$O(n \log n)$	Linearithmic	Merge sort
$O\left(n^2\right)$	Quadratic	Bubble sort
$O\left(n^2\right)$	Cubic	Matrix–vector multiplication
$O\left(n^3\right)$	Cubic	Matrix–matrix multiplication
$O\left(n^c\right)$	Polynomial	
$O\left(c^n\right), c > 1$	Exponential	Recursive Fibonacci
$O(n!)$	Factorial	Permutations of a set

$n + (n-1) + \cdots + 2 + 1$ times, which equals $n(n-1)/2 = n^2/2 - n/2$ times. This is the time complexity of the selection sorting algorithm. Now, since the dominant term is $n^2/2$, ignoring the coefficient, we get $O(n^2)$.

Table 1.1 lists common time complexities and the corresponding examples, from the best to the worst in terms of complexity:

The formal definition of Big O is as follows:

Definition 1.1 Given two functions $f(n)$ and $g(n)$, then $f(n) = O(g(n))$ if there exist constants $c > 0$ and $n_0 \leq 0$ such that $f(n) \leq c \cdot g(n)$ for all $n \leq n_0$. Observe that in this case, $f(n)$ does not grow faster than $c \cdot g(n)$ for all $n \leq n_0$.

1.8 Historical Milestones in Channel Coding

The field of error correction codes and channel coding has been pivotal in enabling reliable communication over noisy channels, transforming theoretical insights into practical engineering marvels. Rooted in Claude Shannon's 1948 groundbreaking work on information theory, this field has evolved through decades of innovation, driven by the quest to approach the channel capacity—the theoretical limit of reliable communication. Early milestones, such as Hamming codes and Reed–Solomon codes, provided foundational methods for error detection and correction. The introduction of convolutional codes and the Viterbi algorithm revolutionized real-time communications, while advances like turbo codes and low-density parity-check (LDPC) codes approached channel capacity with iterative decoding. More recently, the advent of polar codes has offered a capacity-achieving solution with efficient implementations, finding applications in modern standards like 5G. Each milestone reflects the ingenuity of researchers in tackling the challenges of noise, bandwidth, and complexity, shaping the technologies that underpin modern communication and data storage systems. Here is a timeline of major historical milestones in channel coding, highlighting key contributions, authors, and publications:

1948: Foundations of Information Theory
- **Author**: Claude E. Shannon
- **Publication**: *A Mathematical Theory of Communication* (Bell System Technical Journal)
- **Contribution**: Shannon's seminal work established the theoretical foundations of information theory, including the concept of channel capacity and the idea that error-free communication is possible with proper coding.

1950: Hamming Codes
- **Author**: Richard W. Hamming
- **Publication**: *Error Detecting and Error Correcting Codes* (Bell System Technical Journal)
- **Contribution**: Hamming introduced the first practical error-correcting code, known as Hamming codes, which could detect and correct single-bit errors in data.

1954: Reed–Muller Codes
- **Authors**: Irving S. Reed and David E. Muller
- **Publication**: *A Class of Multiple-Error-Correcting Codes and Their Decoding Scheme* (IRE Transactions on Information Theory)
- **Contribution**: Reed–Muller codes provided a flexible class of error-correcting codes that could correct multiple errors and were used in early space communication systems.

1954: Golay Codes
- **Author**: Marcel Golay
- **Publication**: *Notes on Digital Coding*
- **Contribution**: Golay introduced the Golay code, a perfect error-correcting code that has been used in various applications, including space exploration.

1955: Convolutional Codes
- **Author**: Peter Elias
- **Publication**: *Coding for Noisy Channels* (IRE Convention Record)
- **Contribution**: Elias introduced convolutional codes, which encode data using a sliding window and are widely used in wireless communication systems.

1957: Cyclic Codes

- **Author**: W. Wesley Peterson
- **Publication**: *Encoding and Error-Correction Procedures for the Bose–Chaudhuri Codes* (IRE Transactions on Information Theory)
- **Contribution**: Peterson developed cyclic codes, including BCH codes, which are efficient for correcting multiple errors and are widely used in digital communication and storage systems.

1960: Bose–Chaudhuri–Hocquenghem (BCH) Codes

- **Authors**: Raj Chandra Bose, Dwijendra K. Ray-Chaudhuri, and Alexis Hocquenghem
- **Publication**: Independent papers in 1959 and 1960
- **Contribution**: BCH codes are a class of cyclic codes that can correct multiple errors and are widely used in applications like satellite communications and QR codes.

1962: Low-Density Parity-Check (LDPC) Codes

- **Author**: Robert G. Gallager
- **Publication**: Low-Density Parity-Check Codes (*IRE Transactions on Information Theory*)
- **Contribution**: Gallager introduced LDPC codes in his doctoral dissertation at MIT, which are capacity-approaching codes. Although initially overlooked, they became widely used in modern communication systems (e.g. Wi-Fi, 5G).

1963: Fano Algorithm

- **Author**: Robert Fano
- **Publication**: A Heuristic Discussion of Probabilistic Decoding (*IEEE Transactions on Information Theory*)
- **Contribution**: The Fano algorithm is a sequential decoding algorithm for convolutional codes, providing an efficient way to decode long constraint-length codes.

1967: Reed-Solomon Codes

- **Authors**: Irving S. Reed and Gustave Solomon
- **Publication**: Polynomial Codes over Certain Finite Fields (*Journal of the Society for Industrial and Applied Mathematics*)
- **Contribution**: Reed–Solomon codes are highly effective for correcting burst errors and are used in CDs, DVDs, and deep-space communication.

1969: Berlekamp–Massey Algorithm

- **Authors**: Elwyn Berlekamp and James Massey
- **Publication**: On the Solution of Algebraic Equations over Finite Fields (*Information and Control*)
- **Contribution**: The Berlekamp–Massey algorithm is an efficient method for decoding BCH and Reed–Solomon codes by solving the key equation in error correction.

1970: Goppa Codes

- **Author**: Valerii D. Goppa
- **Publication**: A New Class of Linear Error-Correcting Codes (*Problemy Peredachi Informatsii*)
- **Contribution**: Goppa codes are used in algebraic coding theory and cryptography, particularly in the McEliece cryptosystem.

1972: Viterbi Algorithm

- **Author**: Andrew J. Viterbi
- **Publication**: *Principles of Coherent Communication* (McGraw-Hill)
- **Contribution**: The Viterbi algorithm provides an efficient method for decoding convolutional codes and is widely used in digital communication systems.

1974: BCJR Algorithm

- **Authors**: Bahl, Cocke, Jelinek, and Raviv
- **Publication**: Optimal Decoding of Linear Codes for Minimizing Symbol Error Rate (*IEEE Transactions on Information Theory*)
- **Contribution**: The BCJR algorithm is a forward–backward algorithm for maximum a posteriori (MAP) decoding of convolutional codes, later used in turbo codes.

1976: Concatenated Codes

- **Author**: David Forney
- **Publication**: *Concatenated Codes* (MIT Press)
- **Contribution**: Forney introduced concatenated codes, which combine two or more error-correcting codes to achieve high performance. These were used in deep-space communication.

1982: Trellis-Coded Modulation (TCM)

- **Author**: Gottfried Ungerboeck
- **Publication**: *Channel Coding with Multilevel/Phase Signals* (IEEE Transactions on Information Theory)
- **Contribution**: TCM integrates coding and modulation to improve error performance without increasing bandwidth, revolutionizing digital communication.

1993: Turbo Codes

- **Authors**: Claude Berrou, Alain Glavieux, and Punya Thitimajshima
- **Publication**: *Near Shannon Limit Error-Correcting Coding and Decoding: Turbo Codes* (Proceedings of ICC'93)
- **Contribution**: Turbo codes introduced iterative decoding and achieved near-Shannon-limit performance, revolutionizing wireless communication (e.g. 3G and 4G).

1996: Rediscovery of LDPC Codes

- **Authors**: David MacKay and Radford Neal

- **Publication**: *Near Shannon Limit Performance of Low Density Parity Check Codes* (IEEE Transactions on Information Theory)
- **Contribution**: MacKay and Neal rediscovered and modernized LDPC codes, demonstrating their near-Shannon-limit performance and leading to practical implementations in modern communication systems.

2002: Fountain Codes

- **Author**: Michael Luby
- **Publication**: *LT Codes* (Proceedings of the 43rd Annual IEEE Symposium on Foundations of Computer Science)
- **Contribution**: Fountain codes, such as LT (Luby Transform) codes and Raptor codes, introduced rateless coding, which is essential for efficient data streaming and file delivery systems.

2009: Polar Codes

- **Author**: Erdal Arıkan
- **Publication**: *Channel Polarization: A Method for Constructing Capacity-Achieving Codes for Symmetric Binary-Input Memoryless Channels* (IEEE Transactions on Information Theory)
- **Contribution**: Polar codes are the first class of codes proven to achieve channel capacity and are used in 5G networks.

1.9 Why Study Polar Codes?

Polar codes, introduced by Arıkan in 2009, represent a landmark in coding theory as the first class of codes proven to achieve the capacity of symmetric B-DMCs with an explicit construction. This breakthrough has sparked immense interest in both academia and industry due to their theoretical elegance and practical potential.

The primary reasons to study polar codes include:

- Capacity achievement: Polar codes achieve channel capacity asymptotically under successive cancellation (SC) decoding, making them a foundational element in information theory.
- Superior performance for short code-lengths: The error correction performance of short and medium-length polar codes concatenated with cyclic codes or rate-1 convolutional codes under SC list decoding is superior to that of other modern codes.
- Adoption in standards: Polar codes are used in 5G New Radio (NR) standards for control channel communications, underscoring their relevance in practical applications.
- Flexibility: Polar codes support rate adaptation, which makes them suitable for a wide range of communication scenarios, from ultra-reliable low-latency communications (URLLCs) to massive machine-type communications (mMTCs).

- Theoretical insights: Polar codes provide a new perspective on channel polarization, which has led to further developments in coding theory, including polar-like code constructions.

1.9.1 How Do Polar Codes Compare with Other Codes?

Polar codes have unique features, but they also have limitations when compared to other coding schemes, such as LDPC codes, turbo codes, and convolutional codes. Below is a comparison based on various attributes:

Table 1.2 Comparison of polar codes with other coding schemes.

Feature	Polar codes	LDPC codes	Turbo codes	Convolutional codes
Capacity achievement	Achieves channel capacity asymptotically	Approaches capacity with iterative decoding	Approaches capacity for large block lengths	Not capacity achieving
Complexity	$O(L \cdot n \cdot \log n)$	$O(I \cdot n \cdot d_c)$	$O(I \cdot n \cdot 2^m)$	$O(n \cdot 2^m)$
Decoding algorithm	SC list (SCL) decoding	Message passing (belief propagation) algorithm	SISO iterative decoding (BCJR)	Viterbi algorithm
Performance at short lengths	Excellent	Inferior, widely used for long codes	Inferior, used for long codes	Good
Adoption in standards	5G control channels	5G traffic channels, Wi-Fi, DVB-S2, etc.	3G, 4G LTE	Legacy systems like 3G GSM
Design complexity	Explicit construction with fixed parameters	Requires optimization of parity-check matrix	Requires careful interleaver design	Simple design, no interleaver needed
Flexibility in code rate	Requires puncturing/shortening	Highly flexible	Flexible with interleavers	Limited
Error floor	No inherent error floor	May suffer from error floors at high SNR	May suffer from error floors at high SNR	No inherent error floor
Hardware efficiency	Efficient for SC decoding, low for SCL decoding	Moderate, depends on iterative decoder design	Moderate to high	Highly efficient

The parameters used in Table 1.2 are as follows. The code length is denoted by n, the number of iterations in interactive decoding is denoted by I (typically ranges from 10 to 50, depending on the target performance and latency constraints), L is the list size in list decoding algorithm (number of paths retained during decoding, typically small in practical systems; e.g. $L = 8, 16, 32$, to balance complexity and performance), d_c is the average column weight of the parity-check matrix (average number of connections per variable node, or the average degree of variable nodes), m is the constraint length (number of memory elements in the component convolutional codes). The complexity of decoding convolutional codes (standalone or as component codes in Turbo codes) grows exponentially with the constraint length m, making Viterbi decoding suitable for short constraint lengths (≤ 10). However, this makes turbo decoding more computationally expensive than LDPC decoding for large block lengths due to higher complexity in component decoders. Furthermore, hardware optimizations such as parallelism in belief propagation algorithm can significantly reduce the decoding latency of LDPC codes. It is worth noting that the decoding complexity of polar codes in Table 1.2 does not consider simplified and fast decoding techniques using special nodes, described in Chapter 9.

A comprehensive performance comparisons under various decoding algorithms and code parameters can be observed in Chapter 13.

Exercises

1 Using the relation $p(x,y) = p(x)p(y|x)$, show the following chain rule:

$$H(X,Y) = H(X) + H(Y|X).$$

2 Given a BSC with crossover probability $p = 0.01$. If an independent-symbol binary source with probabilities 0.25 and 0.75 is transmitted over this channel, calculate $H(X/Y)$ and $I(X,Y)$.

3 Given the joint distribution of (X, Y) as follows:

Y	1	2
1	0	$\frac{3}{4}$
2	$\frac{1}{8}$	$\frac{1}{8}$

Calculate $H(X)$ and $H(X|Y = 1)$.

4 Suppose that we have two random variables, X and Y, where X is the outcome of a fair coin toss (heads or tails), and Y is the outcome of a roll of a fair die (1–6).

a) Calculate the mutual information between X and Y.
b) Change the random variable X to the outcome of a biased coin toss (heads with probability 0.7 and tails with probability 0.3) and recalculate mutual information between X and Y.

5 Using the information inequality, show that $I(X; Y) = 0$ only if X and Y are independent.

6 A binary erasure channel has an erasure probability of 0.2. What is the capacity of the channel?

7 A channel with a bandwidth $B = 25\,\text{kHz}$ and an SNR of 18 dB is perturbed by AWGN. What is the capacity of this channel in bits per second?

8 What is the Shannon limit if we employ an ideal 2-PAM modulation scheme?

Bibliographic Notes

In his 1948 paper, *A Mathematical Theory of Communication* Shannon [1948], Claude E. Shannon laid the foundation for modern information theory. He introduced the concept of channel capacity and developed the channel coding theorem. In addition to his work on channel capacity and channel coding, Shannon also made significant contributions to the theory of entropy and mutual information. Shannon's work has had a profound impact on the design of communication systems. Gallager in his classic text, Information Theory and reliable communication Gallager [1968], delves into the fundamentals of information theory, discrete entropy, and various coding techniques. It provides a comprehensive treatment of the mathematical aspects of information theory. Furthermore, Cover and Thomas' book, *Elements of Information Theory* Cover and Thomas [2006], is widely used as a reference for understanding information theory. It covers topics such as entropy, mutual information, and channel capacity, providing both theoretical insights and practical applications.

Investigating the lower bound for the error performance in channel coding dates back to the work of Shannon et al. [1967] and earlier. Among recent results available in the literature, an approach called normal approximation (NA) is commonly used to identify the limit of reliable communication for a block code of length n. This limit for the code rate R_{NA} for a wide range of channels is stated as Polyanskiy et al. [2010]. Unlike Shannon [1959] based on the geometric construction associated with the sphere packing, the normal approximation (NA) limit of Polyanskiy et al. [2010] is based on statistical properties and can then be applied to any channel model. More refined and simple evaluations of this approach based on integral expressions for various channels are provided by Erseghe [2016].

Jacobs and Berlekamp [1967], Jacobs and Wozencraft [1965], and Forney [1968] have shown the relationship between cutoff rate, code rate, error probability, and complexity of sequential and list decoding.

For a discussion of the Bhattacharyya parameter on the AWGN channel, the readers can refer to Li and Yuan [2013].

Bibliography

Thomas M Cover and Joy A Thomas. *Elements of information theory*. John Wiley & Sons, 2006.

Tomaso Erseghe. Coding in the finite-blocklength regime: Bounds based on Laplace integrals and their asymptotic approximations. *IEEE Transactions on Information Theory*, 62(12):6854–6883, 2016.

G Forney. Exponential error bounds for erasure, list, and decision feedback schemes. *IEEE Transactions on Information Theory*, 14(2):206–220, 1968.

Robert G Gallager. *Information theory and reliable communication*, volume 588. Springer, 1968.

I Jacobs and E Berlekamp. A lower bound to the distribution of computation for sequential decoding. *IEEE Transactions on Information Theory*, 13(2):167–174, 1967.

Irwin Mark Jacobs and John M Wozencraft. *Principles of communication engineering*, 1965.

Huijun Li and Jinhong Yuan. A practical construction method for polar codes in AWGN channels. In *IEEE 2013 Tencon-Spring*, pages 223–226. IEEE, 2013.

Yury Polyanskiy, H Vincent Poor, and Sergio Verdú. Channel coding rate in the finite blocklength regime. *IEEE Transactions on Information Theory*, 56(5):2307–2359, 2010.

Claude Elwood Shannon. A mathematical theory of communication. *The Bell System Technical Journal*, 27(3):379–423, 1948.

Claude E Shannon. Probability of error for optimal codes in a Gaussian channel. *Bell System Technical Journal*, 38(3):611–656, 1959.

Claude E Shannon, Robert G Gallager, and Elwyn R Berlekamp. Lower bounds to error probability for coding on discrete memoryless channels. I. *Information and Control*, 10(1):65–103, 1967.

2

Linear Codes

Nature doesn't create perfect channels, but mathematics gives us the tools to overcome their imperfections.

— Claude Berrou

In this chapter, we introduce the fundamental concepts of binary linear codes, a crucial class of codes in coding theory. We begin by defining key terms such as codewords, generator matrices, parity-check matrices, syndrome, and Hamming distance. These definitions provide a solid foundation for understanding the structure and performance of linear block codes. We also discuss important bounds for the parameters of these codes, which play a significant role in their analysis and application.

The chapter then explores the decoding of linear block codes, with a focus on syndrome decoding as an effective technique. We follow this by examining several well-known linear codes, including repetition codes, single-parity check codes, Reed–Muller codes, cyclic codes, and convolutional codes. For each of these, we describe their structure, properties, and encoding methods, highlighting their distinct features and practical uses.

Finally, we address the concept of code concatenation, a powerful method for constructing new, more efficient codes from existing shorter ones. This technique is essential for improving the performance of coding schemes in various communication systems. By the end of the chapter, you will have a solid understanding of the basics of linear block codes, their encoding and decoding processes, and how to implement code concatenation effectively.

2.1 Generator and Parity-Check Matrices

From an abstract point of view, a (n, k) *linear code* is a k-dimensional subspace of the n-dimensional vector space $\mathbb{F}_2^n$ of binary vectors of length n. The *codebook* C

Polar Codes: From Theory to Practice, First Edition. Mohammad Rowshan and Emanuele Viterbo.

is the set of 2^k *codewords* out of the 2^n vectors in $\mathbb{F}_2^n$. The code rate is defined as $R = k/n$.

All linear codes can be generated by a $k \times n$ binary *generator matrix* $\mathbf{G}$ as

$$C = \{\mathbf{c} = \mathbf{uG} \mid \mathbf{u} \in \mathbb{F}_2^k\}.$$

The k-dimensional vectors $\mathbf{u}$ are the so-called *information vectors* and can be thought of as the 2^k binary labels for all the codewords in the codebook. The generator matrix obtained by stacking k row vectors $\mathbf{g}_1, \ldots, \mathbf{g}_k$

$$\mathbf{G} = \begin{pmatrix} \mathbf{g}_1 \\ \vdots \\ \mathbf{g}_k \end{pmatrix}$$

defines a linear map from $\mathbb{F}_2^k$ to $\mathbb{F}_2^n$ which implies that the sum of two codewords is still a codeword and that the all-zero codeword belongs to the codebook. The k row vectors $\mathbf{g}_1, \ldots, \mathbf{g}_k$ of $\mathbf{G}$ are codewords labeled by the k unit vectors $\mathbf{u} = (1, 0, \ldots, 0)$ to $\mathbf{u} = (0, \ldots, 0, 1)$ (of length k) and are known as the *basis vectors* of the k-dimensional vector subspace of $\mathbb{F}_2^n$. Equivalently, all the codewords can be constructed as binary linear combinations of the basis vectors

$$\mathbf{c} = \sum_{i=1}^{k} u_i, \mathbf{g}_i \quad u_i \in \mathbb{F}_2.$$

We say that the information vectors of length k are *encoded* by the generator matrix to the codewords of length $n > k$. This operation implies that $n - k$ additional (*redundant*) bits will be used to transmit k bits of information. If we specifically choose the generator matrix as $\mathbf{G} = (\mathbf{I}_k | \mathbf{P}_{k \times (n-k)})$, where $\mathbf{I}_k$ is a $k \times k$ identity matrix, then the encoded codewords have the so-called systematic form $\mathbf{c} = (\mathbf{u}|\mathbf{uP})$, where the first k bits are the *information bits* and the remaining $n - k$ bits are the *parity bits*. A code in this form is known as a *systematic code*. The above definition of a binary linear code can easily be extended to q-ary linear codes if we replace the base field of the vector space $\mathbb{F}_2$ with $\mathbb{F}_q$.

The linear structure of the codes enables encoding with simple matrix–vector multiplication. In contrast, a *nonlinear code* does not have this feature and would generally require storing a lookup table with a list of all codewords and their labels. For nonlinear codes, the memory requirement scales exponentially in k, while for linear codes, the encoding operation generally requires $O(nk)$ multiplications over $\mathbb{F}_q$ (or $O((n-k)k)$ for systematic codes). The particular structures of the generator matrix enable us to reduce the number of multiplications, as we will see in the case of cyclic codes and polar codes.

Let us now consider the problem of checking whether a given sequence of n bits $\mathbf{x}$ belongs to the codebook C or not. Clearly, for nonlinear codes, this would require scanning the entire lookup table. On the other hand, for linear codes, this can be easily performed by using the $(n-k) \times n$ *parity check matrix*

$$\mathbf{H} = \begin{pmatrix} \mathbf{h}_1 \\ \vdots \\ \mathbf{h}_{n-k} \end{pmatrix}.$$

We first notice that the $(n-k) \times n$ matrix $\mathbf{H}$ generates a $(n, n-k)$ linear code. If the vectors $\{\mathbf{h}_j\}_{j=1}^{n-k}$ span the dual subspace of the subspace spanned by the rows $\{\mathbf{g}_i\}_{i=1}^{k}$ of $\mathbf{G}$, that is,

$$\langle \mathbf{g}_i, \mathbf{h}_j \rangle = \sum_{k=1}^{n} g_{ik} h_{jk} \mod 2 = 0 \quad \text{for all } i, j, \tag{2.1}$$

or[1] more compactly

$$\mathbf{G} \cdot \mathbf{H}^T = \mathbf{0}_{k \times (n-k)}, \tag{2.2}$$

then $\mathbf{H}$ is the generator matrix of the so-called *dual code* $C^\perp$ of C. We may think of (2.2) as the orthogonality relations between the two subspaces C and $C^\perp$, which imply that any codeword of C is orthogonal to any vector in $C^\perp$ as shown below. Given any

$$\mathbf{c} = \sum_{i=1}^{k} u_i \mathbf{g}_i \in C \quad \text{and} \quad \mathbf{c}' = \sum_{j=1}^{n-k} u_j' \mathbf{h}_j \in C^\perp,$$

it is straightforward to prove using (2.1) that $\langle \mathbf{c}, \mathbf{c}' \rangle = 0$ and

$$\mathbf{c} \cdot \mathbf{H}^T = \mathbf{0}_{n-k} \quad \text{for all } \mathbf{c} \in C. \tag{2.3}$$

The $n-k$ equations in (2.3) are known as the *parity check equations*, which are satisfied by all the codewords. It can be easily shown that C and $C^\perp$ only share the all-zero codeword, i.e. $C \cap C^\perp = \{\mathbf{0}_n\}$, and all the $2^n = 2^k \times 2^{n-k}$ vectors of $\mathbb{F}_2^n$ are obtained as

$$\mathbb{F}_2^n = C \oplus C^\perp = \{\mathbf{x} = \mathbf{c} + \mathbf{c}' \mid \mathbf{c} \in C, \mathbf{c}' \in C^\perp\}. \tag{2.4}$$

Using (2.3), it is easy to verify if $\mathbf{x} = \mathbf{c} \in C$ or not by computing the *syndrome* $\mathbf{s} = \mathbf{x} \cdot \mathbf{H}^T$ and checking if $\mathbf{s} = \mathbf{0}_{n-k}$ or $\mathbf{s} \neq \mathbf{0}_{n-k}$. This operation is known as *error detection* and amounts to verifying that the $n-k$ parity check equations $\mathbf{x} \cdot \mathbf{H}^T = \mathbf{0}_{n-k}$ are satisfied for a given vector $\mathbf{x} = \mathbf{c} + \mathbf{e}$, where $\mathbf{e}$ is an error pattern that perturbs a given codeword $\mathbf{c}$. We note that passing the error detection test does not guarantee $\mathbf{x} = \mathbf{c}$, the codeword that was selected before the errors occurred. It may fail whenever $\mathbf{e} = \mathbf{c}_1$ is a nonzero codeword of C since $\mathbf{c}_2 = \mathbf{c} + \mathbf{c}_1$ is also a codeword in C.

Finding the parity check matrix given the generator is very simple for systematic codes, where $\mathbf{G} = [\mathbf{I}_k | \mathbf{P}]$ and $\mathbf{H} = [\mathbf{P}^T | \mathbf{I}_{n-k}]$.

---

1 Note that the scalar product (2.1) does not define a norm since $\langle \mathbf{x}, \mathbf{x} \rangle = 0$ for some nonzero vectors (in fact for all the vectors with an even number of ones).

Example 2.1 For the (6, 3) code with **G** given by:

$$\mathbf{G} = \begin{pmatrix} 1 & 0 & 0 & 1 & 0 & 1 \\ 0 & 1 & 0 & 0 & 1 & 1 \\ 0 & 0 & 1 & 1 & 1 & 0 \end{pmatrix}$$

we follow these steps:

1) Extract **P**: The matrix **G** is in systematic form, so the left 3×3 submatrix is $\mathbf{I}_3$, and the right 3×3 submatrix is the parity matrix **P**:

$$\mathbf{P} = \begin{pmatrix} 1 & 0 & 1 \\ 0 & 1 & 1 \\ 1 & 1 & 0 \end{pmatrix}.$$

2) Form **H**: Now we construct **H** by taking $\mathbf{P}^T$ and appending $\mathbf{I}_3$:

$$\mathbf{H} = \begin{pmatrix} 1 & 0 & 1 & 1 & 0 & 0 \\ 0 & 1 & 1 & 0 & 1 & 0 \\ 1 & 1 & 0 & 0 & 0 & 1 \end{pmatrix}.$$

2.2 Distance, Weight, and Bounds

Let us first define the Hamming distance and weight. Given two codewords $\mathbf{c} = (c_0, c_1, \dots, c_{n-1})$ and $\mathbf{c}' = (c'_0, c'_1, \dots, c'_{n-1})$ in C over $\mathbb{F}_2$, the Hamming distance between the two codewords, denoted by $d(\mathbf{c}, \mathbf{c}')$, is defined as the number of coordinates i where $\mathbf{c}$ and $\mathbf{c}'$ differ, that is, $c_i \neq c'_i$. Hamming distance as a metric satisfies the following properties for a given $\mathbf{c}, \mathbf{c}', \mathbf{c}''$:

1) Positive definiteness: $d(\mathbf{c}, \mathbf{c}') \geq 0$, with $d(\mathbf{c}, \mathbf{c}') = 0$ if and only if $\mathbf{c} = \mathbf{c}'$.
2) Symmetry: $d(\mathbf{c}, \mathbf{c}') = d(\mathbf{c}', \mathbf{c})$.
3) Triangle inequality: $d(\mathbf{c}, \mathbf{c}'') \leq d(\mathbf{c}, \mathbf{c}') + d(\mathbf{c}', \mathbf{c}'')$.

Then, the minimum distance of the code C is defined as

$$d_{\min} = d(C) \triangleq \min_{\mathbf{c},\mathbf{c}' \in C, \mathbf{c} \neq \mathbf{c}'} d(\mathbf{c}, \mathbf{c}'), \tag{2.5}$$

which is the smallest Hamming distance between two different codewords in C. The Hamming weight of a codeword or any n-tuple $\mathbf{c}$, denoted by $\mathrm{w}(\mathbf{c})$, is the number of nonzero elements in $\mathbf{c}$, i.e.

$$\mathrm{w}(\mathbf{c}) \triangleq |\mathrm{supp}(\mathbf{c})|. \tag{2.6}$$

Similar to $d_{\min}$, we can find $w_{\min}$, which is the minimum non-zero weight among all codewords of the code C. The Hamming distance and Hamming weight are usually referred to as distance and weight. From the definitions of distance and weight, it follows that the distance between two codewords is equal to the weight of the vector sum of codewords, i.e.

$$d(\mathbf{c}, \mathbf{c}') = \mathrm{w}(\mathbf{c} + \mathbf{c}'). \tag{2.7}$$

Using (2.7), we can easily show that the minimum distance $d_{\min}$ of C is equal to the minimum weight $w_{\min}$ of C, i.e.

$$\begin{aligned} d_{\min}(C) &= \min_{\mathbf{c},\mathbf{c}'\in C, \mathbf{c}\neq\mathbf{c}'} d(\mathbf{c}, \mathbf{c}') \\ &= \min_{\mathbf{c},\mathbf{c}'\in C, \mathbf{c}\neq\mathbf{c}'} \mathrm{w}(\mathbf{c} + \mathbf{c}') \\ &= \min_{\mathbf{c}''\in C, \mathbf{c}''\neq\mathbf{0}} \mathrm{w}(\mathbf{c}'') \\ &= w_{\min}(C). \end{aligned} \tag{2.8}$$

We may now consider the nearest-neighbor decoding scheme. Suppose that C is a code of length n. The codewords that are closest (in Hamming distance) to a given received word $\mathbf{r}$ are called *nearest-neighbor codewords*.

The *nearest neighbor decoding*, also known as *minimum distance decoding*, is a decoding scheme in which a received sequence $\mathbf{r}$ is decoded to its nearest neighbor codeword $\mathbf{c}$. Thus, the distance $d(\mathbf{c}, \mathbf{r})$ indicates the number of erroneous coordinates. The error detection and correction capabilities of a code C can be expressed in terms of the minimum distance between its codewords as follows. Let us denote the minimum distance as d, where we omit the "min" subscript for brevity. Then, C can detect any t errors if and only if $d \geq t + 1$. Likewise, C can correct any t errors if and only if $d \geq 2t + 1$.

Now, let $\mathbf{r}$ be an n-tuple in $\mathbb{F}_2^n$. The sphere $\mathcal{S}_2^n(\mathbf{r}, \rho)$ (also known as *Hamming ball*) with center $\mathbf{r}$ and radius ρ is the set of all n-tuples in $\mathbb{F}_2^n$ whose distance from $\mathbf{r}$ is at most ρ, i.e.

$$\mathrm{S}_2^n(\mathbf{r}, \rho) = \left\{ \mathbf{c} \in \mathbb{F}_2^n \, : \, d(\mathbf{r}, \mathbf{c}) \leq \rho \right\}. \tag{2.9}$$

Here, we give a packing interpretation of the codes. If a code C has distance d, then we can construct disjoint Hamming balls $\mathrm{S}_2^n(\mathbf{c}, t)$ of radius $t = \lfloor (d-1)/2 \rfloor$ around every codeword $\mathbf{c} \in C$ (considered as points in n-dimensional space). Each such sphere contains

$$\mathrm{V}_2^n(t) = \sum_{i=0}^{t} \binom{n}{i} \tag{2.10}$$

points in $\mathbb{F}_2^n$. Since 2^n points in total exist in the entire space $\mathbb{F}_2^n$, we conclude that

$$|C| \leq \frac{2^n}{\mathrm{V}_2^n(t)}, \tag{2.11}$$

which is called the *binary Hamming bound* or *sphere packing bound*. This bound establishes an upper bound for $M = |C|$ such that the code $(n, M, d)_2$ exists.

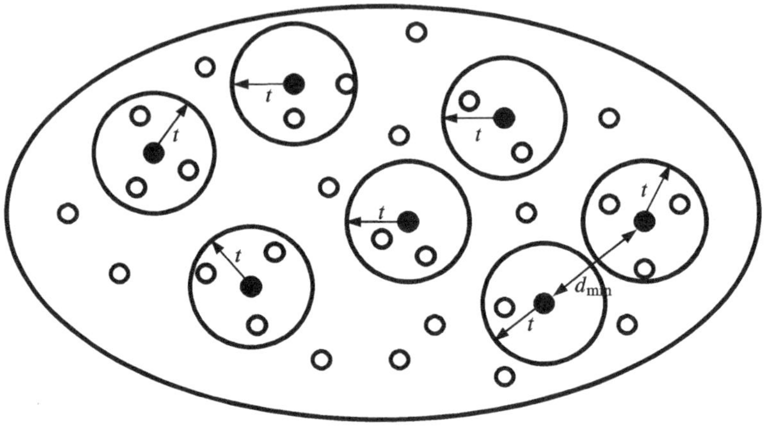

Figure 2.1 An n-dimensional space containing all n-tuples, including codewords.

If the equality in (2.11) holds for a code, i.e. $M = 2^n/V_2^n(t)$, then the code is called *perfect code*.

Figure 2.1 illustrates a code of size $M = |C|$ as a collection of M spheres all with radius t, where each sphere is centered around one of M codewords. In this figure, the open dots represent non-codewords of length n.

Let us introduce another bound on the minimum distance of a code, called the *Singleton bound* as follows:

$$d \leq n - k + 1. \tag{2.12}$$

Codes that satisfy this bound with equality are called *maximum distance separable (MDS) codes.*

Now, we turn our attention to the probability of error. Assuming that error patterns $\mathbf{e} \in \mathbb{F}_2^n$ are equally likely, i.e. $P_e = 2^{-n}$, the probability of error detection failure is

$$P_{\mathrm{df}} = \frac{2^k - 1}{2^n} \approx 2^{-(n-k)},$$

which decreases exponentially with increasing redundancy $n - k$.

In most applications, the error patterns $\mathbf{e}$ are not equally likely. In particular, if the error patterns $\mathbf{e} = (e_1, \dots, e_n)$ are generated by a biased coin with $P(e_i = 1) = p$ and $P(e_i = 0) = 1 - p$ then

$$P_e = (1 - p)^{n-w} p^w,$$

is the probability of occurrence of an error pattern $\mathbf{e}$ of weight $w = \mathrm{w}_H(\mathbf{e})$, i.e. the number of nonzero entries of $\mathbf{e}$. This error pattern distribution results from transmission over a binary symmetric channel (BSC) with a transition probability p, see Section 1.2.1. If $p = 0.5$, the error pattern distribution reduces to uniform but for $p < 0.5$, some error patterns are more likely than others. In particular, the error patterns with higher Hamming weight are less likely to occur, i.e.

$$(1-p)^{n-w_1}p^{w_1} > (1-p)^{n-w_2}p^{w_2} \quad \text{for } w_1 < w_2.$$

In this case, the probability of an error detection failure is

$$P_{\text{df}} = \sum_{w \in \mathcal{W}_C} A_w (1-p)^{n-w} p^w, \tag{2.13}$$

where $\mathcal{W}_C$ is the set of Hamming weights of the non-zero codewords of C and A_w the number of codewords of weight w.

We define the *weight enumerating function* (WEF) of a code C as the polynomial in the variable D

$$A_C(D) = \sum_{w=0}^{n} A_w D^w,$$

where A_w is the multiplicity of the codewords of weight w. The term $A_{w_{\text{min}}}$ is the multiplicity of the nonzero minimum weight codewords and is also known as the *error coefficient*. In the error detection failure expression (2.13), for $p < 0.5$, the dominant term is the one corresponding to w_{min}.

It is trivial to see that for the (n, n) rate 1 code $C = \mathbb{F}_2^n$

$$A_C(D) = \sum_{w=0}^{n} \binom{n}{w} D^w.$$

The WEF of a code is only known explicitly in a few rare cases (e.g. for the Hamming code) and can be evaluated in general by exhaustive enumeration of all codewords. Evaluating the WEF at $D = \frac{p}{1-p}$ we can rewrite (2.13) as

$$P_{\text{df}} = \frac{1}{1-p} A_C \left(\frac{p}{1-p} \right).$$

2.3 Syndrome Decoding

Redundancy enables not only the detection of errors but also the correction of some error patterns, thus avoiding the need for retransmission of the codeword. In the following, we will show using an example how this feature can be realized, specifically using the actual value of the non-zero syndrome $\mathbf{s} = \mathbf{r} \cdot \mathbf{H}^T$, which signaled the presence of an error in the received vector $\mathbf{r} = \mathbf{c} + \mathbf{e}$.

Example 2.2 Let us revisit Example 2.1 and consider the (6,3) code generated by

$$\mathbf{G} = \begin{pmatrix} 1 & 0 & 0 & 1 & 0 & 1 \\ 0 & 1 & 0 & 0 & 1 & 1 \\ 0 & 0 & 1 & 1 & 1 & 0 \end{pmatrix}. \tag{2.14}$$

We list the $2^3 = 8$ codewords in the table below

$\mathbf{u}$	$\mathbf{c} = \mathbf{uG}$
000	000000
001	001110
010	010011
011	011101
100	100101
101	101011
110	110110
111	111000

Figure 2.2 illustrates the standard array. This array is used to list all possible received vectors $\mathbf{x}$ and their corresponding syndromes, organized by increasing error likelihood. The first column contains all eight codewords (valid $\mathbf{c} = \mathbf{uG}$), and each subsequent column represents received vectors that differ from codewords by single or multiple-bit errors.

The six more likely single-error patterns (with dark-gray background) have distinct syndromes. The 15 less-likely double error patterns (with light-gray background) cannot be distinguished by their syndrome. Syndrome 111 does not correspond to any single error pattern and cannot be used to decide among the three possible double-error patterns 001001, 010100, 100010, which have the same probability of occurrence.

	$\mathbf{s} = \mathbf{rH}^T = (\mathbf{c}+\mathbf{e})\mathbf{H}^T = \mathbf{eH}^T$						
000	001	010	011	100	101	110	111
000000	000001	000010	000011	000100	000101	000110	000111
001110	001111	001100	001101	001010	001011	001000	001001
010011	010010	010001	010000	010111	010110	010101	010100
011101	011100	011111	011110	011001	011000	011011	011010
100101	100100	100111	100110	100001	100000	100011	100010
101011	101010	101001	101000	101111	101110	101101	101100
110110	110111	110100	110101	110010	110011	110000	110001
111000	111001	111010	111011	111100	111101	111110	111111
$\mathbf{c} = \mathbf{uG}$	$\mathbf{r} = \mathbf{c} + \mathbf{e} \rightarrow \hat{\mathbf{c}} = \mathbf{r} + \hat{\mathbf{e}}$						

Figure 2.2 The standard array of the (6,3) code.

2.4 Repetition and Single Parity Check Codes

Repetition (REP) codes and single parity check (SPC) codes are two simple types of linear block codes. A binary repetition code C_{REP} of length n, denoted by $(n, 1)$, is a linear code with a single information bit in $\{0, 1\}$, which is obtained by repeating this information bit n times, consisting of one information bit and $n - 1$ identical parity check bits, which are copies of the information bit. As a result, it contains two codewords: the all-zero codeword, $\mathbf{0} = (00 \ldots 0)$, and the all-one codeword $\mathbf{1} = (11 \ldots 1)$. Therefore, the generator matrix of a repetition code of length n is a single row as

$$\mathbf{G}_{\text{REP}} = [\mathbf{I} \,|\, \mathbf{P}] = [1\ 1\ \ldots\ 1]_{1 \times n}.$$

A SPC code C_{SPC} of length n is a binary $(n, n-1)$ linear code. Each SPC codeword consists of $n - 1$ information bits and a SPC bit. Observe that this arrangement is the opposite of what we have in repetition codes. Given the information bit sequence $\mathbf{u} = \left(u_0\, u_1\ \ldots\ u_{n-2}\right)$, then the modulo-2 sum of the $n - 1$ information bits, $p = u_0 + u_1 + \cdots + u_{n-2}$ as a parity check bit, is appended to the sequence to form an n-bit codeword $\left(u_0\, u_1\ \ldots\ u_{n-2}\, p\right)$. Observe that adding the single-parity bit p results in codewords for C_{SPC} with even weight. Therefore, the codewords consist of all binary n-tuples with even weight, resulting in a minimum distance of $d = 2$ for SPC codes. Therefore, any error pattern of even weight results in another codeword, i.e. given $\mathbf{c} \in C_{\text{SPC}}$ and $\text{w}(\mathbf{e}) \bmod 2 = 0$, then we have $\mathbf{c} + \mathbf{e} \in C_{\text{SPC}}$. This implies that C_{SPC} is not capable of detecting error patterns with odd weight. Nevertheless, the capability of detecting odd-weight error patterns has made SPC codes popular in communications systems.

The generator matrix of an $(n, n-1)$ SPC code in systematic form is given as follows:

$$\mathbf{G}_{\text{SPC}} = [\mathbf{I}_{n-1} \,|\, \mathbf{1}] = \begin{bmatrix} 1 & 0 & \cdots & 0 & 0 & 1 \\ 0 & 1 & \cdots & 0 & 0 & 1 \\ \vdots & \vdots & \ddots & \vdots & \vdots & \vdots \\ 0 & 0 & \cdots & 1 & 0 & 1 \\ 0 & 0 & \cdots & 0 & 1 & 1 \end{bmatrix}_{(n-1) \times n} . \tag{2.15}$$

Note that the $(n, 1)$ REP code and the $(n, n-1)$ SPC code are dual codes, hence, $\mathbf{G}_{\text{REP}}\mathbf{G}_{\text{SPC}}^T = \mathbf{0}$ and $\mathbf{G}_{\text{SPC}}\mathbf{G}_{\text{REP}}^T = \mathbf{0}$.

2.5 Reed–Muller Codes

Reed–Muller (RM) codes are a class of linear codes with an elegant structure that is easy to describe. For a given integer m, we can construct RM codes $\mathcal{R}(r, m)$ of block length $n = 2^m$ and *order* r for $0 \leq r \leq m$. Among these, we have three basic codes:

1) $\mathcal{R}(0, m) = \{\mathbf{0}, \mathbf{1}\}$ is the repetition code of length 2^m, where $\mathbf{0}$ is the all-zeros codeword and $\mathbf{1}$ is the all-ones codeword.
2) $\mathcal{R}(m, m) = \{0, 1\}^{2^m}$ is a (useless) Rate-1 code consisting of all binary sequences of length 2^m.
3) $\mathcal{R}(m-1, m)$ is a SPC code that is between the two extreme orders $r = 0$ and $r = m$.

These codes are known to be easy to decode thanks to their structure. As the order r increases (for fixed m), the minimum distance decreases exponentially. First-order RM codes actually have a better minimum distance than higher-order ones. There are three main approaches for constructing their generator matrices in a non-systematic form. We discuss each one of them in the following sections.

2.5.1 Boolean Functions and Boolean Polynomials

RM codewords can be described by multivariate Boolean polynomials. To define RM codes $\mathcal{R}(r, m)$ of length $n = 2^m$, we use m variables $x_0, \ldots, x_{m-1}$ that can take values in $\mathbb{Z}_2 = \{0, 1\}$ and the m-variate monomials are defined as

$$\mathbf{x}^{\mathbf{i}} \triangleq \prod_{j=0}^{m-1} x_j^{i_j} = x_0^{i_0} \cdots x_{m-1}^{i_{m-1}},$$

where $\mathbf{i} = (i_0, \ldots, i_{m-1})$ for $i_j \in \{0, 1\}, j \in [0, m-1]$ and the monomials are of degrees $\deg(\mathbf{x}^{\mathbf{i}}) = i_0 + i_1 + \cdots + i_{m-1} \leq r$. The set of all m-variable monomials with degrees up to r forms the generating set of code $\mathcal{R}(r, m)$, also known as *basis monomials*.

Example 2.3 Given the code $\mathcal{R}(3, 2)$, the generating set of this code is

$$\{1, x_0, x_1, x_2, x_0x_1, x_0x_2, x_1x_2\}$$

As we will see in Chapter 7, the generating set of polar codes is a subset of the RM code's generating set, satisfying some additional properties.

2.5.1.1 Evaluation of Monomials

The evaluation of monomial $\mathbf{x}^{\mathbf{i}}$, $\mathrm{ev}(\mathbf{x}^{\mathbf{i}})$, is the binary vector with length 2^m obtained by evaluating $\mathbf{x}^{\mathbf{i}}$ over all 2^m binary variables entries in $\{0, 1\}^m$, i.e.

$$\mathrm{ev}(\mathbf{x}^{\mathbf{i}}) \triangleq \left(\mathrm{ev}(\mathbf{x}^{\mathbf{i}})(0, \ldots, 0, 0), \ldots, \mathrm{ev}(\mathbf{x}^{\mathbf{i}})(1, \ldots, 1, 1)\right).$$

We use binary functions to represent this evaluation, where $f_{\mathbf{i}}(x_0, \ldots x_{m-1}) = x_0^{i_0} \cdots x_{m-1}^{i_{m-1}}$ maps a binary m-tuple $(x_0, \ldots, x_{m-1})$ to a binary number $\{0, 1\}$. Then, a binary word $\mathbf{V}$ of length 2^m corresponding to some monomial $f_{\mathbf{i}}$ can be represented as

$$\mathbf{V} = \left(v_0 \; v_1 \; \ldots \; v_{2^m-1}\right),$$

where

$$v_\ell = f_{\mathbf{i}}(x_0, x_1, \ldots, x_{m-1}) = \begin{cases} 1, & \text{if } x_j = 1 \text{ for all } j \in \text{supp}(\mathbf{i}), \\ 0, & \text{otherwise.} \end{cases} \tag{2.16}$$

and $0 \leq \ell \leq 2^m - 1$ is the decimal value corresponding to the binary m-tuple $(x_0, x_1, \ldots, x_{m-1})$.

Example 2.4 Consider the monomial $\mathbf{x}^{\mathbf{i}} = x_0 x_2$ and $m = 3$. Observe that we have $\mathbf{i} = (i_0, i_1, i_2) = (1, 0, 1)$ and $\text{supp}(\mathbf{i}) = \{0, 2\}$. To find the corresponding binary word $\mathbf{V} = (v_0 v_1 \ldots v_{2^m-1})$, we find its elements by following (2.16) as below:

$$\begin{aligned} v_0 &= f(0,0,0) = 0, & v_1 &= f(1,0,0) = 0, \\ v_2 &= f(0,1,0) = 0, & v_3 &= f(1,1,0) = 0, \\ v_4 &= f(0,0,1) = 0, & v_5 &= f(1,0,1) = 1 \\ v_6 &= f(0,1,1) = 0, & v_7 &= f(1,1,1) = 1 \end{aligned}$$

According to (2.16), we have $f(1, i_1, 1) = 1$ for $i_1 \in \{0, 1\}$, otherwise 0.

Alternatively, we can evaluate this monomial by computing $x_0 x_2$ for all possible binary combinations of (x_0, x_1, x_2) as follows:

$$\begin{aligned} (x_0, x_1, x_2) &= (0,0,0) \Rightarrow x_0 x_2 = 0 \cdot 0 = 0 \\ (x_0, x_1, x_2) &= (1,0,0) \Rightarrow x_0 x_2 = 1 \cdot 0 = 0 \\ (x_0, x_1, x_2) &= (0,1,0) \Rightarrow x_0 x_2 = 0 \cdot 0 = 0 \\ (x_0, x_1, x_2) &= (1,1,0) \Rightarrow x_0 x_2 = 1 \cdot 0 = 0 \\ (x_0, x_1, x_2) &= (0,0,1) \Rightarrow x_0 x_2 = 0 \cdot 1 = 0 \\ (x_0, x_1, x_2) &= (1,0,1) \Rightarrow x_0 x_2 = 1 \cdot 1 = 1 \\ (x_0, x_1, x_2) &= (0,1,1) \Rightarrow x_0 x_2 = 0 \cdot 1 = 0 \\ (x_0, x_1, x_2) &= (1,1,1) \Rightarrow x_0 x_2 = 1 \cdot 1 = 1 \end{aligned}$$

Thus, the result of the evaluation is

$$\text{ev}\left(x_0 x_2\right) = (0\;0\;0\;0\;0\;1\;0\;1).$$

Example 2.5 Let $m = 2$. Evaluation of all possible monomials up to degree 2, $\text{ev}(x_0^{i_0} x_1^{i_1})$, for all binary entries results in rows of matrix $\boldsymbol{G}_{2^m} = [\mathbf{g}_0\ \mathbf{g}_1\ \mathbf{g}_2\ \mathbf{g}_3]^T$ as follows.

	$\text{ev}(x_0^{i_0} x_1^{i_1})$	$f(1,1)$	$f(0,1)$	$f(1,0)$	$f(0,0)$	Where is $v_{\mathbf{i}} = 1$?
$\mathbf{g}_0$	$\text{ev}\left(x_0x_1\right)$	1	0	0	0	$f(1\,1) = 1$
$\mathbf{g}_1$	$\text{ev}\left(x_1\right)$	1	1	0	0	$f(i_0\,1) = 1$
$\mathbf{g}_2$	$\text{ev}\left(x_0\right)$	1	0	1	0	$f(1\,i_1) = 1$
$\mathbf{g}_3$	$\text{ev}(1)$	1	1	1	1	$f(i_0\,i_1) = 1$

where, for instance, Boolean function $f(1\,i_1)$ gives 1 for all entries with coordinates $(1, i_1), i_1 \in \{0, 1\}$ and 0 otherwise.

2.5.1.2 Boolean Products

Given the evaluation of monomials of degree 0 as $\mathbf{V}_0 = \text{ev}(\mathbf{1})$, i.e. all-one vector, and degree 1 as $\mathbf{V}_{\ell+1} = \text{ev}(x_\ell), \ell \in [0, m-1]$, the evaluation of monomials of degree 2 and greater is found using Boolean products $\mathbf{V}_1 \cdot \mathbf{V}_2$, $\mathbf{V}_1 \cdot \mathbf{V}_3, \ldots, \mathbf{V}_1 \cdots \mathbf{V}_{m-1}$. The products are performed component-wise by the logic "AND" operation. The set of all these vectors forms the basis vectors. That is, RM codes can be generated using the following basis vectors:

$$\mathcal{R}(r, n) = \left\{\mathbf{V}_0, \mathbf{V}_1, \ldots, \mathbf{V}_m, \mathbf{V}_1 \cdot \mathbf{V}_2, \ldots, \mathbf{V}_{m-1} \cdot \mathbf{V}_m, \ldots, \text{higher products}\right\}.$$

Example 2.6 For $m = 3$, we have three variables x_0, x_1, and x_2, and the length of each evaluation vector will be $2^3 = 8$. Evaluating each single-variable monomial over all inputs yields:

$$\mathbf{V}_1 = \text{ev}\left(x_0\right) = (0\ 0\ 0\ 0\ 1\ 1\ 1\ 1) \tag{2.17}$$

$$\mathbf{V}_2 = \text{ev}\left(x_1\right) = (0\ 0\ 1\ 1\ 0\ 0\ 1\ 1) \tag{2.18}$$

$$\mathbf{V}_3 = \text{ev}\left(x_2\right) = (0\ 1\ 0\ 1\ 0\ 1\ 0\ 1) \tag{2.19}$$

Now, we can construct the evaluation of higher-degree monomials by multiplying evaluation vectors of single variable monomials as follows.

$$\mathbf{V}_1 \cdot \mathbf{V}_2 = \text{ev}\left(x_0x_1\right) = \text{ev}\left(x_0\right) \cdot \text{ev}\left(x_1\right) = (0\ 0\ 0\ 0\ 0\ 0\ 1\ 1) \tag{2.20}$$

$$\mathbf{V}_1 \cdot \mathbf{V}_3 = \text{ev}\left(x_0x_2\right) = \text{ev}\left(x_0\right) \cdot \text{ev}\left(x_2\right) = (0\ 0\ 0\ 0\ 0\ 1\ 0\ 1) \tag{2.21}$$

$$\mathbf{V}_2 \cdot \mathbf{V}_3 = \text{ev}\left(x_1x_2\right) = \text{ev}\left(x_1\right) \cdot \text{ev}\left(x_2\right) = (0\ 0\ 0\ 1\ 0\ 0\ 0\ 1) \tag{2.22}$$

2.5.1.3 Boolean Polynomials

While every monomial is a codeword, the summation of monomials resulting in polynomials is also a codeword. These are equivalent to the single rows and sum of the rows of the code's generator matrix, respectively.

The resulting RM code $\mathcal{R}(r,m)$ of order r is characterized by the code parameters of

$$n = 2^m,$$
$$k = \sum_{r=0}^{r} \binom{m}{r} = 1 + \binom{m}{1} + \binom{m}{2} + \cdots + \binom{m}{r},$$
$$d = 2^{m-r}.$$

where the code dimension k is equal to the number of distinct monomials of degree $\leq r$.

2.5.2 Submatrix of Binary Walsh–Hadamard Matrix

The concept of order r in the $\mathcal{R}(r,m)$ code comes from the Boolean products of the 2^m-tuple $\mathbf{V}_{i+1} = \mathbf{g}_{2^m-2^i-1}, i \in [0, m-1]$ up to degree r, where $\mathbf{g}_j$ is the jth row of the matrix $\boldsymbol{G}_{2^m}$ constructed by the mth Kronecker power of the binary Walsh–Hadamard matrix $\mathbf{G}_2 = \begin{bmatrix} 1 & 0 \\ 1 & 1 \end{bmatrix}$, that is, $\mathbf{G}_{2^m} = \mathbf{G}_2^{\otimes m}$. We can write the $\mathbf{G}_{2^m}$ transform by rows as

$$\mathbf{G}_{2^m} = \begin{bmatrix} \mathbf{g}_0 \\ \mathbf{g}_1 \\ \vdots \\ \mathbf{g}_{2^m-1} \end{bmatrix}. \tag{2.23}$$

By default, $\mathbf{V}_0 = \mathbf{g}_{2^m-1} = [1\ 1\ \ldots\ 1]$. For instance, when $m = 3$, we obtain

$$\mathbf{V}_0 = \mathbf{g}_7 = [1\ 1\ 1\ 1\ 1\ 1\ 1\ 1],$$
$$\mathbf{V}_1 = \mathbf{g}_6 = [1\ 0\ 1\ 0\ 1\ 0\ 1\ 0],$$
$$\mathbf{V}_2 = \mathbf{g}_5 = [1\ 1\ 0\ 0\ 1\ 1\ 0\ 0],$$
$$\mathbf{V}_3 = \mathbf{g}_3 = [1\ 1\ 1\ 1\ 0\ 0\ 0\ 0].$$

The vectors $\mathbf{V}_i, i \in [0, 3]$ in the example above form the generator matrix of $\mathcal{R}(1,3)$. As an example for order 2, the generator matrix for $\mathcal{R}(2,3)$ is given by

$$\mathbf{G}_{\mathrm{RM}(2,3)} = \begin{bmatrix} \mathbf{g}_1 \\ \mathbf{g}_2 \\ \mathbf{g}_3 \\ \mathbf{g}_4 \\ \mathbf{g}_5 \\ \mathbf{g}_6 \\ \mathbf{g}_7 \end{bmatrix} = \begin{bmatrix} \mathbf{V}_2 \cdot \mathbf{V}_3 \\ \mathbf{V}_1 \cdot \mathbf{V}_3 \\ \mathbf{V}_3 \\ \mathbf{V}_1 \cdot \mathbf{V}_2 \\ \mathbf{V}_2 \\ \mathbf{V}_1 \\ \mathbf{V}_0 \end{bmatrix} = \begin{bmatrix} 1&1&0&0&0&0&0&0 \\ 1&0&1&0&0&0&0&0 \\ 1&1&1&1&0&0&0&0 \\ 1&0&0&0&1&0&0&0 \\ 1&1&0&0&1&1&0&0 \\ 1&0&1&0&1&0&1&0 \\ 1&1&1&1&1&1&1&1 \end{bmatrix}, \tag{2.24}$$

which has rows with minimum Hamming weight $2^{m-r} = 2^{3-2} = 2$. It can be seen that the generator matrix of $\mathcal{R}(m,m)$ is $\boldsymbol{G}_{2^m}$. In the above example, only $\mathbf{g}_0$ of $\mathbf{G}_8$ that has weight 1 is not included in (2.24).

The basis vectors for the rth order RM code $\mathcal{R}(r, 3)$ are:

Order r	Rows of $\boldsymbol{G}_{2^m}$	Minimum row weight, 2^{3-r}
0	$\mathbf{g}_7$	8
1	$\mathbf{g}_3, \mathbf{g}_5, \mathbf{g}_6, \mathbf{g}_7$	4
2	$\mathbf{g}_1, \dots, \mathbf{g}_7$	2
3	all rows of $\boldsymbol{G}_{2^m}$	1

Therefore, to form the generator matrix of the $\mathcal{R}(r,m)$ code based on the Walsh–Hadamard binary matrix, we can select the rows $\mathbf{g}_i$ of $\mathbf{G}_2^{\otimes m}$ with weights $w(\mathbf{g}_i) \geq 2^{m-r}$. This approach based on row weights is called the *RM rule*. We will learn later that the generator matrix of polar codes is also formed based on the binary Walsh–Hadamard matrix, but the row selection rule is different.

2.5.3 Plotkin's Construction

RM codes of length 2^{m+1} can be simply obtained from RM codes of length 2^m using the concatenation of u and $u + v$ known as *$u|u + v$ construction*, which is defined as follows:

$$\begin{aligned}\mathcal{R}(r,m) &= \{u|u+v : u \in \mathcal{R}(r,m-1), v \in \mathcal{R}(r-1,m-1)\}\\ &= \mathcal{R}(r,m-1) * \mathcal{R}(r-1,m-1).\end{aligned}$$

For the recursive construction of RM codes, we need to define the first few RM codes as follows:

$\mathcal{R}(0,0) = \{0,1\}^1 = \{0,1\}$, a rate-1 code including all sequences of length $2^0 = 1$,
$\mathcal{R}(0,1) = \{00,11\}$, a repetition code of length $2^1 = 2$,
$\mathcal{R}(1,1) = \{0,1\}^2 = \{00,01,10,11\}$, a rate-1 code including all sequences of length $2^1 = 2$,
$\mathcal{R}(0,2) = \{0000,1111\}$, a repetition code of length $2^2 = 4$.
$\mathcal{R}(2,2) = \{0,1\}^4$, a rate-1 code including all sequences of length $2^2 = 4$,

Then, by Plotkin's construction, we can recursively obtain the following codes:

$\mathcal{R}(1,2) = \mathcal{R}(1,1) * \mathcal{R}(0,1)$,
$\mathcal{R}(1,3) = \mathcal{R}(1,2) * \mathcal{R}(0,2)$,
$\mathcal{R}(2,3) = \mathcal{R}(2,2) * \mathcal{R}(1,2) = \mathcal{R}(2,2) * (\mathcal{R}(1,1) * \mathcal{R}(0,1))$, a SPC code.

and this can continue.

In Table 2.1, the rows with the minimum distance $d = 1$ and 2 include the Rate-1 codes and the SPC codes, respectively. The repetition codes have the dimension $k = 1$, which are located on the diagonal of the table.

Table 2.1 The $(2^m, k, d)$-RM codes defined over $m \in [0, 9]$ with all possible dimensions k.

m	0	1	2	3	4	5	6	7	8	9
Length 2^m	1	2	4	8	16	32	64	128	256	512
Distance d					Dimension k					
1	1	2	4	8	16	32	64	128	256	512
2		1	3	7	15	31	63	127	255	511
4			1	4	11	26	57	120	247	502
8				1	5	16	42	99	219	466
16					1	6	22	64	163	382
32						1	7	29	93	256
64							1	8	37	130
128								1	9	46
255									1	10
512										1

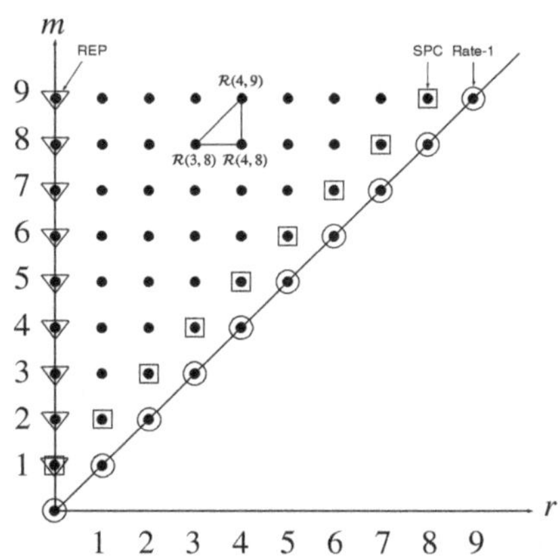

Figure 2.3 RM codes of $\mathcal{R}(r, m)$ for $r, m \in [0, 9]$. By forming right triangles, such as $\mathcal{R}(4, 9)$, we can find the corresponding component codes in Plotkin's construction.

Figure 2.3 illustrates these codes in terms of order r and length m. As we will see in Chapter 5, these component codes have been used in the simplification of successive cancellation decoding of polar codes.

Example 2.7 Give the recursive construction of the $\mathcal{R}(3, 5)$ code down to trivial codes: REP codes and SPC codes.

$$\begin{aligned}\mathcal{R}(3,5) &= \mathcal{R}(3,4) * \mathcal{R}(2,4)\\ &= \mathcal{R}(3,4) * (\mathcal{R}(2,3) * \mathcal{R}(1,3))\\ &= \mathcal{R}(3,4) * (\mathcal{R}(2,3) * (\mathcal{R}(1,2) * \mathcal{R}(0,2))),\end{aligned}$$

Equivalently, this construction can be expressed in terms of generator matrices. Let $\boldsymbol{G}_{\mathcal{R}(r,\ m)}$ be a generator matrix for $\mathcal{R}(r,m)$. This generator matrix is obtained from the generator matrices of codes $\mathcal{R}(r,m-1)$ and $\mathcal{R}(r-1,m-1)$ as follows:

$$\boldsymbol{G}_{\mathcal{R}(r,m)} = \begin{pmatrix} \boldsymbol{G}_{\mathcal{R}(r,\ m-1)} & \boldsymbol{G}_{\mathcal{R}(r,\ m-1)} \\ 0 & \boldsymbol{G}_{\mathcal{R}(r-1,\ m-1)} \end{pmatrix}.$$

Example 2.8 Example: Construct the generator matrix of the $\mathcal{R}(1,3)$ code.
$\mathcal{R}(1,3) = (\mathcal{R}(1,2) * \mathcal{R}(0,2))$

$$\boldsymbol{G}_{\mathcal{R}(1,\ 3)} = \begin{pmatrix} \boldsymbol{G}_{\mathcal{R}(1,\ 2)} & \boldsymbol{G}_{\mathcal{R}(1,\ 2)} \\ 0 & \boldsymbol{G}_{\mathcal{R}(0,\ 2)} \end{pmatrix}.$$

where

$$\boldsymbol{G}_{\mathcal{R}(0,\ 2)} = \mathbf{G}_{\text{REP}} = [\ 1\ 1\ 1\ 1\ ],$$

and

$$\boldsymbol{G}_{\mathcal{R}(1,\ 2)} = \mathbf{G}_{\text{SPC}} = [\mathbf{I}_3 \mid \mathbf{1}] = \begin{bmatrix} 1\ 0\ 0\ 1 \\ 0\ 1\ 0\ 1 \\ 0\ 0\ 1\ 1 \end{bmatrix}. \tag{2.25}$$

Therefore, we have

$$\boldsymbol{G}_{\mathcal{R}(1,\ 3)} = \begin{bmatrix} 1\ 0\ 0\ 1\ 1\ 0\ 0\ 1 \\ 0\ 1\ 0\ 1\ 0\ 1\ 0\ 1 \\ 0\ 0\ 1\ 1\ 0\ 0\ 1\ 1 \\ 0\ 0\ 0\ 0\ 1\ 1\ 1\ 1 \end{bmatrix}. \tag{2.26}$$

Observe that the matrix $\boldsymbol{G}^T_{\mathcal{R}(1,\ 3)}$ is a submatrix of $\left(\boldsymbol{G}_2^{\otimes 3}\right)^T$ that includes the rows $\mathbf{g}_i, i \in [0,7]$ with weight $\mathrm{w}(\mathbf{g}_i) \geq 2^{3-1}$.

Note that we can also use $\boldsymbol{G}_{\mathcal{R}(r,\ m)}$ with permuted rows. Row permutation simply reorders the basis vectors of the code. This changes the order in which codewords appear when forming linear combinations of rows, but it does not alter the linear span of the rows. Consequently, the set of codewords in the code remains unchanged; what changes is only the correspondence between the k information bits and the particular basis vectors (rows) of the generator matrix.

A decoding algorithm for RM codes called recursive decoding will be discussed in Chapter 3.

2.6 Cyclic Codes

Cyclic codes are a class of linear block codes that can be easily implemented using sequential logic or shift registers and are commonly used as an error detection tool. Given a vector with n components,

$$\mathbf{c} = \left(c_0, c_1, \ldots, c_{n-1}\right),$$

performing a cyclic right-shift rotation by i positions results in the new vector:

$$\mathbf{c}^{(i)} = \left(c_{n-i}, c_{n-i+1}, \ldots, c_{n-1}, c_0, c_1, \ldots, c_{n-i-1}\right).$$

A linear block code is categorized as a cyclic code when every code vector it contains, upon undergoing the ith cyclic rotation, yields another valid code vector within the same code. Additionally, recall since it is a linear block code, the sum of any two code vectors from a cyclic code is still a code vector.

In a given cyclic code $C_{\text{cyc}}(n, k)$, codewords can be efficiently represented using polynomials. These polynomials are defined within a Galois field $\mathbb{F}_2$. In this book, we focus on the fields of characteristic 2 as $\mathbb{F}_2$.

A polynomial is an expression that involves a variable X, comprising monomials of the form $c_i X^i$, where the coefficients c_i belong to a field $\mathbb{F}_2$. The polynomial representation $c(X)$ of a code vector $\mathbf{c} = \left(c_0, c_1, \ldots, c_{n-1}\right)$ follows the form:

$$c(X) = c_0 + c_1 X + \cdots + c_{n-1} X^{n-1} \quad c_i \in \mathbb{F}_2.$$

2.6.1 Arithmetic of Polynomials

Two polynomials defined over GF(2), $c_1(X)$ and $c_2(X)$, are added using the modulo-2 addition of the coefficients corresponding to the same term X^i. Given

$$c_1(X) = c_{01} + c_{11} X + \cdots + c_{n-1,1} X^{n-1}$$

and

$$c_2(X) = c_{02} + c_{12} X + \cdots + c_{n-1,2} X^{n-1},$$

then the addition of these two polynomials equals

$$c_1(X) \oplus c_2(X) = \left(c_{01} \oplus c_{02}\right) + \left(c_{11} \oplus c_{12}\right) X + \cdots + \left(c_{n-1,1} \oplus c_{n-1,2}\right) X^{n-1}. \tag{2.27}$$

When the degrees of the polynomials are not identical, we also use (2.27) for the addition. Similarly, the multiplication of two polynomials $c_1(X)$ and $c_2(X)$, $c_1(X) \bullet c_2(X)$, is performed as

$$c_1(X) \bullet c_2(X) = c_{01} \bullet c_{02} + \left(c_{01} \bullet c_{12} \oplus c_{02} \bullet c_{11}\right) X + \cdots + \left(c_{n-1,1} \bullet c_{n-1,2}\right) X^{2(n-1)}, \tag{2.28}$$

where all additions and multiplications are modulo-2 and they follow the commutative, associative and distributive laws. The division of two polynomials $c_1(X)$ and $c_2(X)$, where $c_2(X) \neq 0$ is a nonzero polynomial, is defined as:

$$c_1(X) = q(X)\ c_2(X) + r(X),$$

where $q(X)$ and $r(X)$ are two unique polynomials, called the *quotient* and the *remainder*, respectively.

2.6.2 Generator Polynomial and Encoding

The cyclic right-shift of a code vector c by i positions is given by the following polynomial expression:

$$c^{(i)}(X) = c_{n-i} + c_{n-i+1}X + \cdots + c_{n-i-1}X^{n-1}. \tag{2.29}$$

It can be easily verified that the relationship between $c^{(i)}(X)$ and the original code $c(X)$ is of the form:

$$X^i c(X) = q(X)\left(X^n + 1\right) + c^{(i)}(X) \tag{2.30}$$

for some $q(X)$, or equivalently

$$c^{(i)}(X) = X^i c(X) \bmod \left(X^n + 1\right), \tag{2.31}$$

where the mod operator gives the remainder of the division of $X^i c(X)$ and $X^n + 1$.

The nonzero minimum-degree code polynomial $g(X)$ of a given cyclic code $C_{cyc}(n, k)$ is unique because the addition of two code polynomials of some minimum degree r would result in a polynomial of lower degree $\leq (r - 1)$, which contradicts the statement that the minimum degree is r. We observe that polynomials of the form $Xg(X), X^2g(X), \ldots, X^{n-r-1}g(X)$ all have a degree less than n. Therefore, the division of these polynomials by the polynomial $X^n + 1$ will result in a quotient equal to zero, $q(X) = 0$ in (2.30). That is, the cyclic right-shift of $g(X)$ is code polynomials $g^{(i)}(X)$ as

$$X^i g(X) = g^{(i)}(X), \quad 1 \leq i \leq n - r - 1. \tag{2.32}$$

Furthermore, linear combinations of code polynomials $g^{(i)}(X)$ are also code polynomials. That is,

$$\begin{aligned} c(X) &= m_0 g(X) + m_1 X g(X) + \cdots + m_{n-r-1} X^{n-r-1} g(X), \\ c(X) &= \underbrace{\left(m_0 + m_1 X + \cdots + m_{n-r-1} X^{n-r-1}\right)}_{m(X)} g(X), \end{aligned} \tag{2.33}$$

where $m_i \in \{0, 1\}$ for $i = 0, 1, \ldots, n - r - 1$. Observe that a code polynomial $c(X)$ is a multiple of the nonzero minimum-degree polynomial $g(X)$. We use this property for the encoding and decoding of a cyclic code $C_{cyc}(n,k)$. The polynomial $g(X)$ is called the *generator polynomial* since it fully determines and generates the linear cyclic code $C_{cyc}(n,k)$. As a linear cyclic code $C_{cyc}(n,k)$ generated by $g(X)$ is spanned by the k code polynomials, $g(X), Xg(X), \dots, X^{n-k}g(X)$, which can be represented as row vectors of a generator matrix $\mathbf{G}$ of dimension $k \times n$ as

$$\mathbf{G} = \begin{bmatrix} g_0 & g_1 & g_2 & \cdots & g_{n-k} & 0 & 0 & \cdots & 0 \\ 0 & g_0 & g_1 & \cdots & g_{n-k-1} & g_{n-k} & 0 & \cdots & 0 \\ \vdots & \vdots & & & \vdots & \vdots & & & \vdots \\ 0 & 0 & & \cdots & g_0 & g_1 & g_2 & \cdots & g_{n-k} \end{bmatrix}, \tag{2.34}$$

where $g_0 = g_{n-k} = 1$.

Observe that there are 2^{n-r} possible linear combinations or possible polynomials of degree $n-1$ or less as multiples of $g(X)$. On the other hand, to have a bijective relationship between the message and the coded vector spaces, we need 2^k possible linear combinations. This leads to $2^{n-r} = 2^k$ or $r = n-k$, which means the degree of the nonzero minimum degree polynomial r is also the number of redundancy bits that the code $C_{cyc}(n,k)$ adds to the message vector $\mathbf{m} = [m_0 m_1 \dots m_k]$. As $r = n-k$, we have

$$g(X) = 1 + g_1 X + \cdots + g_{n-k-1} X^{n-k-1} + X^{n-k}.$$

Therefore, to encode k-bit message vector $\mathbf{m}$ or $m(X)$, we multiply it by the generator polynomial $g(X)$ as

$$c(X) = m(X)g(X) = \left(m_0 + m_1 X + \cdots + m_{k-1} X^{k-1}\right) g(X).$$

Note that this encoding procedure is non-systematic.

Example 2.9 Given the generator polynomial $g(X) = 1 + X + X^3$ for the linear cyclic code $C_{cyc}(7,4)$, encode the message vectors $\mathbf{m} = [0\,0\,0\,0], [1\,0\,0\,0], [0\,1\,0\,0]$ and $[1\,1\,0\,0]$ in two forms of codewords and code polynomials.

Ans: Table 2.2 shows the code vectors and the code polynomials resulting from the encoding of the message vectors.

The systematic version of the linear cyclic code $C_{cyc}(n,k)$ is obtained by first forming the polynomial $X^{n-k}m(X) = m_0 X^{n-k} + m_1 X^{n-k+1} + \cdots + m_{k-1} X^{n-1}$, dividing it by the generator polynomial $g(X)$ and taking the reminder $p(X)$ as:

$$X^{n-k}m(X) = q(X)g(X) + p(X). \tag{2.35}$$

Since $p(X)$ is the remainder polynomial of the division by $g(X)$ of degree $n-k$, it has degree $n-k-1$ or less. Rearranging (2.35) yields

$$X^{n-k}m(X) + p(X) = q(X)g(X), \tag{2.36}$$

Table 2.2 Code polynomials of a cyclic code $C_{\text{cyc}}(7,4)$.

Message m	**Code vectors** c	**Code polynomials** $c(X)$
0000	0000000	$0 = 0\,g(X)$
1000	1101000	$1 + X + X^3 = 1\,g(X)$
0100	0110100	$X + X^2 + X^4 = X\,g(X)$
1100	1011100	$1 + X^2 + X^3 + X^4 = (1 + X)\,g(X)$

where the polynomial $X^{n-k}m(X) + p(X)$ is a code polynomial $c(X)$ as it is a factor of $g(X)$. Observe that $X^{n-k}m(X)$ represents the message polynomial right-shifted by $n - k$ positions, whereas $p(X)$ acts as the redundancy polynomial, occupying lower-degree terms up to $n - k + 1$ of the polynomial expression. Therefore, the systematic form of the code polynomial is

$$\begin{aligned} c(X) &= X^{n-k}m(X) + p(X) \\ &= p_0 + p_1X + \cdots + p_{n-k-1}X^{n-k-1} + m_0X^{n-k} + m_1X^{n-k+1} + \cdots + m_{k-1}X^{n-1}, \end{aligned}$$

the codeword form is

$$c = \left(p_0, p_1, \ldots, p_{n-k-1}, m_0, m_1, \ldots, m_{k-1}\right),$$

and the parity polynomial is found using

$$p(X) = \text{Rem}\left(\frac{X^{n-k}m(X)}{g(X)}\right). \tag{2.37}$$

Example 2.10 Suppose we have the message polynomial $m(X) = 1 + X^2$ (with binary vector $(1, 0, 1)$, $k = 3$) and the generator polynomial $g(X) = 1 + X + X^2$ (of degree $r = 2$, so $n - k = 2$). We find the codeword polynomial of length $n = 5$ (since $r = n - k = 2$) as follows. First, we shift the message polynomial: multiply $m(X)$ by $X^{n-k} = X^2$ and we get

$$X^2 \cdot m(X) = X^2 + X^4.$$

Then, we perform polynomial division of $X^2 + X^4$ by $g(X) = 1 + X + X^2$ to find the remainder $p(X)$; i.e.

$$p(X) = \text{Rem}\left(\frac{X^{n-k}m(X)}{g(X)}\right) = \text{Rem}\left(\frac{X^2 + X^4}{1 + X + X^2}\right).$$

The polynomial division is performed as follows: first step: Since $X^4/X^2 = X^2$, we multiply X^2 by $g(X)$:

$$X^2 \cdot \left(1 + X + X^2\right) = X^4 + X^3 + X^2,$$

and subtract it from $m(X)$:

$$\left(X^4 + X^2\right) - \left(X^4 + X^3 + X^2\right) = -X^3 = X^3.$$

Second step: as $X^3/X^2 = X$, we multiply X by $g(X)$:

$$X \cdot \left(1 + X + X^2\right) = X^3 + X^2 + X,$$

followed by subtraction from the remainder of the previous step, X^3:

$$X^3 - \left(X^3 + X^2 + X\right) = -X^2 - X = X^2 + X.$$

Third step: since $X^2/X^2 = 1$, we multiply 1 by $g(X)$:

$$1 \cdot \left(1 + X + X^2\right) = X^2 + X + 1,$$

and then subtract it from the remainder of the previous step, $X^2 + X$:

$$\left(X^2 + X\right) - \left(X^2 + X + 1\right) = -1 = 1.$$

The remainder or the parity polynomial is $p(X) = 1$ and the quotient polynomial is $q(X) = 1 + X + X^2$. Therefore, we have the codeword polynomial

$$c(X) = X^2 m(X) + p(X) = 1 + X^2 + X^4,$$

and the codeword's binary vector

$$c = (1, 0, 1, 0, 1).$$

We can also write the generator matrix for systematic cyclic codes as

$$\mathbf{G} = [\mathbf{I}_k \mid \mathbf{P}_{k\times r}],$$

where r is the degree of the generator polynomial $g(X)$ and the rows $\mathbf{p}_i, i \in [0, k-1]$ of the submatrix $\mathbf{P}_{k\times r} = [\mathbf{p}_{r-1}, \mathbf{p}_{r-2}, \dots, \mathbf{p}_0]^T$ are the coefficients of the polynomial

$$p_i(X) = \text{Rem}\left(\frac{X^{(k+r)-1-i}}{g(X)}\right), \quad i \in [0, k-1], \tag{2.38}$$

where Rem[·] gives the remainder of the polynomial division. Note that the resulting codeword will be in the following form:

$$c = \left(m_{k-1}, \dots, m_1, m_0, p_{n-k-1}, \dots, p_1, p_0\right).$$

In this form, the parity check bits are appended to the message. We can alternatively write $\mathbf{G} = [\mathbf{P}_{k\times r} \mid \mathbf{I}_k]$ to obtain codewords with a bit order similar to the polynomial form.

2.6.3 Cyclic Redundancy Check (CRC) for Error Detection

When a cyclic code is used for error detection, the redundancy bits $p_0, p_1, \dots, p_{n-k-1}$ are called *cyclic redundancy check* bits. The example below demonstrates how to find the CRC bits of a message.

Example 2.11 Consider the message vector $m = [1, 0, 1, 1]$ and the generator polynomial $g(X) = X^3 + X^2 + 1$, represented as $g = [1, 1, 0, 1]$. We calculate the 3 CRC bits that should be appended to the message m as follows:

1) Define the polynomial lengths: The degree of $g(X)$ is 3, so the length of g is $n = 4$. The number of CRC bits, r, is $n - 1 = 3$.
2) Append zeros to the message: Extend m by adding $r = 3$ zeros to create a new vector M:

$$M = [1, 0, 1, 1, \mathbf{0}, \mathbf{0}, \mathbf{0}].$$

3) Perform binary division: We divide M by g using XOR (binary division), working through each bit of M as follows:
   - Step 1: Initial $M = [1, 0, 1, 1, \mathbf{0}, \mathbf{0}, \mathbf{0}]$. Since the first bit is 1, align $g = [1, 1, 0, 1]$ with the first bit of M and XOR:

$$[1, 0, 1, 1, \mathbf{0}, \mathbf{0}, \mathbf{0}] \oplus [1, 1, 0, 1] = [0, 1, 1, 0, 0, 0, 0].$$

   - Step 2: Updated $M = [0, 1, 1, 0, \mathbf{0}, \mathbf{0}, \mathbf{0}]$. The first bit is now 0, so skip to the next bit. This is, shifting to the right to align g with the second 1 in M:

$$[1, 1, 0, \mathbf{0}, \mathbf{0}, \mathbf{0}] \oplus [1, 1, 0, 1] = [0, 0, 0, \mathbf{1}, \mathbf{0}, \mathbf{0}].$$

   - Step 3: Updated $M = [0, 0, 0, 0, \mathbf{1}, \mathbf{0}, \mathbf{0}]$. Observe that there is no 1 within the length of the message m to align g with. Hence, extract the CRC remainder: The final $r = 3$ bits, $[1, 0, 1]$, are the CRC bits for the original message.

$$\begin{array}{r} 1\,0\,1\,1\;\;0\,0\,0 \\ +\;1\,1\,0\,1 \\ \hline 0\,1\,1\,0\;\;0\,0\,0. \\ +\;\;\;1\,1\,0\;\;1 \\ \hline 0\,0\,0\;\;\mathbf{1}\,\mathbf{0}\,\mathbf{0} \end{array}$$

4) Result: The original message $m = [1, 0, 1, 1]$ with the calculated CRC bits $[1, 0, 0]$ appended is:

$$[1, 0, 1, 1, \mathbf{1}, \mathbf{0}, \mathbf{0}].$$

We have provided a Python script to perform the CRC calculation in Appendix C.5.

Note that appending the CRC bits to the message vector can be performed by a systematic form of the generator matrix that can be obtained by operating on the rows of the matrix (2.34).

Now we turn our focus to error detection. Given a vector $r = \left(r_0, r_1, \ldots, r_{n-1}\right)$ received through a noisy channel containing possible errors, its polynomial representation is

$$r(X) = r_0 + r_1X + r_2X^2 + \cdots + r_{n-1}X^{n-1}.$$

To detect a possible error is equivalent to detecting a polynomial that does not belong to the code. To this end, we divide the received polynomial $r(X)$ by $g(X)$ as

$$r(X) = q(X)g(X) + S(X),$$

where the remainder of this division $S(X)$ is a polynomial of degree $n - k - 1$ or less. Since a code polynomial is a multiple of $g(X)$, the received polynomial is a code polynomial if the remainder of the division is zero. Otherwise, the received polynomial does not belong to the code, or the received vector is in error. It is important to recall that error detection will fail whenever the error pattern coincides with one of the 2^k codewords. This procedure is commonly used for detection in the communication system. We will observe in Chapter 6 how a message is cyclically encoded using a generator polynomial before polar encoding.

Example 2.12 Consider the codewords $[1, 0, 1, 1, \mathbf{1}, \mathbf{0}, \mathbf{0}]$ in Example 2.11. This codeword is transmitted over a noisy channel, and the following vector is received, where the fourth element is flipped:

$$[1, 0, 1, 0, \mathbf{1}, \mathbf{0}, \mathbf{0}].$$

Now, to check whether the received vector is a codeword, we repeat the division in Example 2.11 on this vector.

$$\begin{array}{l} \ 1\,0\,1\,0\ \ 1\,0\,0 \\ +\ 1\,1\,0\,1 \\ \hline \ 0\,1\,1\,1\ \ 0\,0\,0 \\ +\ \ \ 1\,1\,0\ \ 1 \\ \hline \ \ \ \ 0\,0\,1\ \ 1\,0\,0 \\ +\ \ \ \ \ \ \ \ 1\ \ 1\,0\,1 \\ \hline \ \ \ \ \ \ \ \ \ \ \ \ \ \mathbf{0\,0\,1} \end{array}$$

Observe that the remainder is nonzero. Hence, there exist an error in the received vector.

Table 2.3 Popular CRC generator polynomials used in standards.

	Generator polynomials $g(x)$
CRC-4	$g(x) = x^4 + x^3 + x^2 + x + 1$
CRC-6 5G	$g(x) = x^6 + x^5 + 1$
CRC-7	$g(x) = x^7 + x^6 + x^4 + 1$
CRC-8	$g(x) = x^8 + x^7 + x^6 + x^4 + x^2 + 1$
CRC-11 5G	$g(x) = x^{11} + x^{10} + x^9 + x^5 + 1$
CRC-12	$g(x) = x^{12} + x^{11} + x^3 + x^2 + x + 1$
CRC-16 ANSI	$g(x) = x^{16} + x^{15} + x^2 + 1$
CRC-16 CCITT	$g(x) = x^{16} + x^{12} + x^5 + 1$
CRC-16 SDLC	$g(x) = x^{16} + x^{15} + x^{13} + x^7 + x^4 + x^2 + x + 1$
CRC-24	$g(x) = x^{24} + x^{23} + x^{14} + x^{12} + x^8 + 1$
CRC-24 5G	$g(x) = x^{24} + x^{23} + x^{21} + x^{20} + x^{17} + x^{13} + x^{12} + x^8 + x^4 + x^2 + x + 1$
CRC-32a	$g(x) = x^{32} + x^{30} + x^{22} + x^{15} + x^{12} + x^{11} + x^7 + x^6 + x^5 + x$
CRC-32b	$g(x) = x^{32} + x^{26} + x^{23} + x^{22} + x^{16} + x^{12} + x^{11} + x^{10} + x^8 + x^7 + x^5 + x^4 + x^2 + x + 1$

Table 2.3 tabulates popular generator polynomials used in standards such as those used for 5G polar codes.

2.7 Convolutional Codes

In linear codes, each codeword depends only on the present k-tuple of information bits, i.e. the encoding is without memory across a long string of bits. *Convolutional codes* (CCs) introduce memory sliding the information bits through a shift register, i.e. every n coded bit, called *code frame*, depends not only on the last k information bits, called *information frame*, but also on the preceding information frames m. This section describes convolutional codes by generator polynomials and generator matrices.

Convolutional codes are a class of linear codes described by a tuple (n, k, m), where k is the number of information bits shifted into the encoder at each time interval (the length of the information frame), n is the number of corresponding outputted coded bits (the length of the code frame), and m is the number of previous information frames stored in a shift-register, a.k.a. *memory size*. The value $m + 1$, named *constraint length*, determines the number of previous information frames plus the current information frame that influences each code frame. A larger constraint length generally provides greater resilience to bit errors.

2.7.1 Encoding of Convolutional Codes

In this section, we review various methods to perform convolutional encoding. Note that we only consider $k = 1$ as this aligns with the convolutional pre-transformation in polar coding, which will be discussed in Chapter 6.

2.7.1.1 Encoding with Sequential Logic

The encoder can be implemented as a sequential circuit with k inputs, n outputs, and m memory elements.

Example 2.13 Consider the convolutional code $C(2, 1, 2)$. Figure 2.4 shows the encoder of this code, comprising a shift register with a length of $m = 2$. It produces $n = 2$ output bits at a time instance, which are formed as linear combinations of the shift register's memory contents and the current input bit. The information bits $\mathbf{u} = \left(u_0, u_1, u_2, \ldots\right)$ are shifted bit by bit (since $k = 1$) to the encoder, where u_t corresponds to the input bit at time t. The $n = 2$ output sequences

$$\mathbf{v}^{(1)} = \left(v_0^{(1)}, v_1^{(1)}, v_2^{(1)}, \ldots\right) \text{ and } \mathbf{v}^{(2)} = \left(v_0^{(2)}, v_1^{(2)}, v_2^{(2)}, \ldots\right)$$

are produced by the circuit. The resulting code sequence comprising code frames $(v_t^{(1)}, v_t^{(2)})$ for $t \geq 1$ is

$$\mathbf{v} = \left(\left(v_0^{(1)} v_0^{(2)}\right), \left(v_1^{(1)} v_1^{(2)}\right), \left(v_2^{(1)} v_2^{(2)}\right), \ldots\right) = \left(\mathbf{v}_0, \mathbf{v}_1, \mathbf{v}_2, \ldots\right).$$

2.7.1.2 Impulse Response and Convolution

Borrowing from system theory, the sequential logic encoder can be represented as a time-discrete linear time-invariant (LTI) system with 1-dimensional input and n-dimensional output.

The output sequences $\mathbf{v}^{(j)}, j \in [1, n]$ are obtained by convolving the input sequence $\mathbf{u}$ with the corresponding system impulse response $\mathbf{g}^{(j)} = (g_0^{(j)}, g_1^{(j)}, \ldots, g_m^{(j)})$, where $g_i^{(j)} \in \{0, 1\}$ and $g_0^{(j)} = g_m^{(j)} = 1$ by default, as

$$\mathbf{v}^{(1)} = \mathbf{u} * \mathbf{g}^{(1)}, \mathbf{v}^{(2)} = \mathbf{u} * \mathbf{g}^{(2)}, \ldots, \mathbf{v}^{(n)} = \mathbf{u} * \mathbf{g}^{(n)}.$$

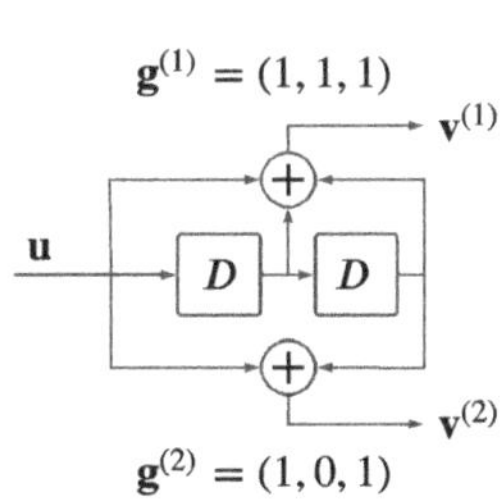

Figure 2.4 Encoder for the code $C(2, 1, 2)$ in Example 2.13.

where $*$ is the convolution operation over $\mathbb{F}_2$. This is the reason why this scheme is named the convolutional coding scheme. The impulse responses are also called generator sequences of the encoder. For example, $\mathbf{g}^{(1)} = (1, 1, 1, 1)$ and $\mathbf{g}^{(2)} = (1, 0, 1, 1)$. A generator sequence can also be represented in the form of a polynomial, similar to the one in cyclic coding.

The convolution of every $v_i^{(j)}$ at time-step i is obtained by

$$v_i^{(j)} = \sum_{\ell=0}^{m} g_\ell^{(j)} u_{i-\ell}. \tag{2.39}$$

By representing bit sequences as polynomials in the delay variable D representing a time-step in the encoder, an output sequence $\mathbf{v}(D)$ is obtained as $\mathbf{g}(D)\mathbf{u}(D)$, where $\mathbf{g}(D) = \sum_{j=0}^{m} g_j D^j$ is the *generator polynomial.*

Example 2.14 Given the input sequence $\mathbf{u} = (1, 0, 1, 1, 1, 0, 0)$ and the generator sequence $g = (1, 1, 0, 1)$, we will perform the convolution $(1, 0, 1, 1, 1, 0, 0) * (1, 1, 0, 1)$ as follows.

$$\begin{array}{rl} & 1\ 0\ 1\ 1\ 1\ 0\ 0 \\ + & \quad\ \ 1\ 0\ 1\ 1\ 1\ 0\ 0 \\ + & \qquad\ \ 1\ 0\ 1\ 1\ 1\ 0\ 0 \\ \hline & 1\ 0\ 0\ 0\ 0\ 0\ 0\ 1\ 0\ 0 \end{array}.$$

Note that the second line is shifted twice to the right to skip the second element of the generator sequence, which is 0.

2.7.1.3 Generator Matrix

Convolution can be described by the multiplication of matrices. To this end, the generator sequences $\mathbf{g}^{(1)}, \ldots, \mathbf{g}^{(n)}$ are organized into a generator matrix $\mathbf{G}$. Since we assumed $k = 1$, the generator matrix becomes

$$\mathbf{G} = \begin{pmatrix} g_0^{(1)} \cdots g_0^{(n)} & g_1^{(1)} \cdots g_1^{(n)} & g_2^{(1)} \cdots g_2^{(n)} & \cdots & g_m^{(1)} \cdots g_m^{(n)} & & \\ & g_0^{(1)} \cdots g_0^{(n)} & g_1^{(1)} \cdots g_1^{(n)} & \cdots & g_{m-1}^{(1)} \cdots g_{m-1}^{(n)} & g_m^{(1)} \cdots g_m^{(n)} & \\ & & g_0^{(1)} \cdots g_0^{(n)} & \cdots & g_{m-2}^{(1)} \cdots g_{m-2}^{(n)} & g_{m-1}^{(1)} \cdots g_{m-1}^{(n)} & g_m^{(1)} \cdots g_m^{(n)} \\ & & & \ddots & & & \ddots \end{pmatrix}. \tag{2.40}$$

The matrix multiplication $\mathbf{v} = \mathbf{u} \cdot \mathbf{G}$ results in a code sequence $\mathbf{v}$ consisting of blocks with n bits as

$$\mathbf{v} = \left(\mathbf{v}_0, \mathbf{v}_1, \mathbf{v}_2, \ldots, \mathbf{v}_t, \ldots\right), \text{ with } \mathbf{v}_t = \left(v_t^{(1)}, v_t^{(2)} \ldots, v_t^{(n)}\right),$$

for the information sequence

$$\mathbf{u} = \left(u_0, u_1, u_2, \ldots, u_t, \ldots\right).$$

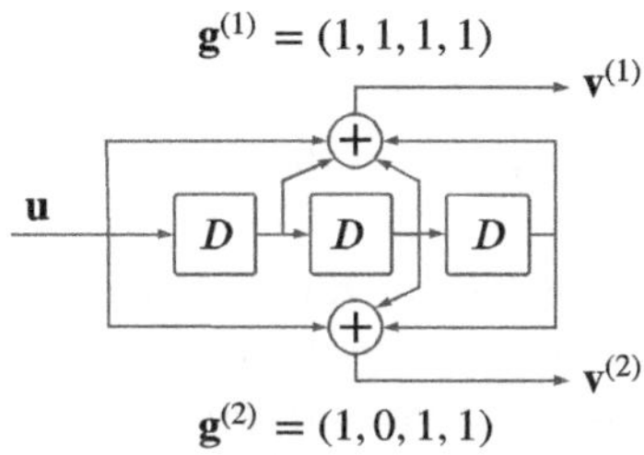

Figure 2.5 Encoder for the code $C(2, 1, 3)$ in Example 2.15.

Example 2.15 Consider the (2, 1, 3) encoder shown in Figure 2.5 where the generator sequences are $\mathbf{g}^{(1)} = (1, 1, 1, 1)$ and $\mathbf{g}^{(2)} = (1, 0, 1, 1)$. The generator matrix is formed as

$$\mathbf{G} = \begin{pmatrix} g_0^{(1)}g_0^{(2)} & g_1^{(1)}g_1^{(2)} & g_2^{(1)}g_2^{(2)} & \cdots & g_m^{(1)}g_m^{(2)} & & & \\ & g_0^{(1)}g_0^{(2)} & g_1^{(1)}g_1^{(2)} & \cdots & g_{m-1}^{(1)}g_{m-1}^{(2)} & g_m^{(1)}g_m^{(2)} & & \\ & & g_0^{(1)}g_0^{(2)} & \cdots & g_{m-2}^{(1)}g_{m-2}^{(2)} & g_{m-1}^{(1)}g_{m-1}^{(2)} & g_m^{(1)}g_m^{(2)} & \\ & & & \ddots & & & & \ddots \end{pmatrix}$$

$$= \begin{pmatrix} 11 & 10 & 11 & 11 & & & & & \\ & 11 & 10 & 11 & 11 & & & & \\ & & 11 & 10 & 11 & 11 & & & \\ & & & 11 & 10 & 11 & 11 & & \\ & & & & 11 & 10 & 11 & 11 & \\ & & & & & \ddots & & & \ddots \end{pmatrix}.$$

Note that every row is a shifted version of an identical sequence by m columns. Now, given the information sequence $\mathbf{u} = (1, 0, 1, 1, 1, 0, \ldots)$, the matrix multiplication yields in the code sequence

$$\mathbf{v} = \mathbf{u} \cdot \mathbf{G} = (1, 0, 1, 1, 1, 0, \ldots) \cdot \mathbf{G} = (11, 10, 00, 10, 10, 10, 00, 11, \ldots).$$

In Chapter 6, we will see that in PAC coding a convolutional encoder with $n = 1$ is used as a pretransformation stage.

2.7.2 Termination, Truncation, and Tail-Biting

The code sequences of a convolutional code could be infinitely long. However, in practice, code sequences of finite length are used. To obtain a finite code length, various techniques are used as follows.

In the *termination* approach for generating finite code sequences, m zero-valued bits are appended to the information sequence to return all the memory elements in the encoder to the zero state at the end of the input sequence. Observe that the code rate R_T of the terminated convolutional code is reduced by the fractional rate loss $L/(L+m)$ compared to the original code R as

$$R_T = \frac{kL}{n(L+m)} = R\frac{L}{L+m},$$

where L is the number of encoded information blocks. When L is chosen sufficiently large, we can assume that $R_T \approx R$.

In the *truncation* approach, the output of the encoder for $t > L$ is stopped regardless of the current state. Therefore, the code rate remains unchanged; however, as the convolution stops, the information bits at the end of a block are not well protected against the error.

Finally, the *tail-biting* approach forces the convolutional encoder to start in the same state in which the encoding will stop after L information blocks. Given the knowledge of the last m blocks, the final state of the encoder can be calculated and used for initialization. This approach makes it possible to have equal protection of all information bits of the entire block.

Example 2.16 Let us take the encoder in Example 2.13 for the $C(2,1,2)$ convolutional code with rate $R = 1/2$ with information sequence of length $L = 5$. The following matrices, from left to right, show the generator matrix for the termination, truncation, and tail-biting approach, respectively.

$$\begin{pmatrix} 11 & 10 & 11 & & & & \\ & 11 & 10 & 11 & & & \\ & & 11 & 10 & 11 & & \\ & & & 11 & 10 & 11 & \\ & & & & 11 & 10 & 11 \end{pmatrix}, \begin{pmatrix} 11 & 10 & 11 & & \\ & 11 & 10 & 11 & \\ & & 11 & 10 & 11 \\ & & & 11 & 10 \\ & & & & 11 \end{pmatrix}, \begin{pmatrix} 11 & 10 & 11 & & \\ & 11 & 10 & 11 & \\ & & 11 & 10 & 11 \\ 11 & & & 11 & 10 \\ 10 & 11 & & & 11 \end{pmatrix}.$$

2.7.3 Tree and Trellis

The set of all codewords, generated by a convolutional encoder, can be represented in graphical forms, namely tree and trellis, where codewords are distinct paths. We will use the following example to describe these graphical forms.

Example 2.17 Consider $(2,1,2)$ convolutional encoder shown in Figure 2.4. This encoder generates the code tree shown in Figure 2.6. In this figure, branches pointing upward represent the coded bits (placed next to the branches) corresponding to the information bit of 0, while branches pointing downward represent the coded bits corresponding to the information bit of 1. To illustrate this, consider the information sequence $\mathbf{u} = (0, 1, 0, 1 \ldots)$, which generates the corresponding code sequence $\mathbf{v} = (11, 10, 00, 01, \ldots)$. This code sequence aligns with the specific path marked within the code tree.

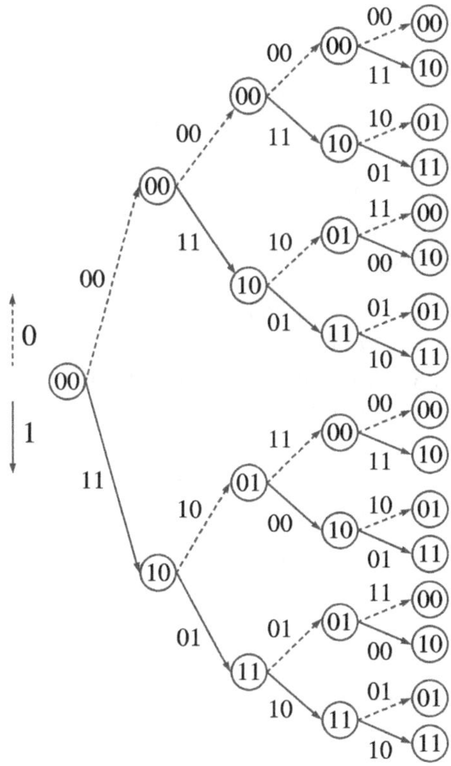

Figure 2.6 Code tree for $C(2,1,3)$ in Example 2.17.

Note that at each depth or time instant $t = 1, 2, \dots$, the tree contains $2^{k \cdot t}$ nodes. Moreover, we observe that sub-trees are periodically repeated. Each node in the tree can be associated with a state σ_t represented by ν-tuples of memory values as

$$\sigma_t = \left(\left(u_{t-1}^{(1)} \dots u_{t-v_1}^{(1)} \right), \left(u_{t-1}^{(2)}, \dots u_{t-v_2}^{(2)} \right) \dots \dots, \left(u_{t-1}^{(k)} \dots u_{t-v_k}^{(k)} \right) \right), \tag{2.41}$$

that can be determined by the path to the node. If we merge the transitions from nodes with identical states for depth $t > m$, this results in the code trellis representation. The trellis contains a constant number of nodes, 2^{v}, at every level $t \geq m$. Figure 2.7 illustrates the trellis for code (2, 1, 2).

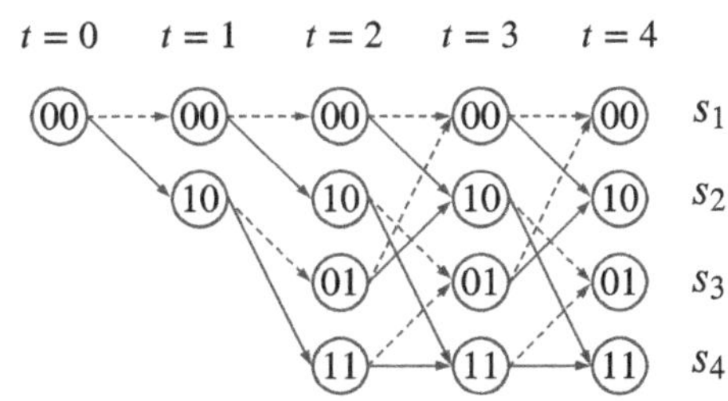

Figure 2.7 Code trellis for $C(2,1,3)$ in Example 2.17.

2.8 Concatenated Codes

The concept of code concatenation was initially explored by Forney in 1966. Forney introduced an *inner code* and paired it with the channel to form a super-channel. Subsequently, a second code, referred to as the *outer code*, was employed to encode the data intended for transmission through this super-channel. Forney highlighted three distinct advantages of concatenated codes:

- Concatenation allows for the construction of considerably long codes by building upon shorter codes.
- Concatenated codes possess the capability to rectify both burst errors and independent single errors simultaneously.
- Decoding concatenated codes involves less complexity compared to decoding a single code of equivalent length. This is because decoding of the shorter component codes can be executed in separate stages.

In 1974, Blokh and Zyablov (and in 1976, Zinov'ev) introduced the concept of generalized concatenated (GC) codes. GC codes extend Forney's concatenation concept by introducing a hierarchical subcode structure for the inner code C_I and multiple outer codes, each corresponding to a specific partition level within this hierarchy. The GC construction amalgamates the concepts of coset decomposition and concatenation.

Diverging from Forney's approach, a generalized concatenated code treats the inner and outer code as a unified entity. These two models are depicted in Figure 2.8. Although the construction of both models is identical, the way the components are grouped presents distinct advantages.

Coded modulation is viewed as a specific instance of generalized concatenation, where the modulation technique is regarded as the inner code. This perspective

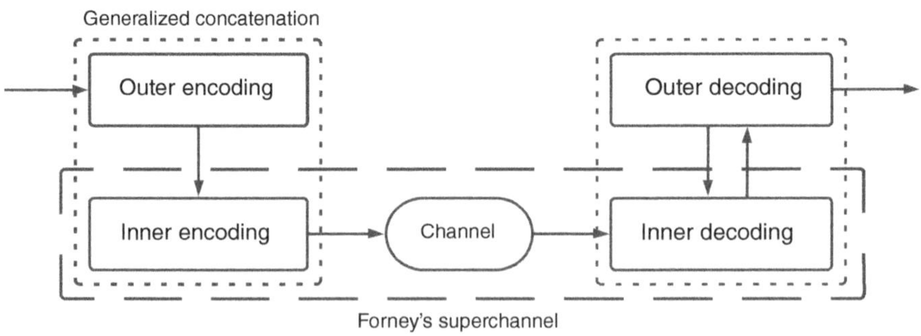

Figure 2.8 Classic concatenation as super-channel versus generalized concatenation.

provides a sophisticated approach to comprehensively define both coded modulation and the corresponding decoding techniques. Here, we describe these two classes of concatenation using two examples.

Example 2.18 Let us use a single parity check (SPC) code $\mathcal{SPC}(4,3,2)$ with length 4 and generator matrix

$$\mathbf{G} = \begin{pmatrix} 1 & 0 & 0 & 1 \\ 0 & 1 & 0 & 1 \\ 0 & 0 & 1 & 1 \end{pmatrix}, \tag{2.42}$$

as the inner code. Our outer code is the repetition code $\mathcal{R}(8,1,8)$ with length 8 and generator matrix

$$\mathbf{G} = \begin{pmatrix} 1 & 1 & 1 & 1 & 1 & 1 & 1 & 1 \end{pmatrix}. \tag{2.43}$$

Therefore, three bits of data $\mathbf{d} = \left(d_0, d_1, d_2\right)$ are encoded with the outer code $\mathcal{R}(8,1,8)$, giving $\mathbf{v} = \left(v_0, v_1, \ldots, v_7\right)$. Since the inner code is $\mathcal{SPC}(4,3,2)$, every three outer coded bits $v_j, j = 0, \ldots, 2$ are encoded with the inner code, resulting in 8 codewords. The length of the concatenated code is $n = 4 \cdot 8$ with dimension $k = 3$ and the minimum distance $d = d_R d_{SPC} = 8 \cdot 2 = 16$. Therefore, the concatenated code can be denoted by (32, 3, 16). This classical concatenation technique is shown in Figure 2.9.

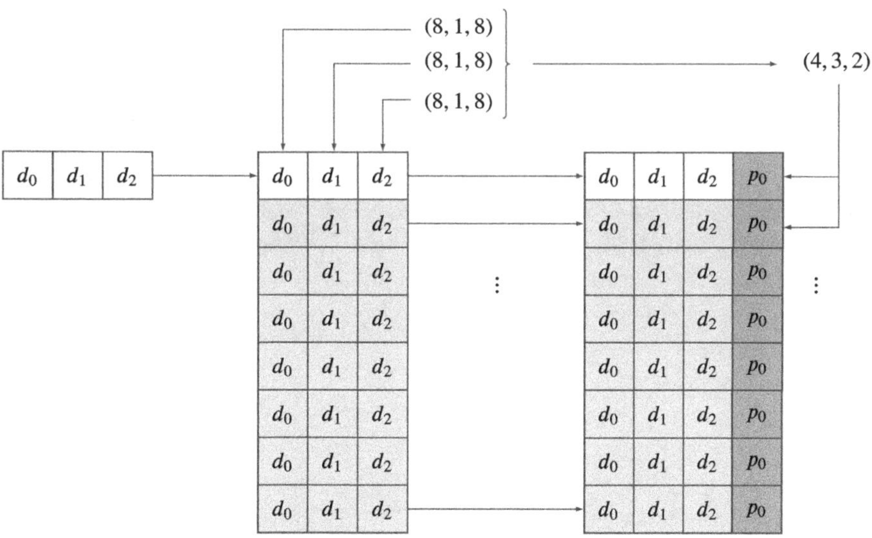

Figure 2.9 The classic concatenation of Example 2.18.

Example 2.19 Let us take the two codes used in Example 2.18. The SPC code $\mathcal{SPC}(4,3,2)$ is denoted by $C^{(1)}$. Clearly, this code has $2^3 = 8$ codewords as listed below:

$$(0000),\ (0011).\ (0101),\ (0110),$$
$$(1001),\ (1010),\ (1100),\ (1111).$$

We partition these codewords into four subsets such that every two codewords belong to a subcode while maximizing the minimum distance within subcodes. They are listed below:

$$C_0^{(2)}(4,1,4) = \{(0000),(1111)\},$$
$$C_1^{(2)}(4,1,4) = \{(0011),(1100)\},$$
$$C_2^{(2)}(4,1,4) = \{(0101),(1010)\},$$
$$C_3^{(2)}(4,1,4) = \{(0110),(1001)\}.$$

We can uniquely determine a particular codeword of $C^{(1)}$ by knowing the subcode label i in $C_i^{(\ell)}$ for $i \in \{0,1,2,3\} = \{(00)_2,(01)_2,(10)_2,(11)_2)_2\}$, and labeling the codewords belonging to each subcode by $j \in \{0,1\}$. Therefore, we need two indices i,j for each codeword. For example, the two binary indices $(10)_2$ and 1 determine the codeword $C_{10.1}^{(3)} = (1010)$, as shown in Figure 2.10 where the three levels of partitioning are illustrated. Note that this is the way we label the constellation points in different signaling/modulation schemes.

Now, we can employ outer codes to protect this enumeration, which is performed by 3-bit sequences. We employ two outer codes with equal length, here it is 2^3; namely $C^{(1)}(8,1,8)$, which is a repetition code, and $C^{(2)}(8,4,4)$, which can be an extended Hamming code. As every codeword of the inner code is determined by 3 bits, one can take two bits (out of three required bits) from two codewords of $C^{(1)}(8,1,8)$ and one bit from a codeword of $C^{(2)}(8,4,4)$. Therefore, we can encode $2 \times 1 + 4 = 6$ information bits in total. On the other hand, since every 3 coded bits (taken from three codewords as illustrated in Figure 2.11) indicate one codeword of $\mathcal{SPC}(4,3,2)$, three codewords of length 8 return eight codewords of length 4. Therefore, the resulting concatenated code C_{GC} has length $n = 8 \times 4 = 32$, representing $k = 6$ information bits.

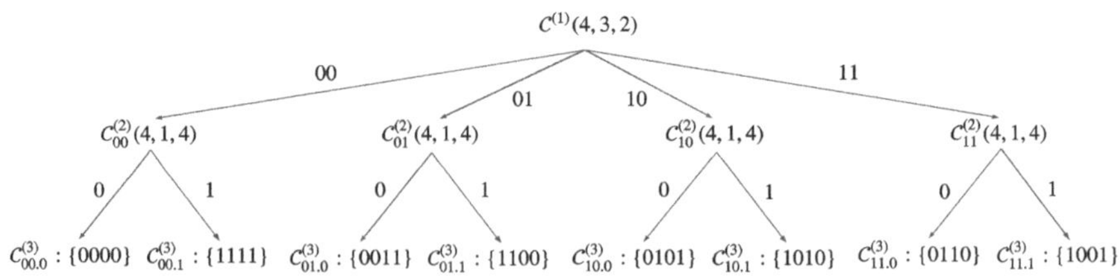

Figure 2.10 The subcodes tree for generalized concatenation in Example 2.19.

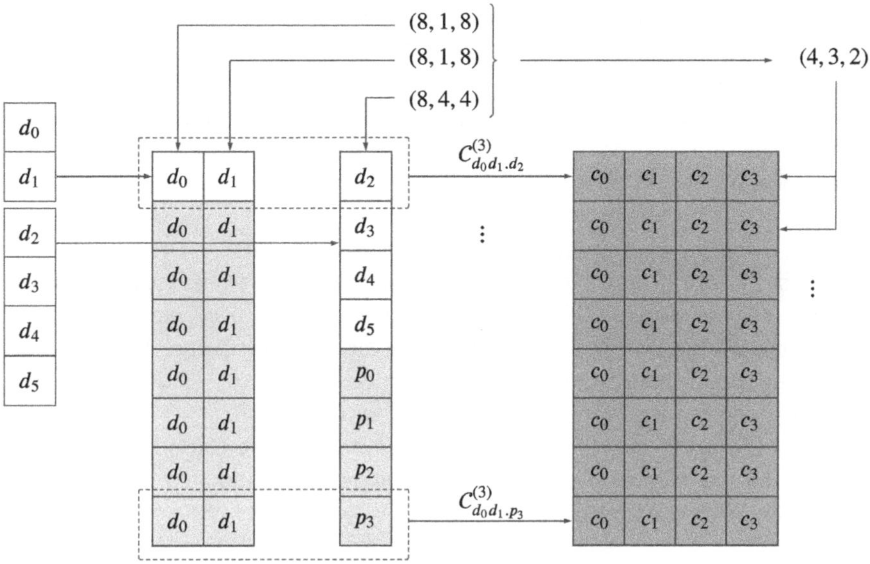

Figure 2.11 The generalized concatenation of Example 2.19.

As we will observe in Chapter 12, the coded modulation is based on the idea of generalized concatenated code, where the coded bits of different codewords are used to label a constellation point. Another well-known example is $u|u+v$-construction, where two binary outer codes U and V of the same length n_o are concatenated through an inner code of length $n_I = 2$, $k_I = 2$ and distance $d_I = 1$ with generator matrix:

$$G_I = \begin{pmatrix} 1 & 1 \\ 0 & 1 \end{pmatrix}.$$

Exercises

1 Show that a binary linear code codebook C has a group structure under the addition of binary vectors of codewords.
Hint: Verify that all four properties in the definition of an Abelian group are satisfied.

2 A sample MATLAB script is given in Appendix C.3 to generate the standard array of the $(6, 3)$ code C generated by (2.14). Modify it to find the standard array of the dual code.

3 The standard array of binary linear code represents the quotient group structure of the error patterns under the addition of binary vectors of codewords. Discuss the relations among the groups $\mathbb{F}_2^n$, C, and $C^{\perp}$.
Hint: The code C is a subgroup of $\mathbb{F}_2^n$ is in the first column. The remaining seven columns are the cosets of C obtained by adding a coset leader to all the codewords.

4 Write a MATLAB script that chooses a random codeword $\hat{\mathbf{c}} \in C$ the code generated by (2.14) and an error pattern with a given $p < 0.5$. Then, use the standard array of C to process the received sequence $\mathbf{r} = \mathbf{c} + \mathbf{e}$ correct single-error patterns when an error is detected. Iterate the above procedure 10^6 times and plot the codeword error rate as a function of $p = 0.001, 0.005, 0.01, 0.05, 0.1, 0.5$ with and without performing the single error correction.
Hint: The codeword error rate is estimated by Monte Carlo simulation by counting the number of cases out of 10^6, where the estimated codeword $\hat{\mathbf{c}}$ is different from the transmitted one $\hat{\mathbf{c}}$.

5 Construct a Reed–Muller code, $\mathcal{R}(r, m)$, with parameters $r = 1$ and $m = 3$. Determine:

- The length n, dimension k, and minimum distance d of the code.
- A generator matrix for $RM(1, 3)$.

Hint: Use the formula $n = 2^m, k = \sum_{i=0}^{r} \binom{m}{i}$, and $d = 2^{m-r}$.

6 For a convolutional code with constraint length $\nu = 2$ and generator polynomials $g_1(D) = 1 + D + D^2$ and $g_2(D) = 1 + D^2$:

- Draw the encoder's state diagram.
- Determine the code rate and the output for the input sequence $\mathbf{u} = (1, 0, 1, 1)$.

Hint: Use the generator polynomials to calculate the output for each bit in the input sequence.

7 Consider a cyclic code with length $n = 7$ and generator polynomial $g(x) = x^3 + x + 1$. Determine:

- The codeword polynomial for the message $\mathbf{m} = (1, 0, 1, 1)$.
- The systematic form of the codeword.

Hint: Multiply $g(x)$ by the message polynomial $m(x)$ modulo $x^n - 1$ to find the codeword polynomial.

8 Design a concatenated code with an inner single parity check code of length $n = 3$ (rate 2/3) and an outer repetition code of length $n = 3$ (rate 1/3).

- Determine the effective rate of the concatenated code.
- Calculate the effective minimum distance of the concatenated code.

Hint: Use the rate multiplication rule for concatenated codes and consider that the minimum distance of a repetition code of length n is n, while for a single parity check code of length n, it is 2.

Bibliographic Notes

For more details on the codes introduced in this chapter, interested readers can refer to any books on linear block codes such as Error Control Coding by Lin and Costello [2004], Error Correction Coding: Mathematical Methods and Algorithms by Moon [2020], and Channel Coding for Telecommunications by Bossert [1999].

The codes known as Reed–Muller codes were proposed by Muller [1954], and Reed's name is associated with the codes due to the majority logic decoder proposed by Reed [1954]. Binary $|u|u + v|$-construction was defined by PLOTKIN in 1951 Plotkin [1960]. The concatenation of codes was first studied by Forney [1966] by defining an inner code together with the channel as a superchannel. This was followed by the concept of generalized concatenated by Blokh and Zyablov [1974] and later by Zinov'ev [1976]. The most method for BCM codes is the multilevel devised by Imai and Hirakawa [1977] and the results from Ungerboeck [1982] have described coded modulation and have resulted in a lot of research activity.

Convolutional codes were first introduced by Elias [1955]. Then, Wozencraft and Reiffen [1961] proposed sequential decoding as an efficient decoding method for convolutional codes with large constraint lengths. Viterbi [1967] proposed a maximum likelihood (ML) decoding algorithm that was relatively easy to implement for soft-decision decoding of convolutional codes with small constraint lengths. The Viterbi algorithm, along with soft-decision versions of sequential decoding, led to the application of convolutional codes to deep-space and satellite communication systems in the 1970s. The other class of codes discussed in this chapter was the cyclic codes introduced by Prange [1957]. We covered it here to give a background on using CRC as outer codes concatenated with polar codes.

Bibliography

È L Blokh and Victor Vasilievich Zyablov. Coding of generalized concatenated codes. *Problemy Peredachi Informatsii*, 10(3):45–50, 1974.

Martin Bossert. *Channel coding for telecommunications*. John Wiley & Sons, Inc., 1999.

Peter Elias. Coding for noisy channels. In *IRE WESCON Convention Record, 1955*, volume 2, pages 94–104, 1955.

G David Forney Jr.," *Concatenated codes*. IT Press, Cambridge, VIA, 1966.

Hideki Imai and Shuji Hirakawa. A new multilevel coding method using error-correcting codes. *IEEE Transactions on Information Theory*, 23(3):371–377, 1977.

S Lin and D Costello. *Error control coding*. Pearson Prentice Hall, Upper Saddle River, NJ, 2004.

Todd K Moon. *Error correction coding: mathematical methods and algorithms*. John Wiley & Sons, 2020.

David E Muller. Application of Boolean algebra to switching circuit design and to error detection. *Transactions of the IRE Professional Group on Electronic Computers*, EC-3(3):6–12, 1954.

Morris Plotkin. Binary codes with specified minimum distance. *IRE Transactions on Information Theory*, 6(4):445–450, 1960.

Eugene Prange. *Cyclic error-correcting codes in two symbols*. Air Force Cambridge Research Center, 1957.

Irving S Reed. A class of multiple-error-correcting codes and the decoding scheme. *IEEE Transactions on Information Theory*, 4(4):38–49, 1954.

Gottfried Ungerboeck. Channel coding with multilevel/phase signals. *IEEE Transactions on Information Theory*, 28(1):55–67, 1982.

Andrew Viterbi. Error bounds for convolutional codes and an asymptotically optimum decoding algorithm. *IEEE Transactions on Information Theory*, 13(2):260–269, 1967.

John M Wozencraft and Barney Reiffen. *Sequential decoding*. Published jointly by the Technology Press of the Massachusetts Institute of Technology and Wiley, New York, 1961.

Victor Alexandrovich Zinov'ev. Generalized cascade codes. *Problemy Peredachi Informatsii*, 12(1):5–15, 1976.

3

Fundamentals of Soft-Decision Decoding

"The art of communication is not only to say the right thing in the right place but to leave unsaid the wrong thing at the tempting moment."

— Peter Elias

In this chapter, we explore the critical role of decoding techniques in ensuring reliable communication over noisy channels. Decoding is central to recovering the transmitted message from the received signal, and understanding how to find the closest codeword to the transmitted codeword is key to minimizing errors. We begin by discussing the role of various cost functions, which are essential for evaluating how closely a received signal corresponds to potential codewords. These functions guide the search for the closest match, directly impacting the error rate of the decoded message.

As we delve deeper into the chapter, you will learn about the factors influencing the error rate of linear codes, providing a comprehensive understanding of how different parameters affect the reliability of communication. A key aspect of this chapter is equipping you with the knowledge to distinguish between different decoding algorithms, enabling you to choose the most suitable one for a given code.

We will explore advanced decoding techniques, including maximum a posteriori (MAP)/maximum-likelihood (ML) decoding, reliability-based universal decoding, recursive decoding, message-passing algorithms, and tree/trellis-based search methods. Each of these approaches leverages different strategies for dealing with the noise and uncertainties in communication systems, and by the end of the chapter, you will be equipped to identify and implement the most effective decoding strategy for a variety of coding scenarios.

Polar Codes: From Theory to Practice, First Edition. Mohammad Rowshan and Emanuele Viterbo.

3.1 MAP/ML Decoding and Likelihood

Suppose we have a generic binary discrete memoryless channel (B-DMC) $W : \mathcal{X} \rightarrow \mathcal{Y}$ with input alphabet $\mathcal{X} = \{0, 1\}$, the arbitrary output alphabet $\mathcal{Y}$, and the arbitrary transition probability $W(x|y) = p_{\mathcal{X}|\mathcal{Y}}(x|y)$ for $x \in \mathcal{X}$ and $y \in \mathcal{Y}$.

The codeword $\mathbf{x} = (x_0, x_1, \dots, x_{N-1}) \in \mathscr{C}$ is transmitted by N uses of the channel W, denoted by W^N, and the word $\mathbf{y}$ is received. The decoder must produce an estimate codeword $\hat{\mathbf{x}}$ based on the received word $\mathbf{y} = (y_0, y_1, \dots, y_{N-1})$. Due to a one-to-one correspondence between the information sequence $\mathbf{u} = (u_0, \dots, u_{K-1})$ and the codeword $\mathbf{x}$, the decoder can produce the corresponding estimate information sequence $\hat{\mathbf{u}}$ of $\mathbf{u}$. Obviously, $\hat{\mathbf{u}} = \mathbf{u}$ if and only if $\hat{\mathbf{x}} = \mathbf{x}$. The error probability of the decoder is $1 - p_{\mathcal{X}|\mathcal{Y}}(\hat{\mathbf{x}}|\mathbf{y})$. To minimize this probability, we need to choose some $\hat{\mathbf{x}}$ that maximizes $p_{\mathcal{X}|\mathcal{Y}}(\hat{\mathbf{x}}|\mathbf{y})$. Hence, the minimum error probability decoding (MED) or more commonly known by the name *maximum a posteriori* (MAP) decoding is mathematically defined as

$$\hat{\mathbf{x}}^{\text{MAP}} = \underset{\mathbf{x} \in \mathscr{C}}{\operatorname{argmax}}\, p_{\mathcal{X}|\mathcal{Y}}(\mathbf{x}|\mathbf{y}). \tag{3.1}$$

For the MAP decoding process, we need to compute the *a posteriori probability* (APP) $p_{\mathcal{X}|\mathcal{Y}}(\mathbf{x}|\mathbf{y})$, that is, the distribution of the random variable $\mathcal{X}$ given the observation random variable $\mathcal{Y}$. Therefore, an MAP decoder can also be called an APP decoder. By Bayes' rule and by omitting the term $p_{\mathcal{Y}}(\mathbf{y})$ which does not depend on the specific codeword $\mathbf{x}$,

$$p_{\mathcal{X}|\mathcal{Y}}(\mathbf{x}|\mathbf{y}) = \frac{p_{\mathcal{Y}|\mathcal{X}}(\mathbf{y}|\mathbf{x}) p_{\mathcal{X}}(\mathbf{x})}{p_{\mathcal{Y}}(\mathbf{y})} \propto p_{\mathcal{Y}|\mathcal{X}}(\mathbf{y}|\mathbf{x}) p_{\mathcal{X}}(\mathbf{x}), \tag{3.2}$$

where $p_{\mathcal{X}}(\mathbf{x})$ is the codeword a priori probability distribution of $\mathcal{X}$. The MAP decoding rule follows as:

$$\hat{\mathbf{x}}^{\text{MAP}} = \underset{\mathbf{x} \in \mathscr{C}}{\operatorname{argmax}}\, p_{\mathcal{Y}|\mathcal{X}}(\mathbf{y}|\mathbf{x}) p_{\mathcal{X}}(\mathbf{x}). \tag{3.3}$$

If all codewords are equally likely, i.e. if $p_{\mathcal{X}}(\mathbf{x}) = \frac{1}{2^K}$ is uniform, then the codeword $\mathbf{x}$ that maximizes $p_{\mathcal{X}|\mathcal{Y}}(\mathbf{x}|\mathbf{y})$ also maximizes $p_{\mathcal{Y}|\mathcal{X}}(\mathbf{y}|\mathbf{x})$, and the MAP reduces to the *maximum likelihood* (ML) rule. That is,

$$\hat{\mathbf{x}}^{\text{MAP}} = \underset{\mathbf{x} \in \mathscr{C}}{\operatorname{argmax}}\, p_{\mathcal{Y}|\mathcal{X}}(\mathbf{y}|\mathbf{x}) p_{\mathcal{X}}(\mathbf{x}) = \underset{\mathbf{x} \in \mathscr{C}}{\operatorname{argmax}}\, p_{\mathcal{Y}|\mathcal{X}}(\mathbf{y}|\mathbf{x}) = \hat{\mathbf{x}}^{\text{ML}}. \tag{3.4}$$

In the following, we will assume that all codewords are equally likely, so that ML decoding can be used as the optimal decoding rule. Note that if the codewords are not equally likely, then the conditional probabilities $p_{\mathcal{Y}|\mathcal{X}}(\mathbf{y}|\mathbf{x})$ must be weighted by a priori probabilities $p_{\mathcal{X}}(\mathbf{x})$ to find which codeword maximizes $p_{\mathcal{X}|\mathcal{Y}}(\mathbf{x}|\mathbf{y})$ according to the MAP decoding rule. Nevertheless, as the a priori probability of the codeword $p_{\mathcal{X}}(\mathbf{x})$ is not known by the receiver in most cases, it becomes a nonoptimal decoding. Then, ML decoding becomes the best feasible decoding rule. Additionally, to find the estimate $\hat{\mathbf{x}}$ with the ML decoding rule, we need to check the conditional probabilities $p_{\mathcal{Y}|\mathcal{X}}(\mathbf{y}|\mathbf{x})$, also known as the *probability function* of all 2^K codewords, which quickly becomes impractical due to the exponential growth in K.

Since the channel is assumed to be memoryless, i.e. each received symbol only depends on the corresponding transmitted symbol, the conditional probability $p_{\mathcal{Y}|\mathcal{X}}(\mathbf{y}|\mathbf{x})$ can be calculated by

$$p_{\mathcal{Y}|\mathcal{X}}(\mathbf{y}|\mathbf{x}) = \prod_{i=0}^{N-1} p_{\mathcal{Y}|\mathcal{X}}(y_i|x_i). \tag{3.5}$$

For a discrete input, continuous output channel, the received word $\mathbf{y}$ is a sequence of real numbers. To compute the conditional probability $p_{\mathcal{Y}|\mathcal{X}}(\mathbf{y}|\mathbf{x})$, we simply need to replace each symbol transition probability $p_{\mathcal{Y}|\mathcal{X}}(y_i|x_i)$ with the corresponding conditional probability density function of the channel, such as the Gaussian conditional probability density function.

Furthermore, since the logarithmic functions are monotonically increasing functions, i.e. $\log x > \log z$ for $x > z$, then maximizing (3.5) is equivalent to maximizing

$$\log\, p_{\mathcal{Y}|\mathcal{X}}(\mathbf{y}|\mathbf{x}) = \sum_{i=0}^{N-1} \log\, p_{\mathcal{Y}|\mathcal{X}}(y_i|x_i), \tag{3.6}$$

which is called *log-likelihood function*. Unless stated otherwise, from now on the logarithm is the natural logarithm.

For each symbol i from a discrete alphabet of size $|\mathcal{X}| = M$, by the ML decision rule, we have

$$p_{\mathcal{Y}|\mathcal{X}}\left(y_i|x_i^{(j^*)}\right) > p_{\mathcal{Y}|\mathcal{X}}\left(y_i|x_i^{(j)}\right), \quad j \in \{1, 2, \ldots, M\}\backslash\{j^*\}. \tag{3.7}$$

So far we have considered a general DMC model. Now, let us consider a particular case, where the channel has a bipolar input $x_i = \pm 1$ and continuous output y_i at the ith channel use. Then, the ML decision rule is

$$p_{\mathcal{Y}|\mathcal{X}}(y_i|x_i = 1) \gtrless p_{\mathcal{Y}|\mathcal{X}}(y_i|x_i = -1). \tag{3.8}$$

We select the value +1 or −1, or the demodulated values 0 or 1, for which the conditional probability density function is higher. The decision rule will obviously remain the same if both sides of (3.8) are replaced by their logarithms, i.e.

$$\log\, p_{\mathcal{Y}|\mathcal{X}}(y_i|x_i = 1) \gtrless \log\, p_{\mathcal{Y}|\mathcal{X}}(y_i|x_i = -1). \tag{3.9}$$

3.1.1 Log-Likelihood Ratio (LLR)

The comparisons in (3.8) and (3.9) can be made more efficiently. Equivalent to (3.9), we can calculate the expression

$$\alpha(y_i) = \log \frac{p_{\mathcal{Y}|\mathcal{X}}(y_i|x_i = 1)}{p_{\mathcal{Y}|\mathcal{X}}(y_i|x_i = -1)}, \tag{3.10}$$

and compare it to a zero threshold. The function $\alpha(y_i)$ is called the *log-likelihood ratio* (LLR) function and is an alternative tool to perform the ML decision rule. The general expression for LLR in (3.10) can be significantly simplified for common probability density functions. For instance, the LLR function for the additive white Gaussian noise (AWGN) channel model reduces to

$$\alpha(y_i) = \log \frac{\frac{1}{\sqrt{2\pi}\sigma} \exp\left[-\frac{(y_i-1)^2}{2\sigma^2}\right]}{\frac{1}{\sqrt{2\pi}\sigma} \exp\left[-\frac{(y_i+1)^2}{2\sigma^2}\right]} = \frac{2}{\sigma^2} y_i, \tag{3.11}$$

where σ^2 is the noise variance. The noise variance for the AWGN channel is $\sigma^2 = \frac{N_0}{2}$, where N_0 denotes the power spectral density of the noise. From this, we can find the relation of noise variance with the signal-to-noise ratio (SNR) $= \frac{E_s}{N_0}$. Knowing that the energy per information bit denoted by E_b is defined as $E_b = E_s/R$, where R is the code rate in information bits per channel use (or the data rate after coding and modulation). Then, by normalizing $E_s = 1$, we can write the SNR per information bit as

$$\frac{E_b}{N_0} = \frac{E_s}{R \cdot N_0} = \frac{1}{2 \cdot R \cdot \sigma^2}, \tag{3.12}$$

or in other words, for a given E_b/N_0 and R, the variance of noise is

$$\sigma^2 = 2R\frac{E_b}{N_0}. \tag{3.13}$$

Furthermore, the LLR of the received symbols $\alpha(y_i)$ can be considered as i.d.d. random numbers that follow the distribution $L_1^{(i)}(y_i) \sim \mathcal{N}(\frac{\pm 2}{\sigma^2}, \frac{4}{\sigma^2})$. As can be seen in (3.11), the ML rule using the LLR function, in this particular case of the AWGN channel with bipolar transition, is reduced to checking if y_i is positive or negative.

Now, we turn our focus to the LLR algebra of binary random variables. Let x be in the binary Galois field $GF(2)$ with the elements $x \in \{+1, -1\}$ and $+1$ as the "null" element under the addition $\oplus$ operation, i.e.

$$1 \oplus 1 = 1 \quad -1 \oplus 1 = -1 \quad 1 \oplus -1 = -1 \quad -1 \oplus -1 = 1, \tag{3.14}$$

and its LLR $\alpha(x)$ in (3.10) as the real number and "soft"-value of x. The sign of $\alpha(x)$ is the hard decision, and the magnitude $|\alpha(x)|$ is the reliability of this hard decision.

We rewrite (3.10) as follows based on Bayes' rule and cancellation of the common terms. Since there is no danger of confusion, we shall henceforth skip the subscripts for the probabilities.

$$\alpha(x_i) = \log \frac{p(x_i = 1|y_i)}{p(x_i = -1|y_i)}, \tag{3.15}$$

Knowing $\alpha(x_i)$, we can estimate x_i by the following hard decision rule:

$$h\left(\alpha(x_i)\right) = \text{sign}(\alpha(x_i)). \tag{3.16}$$

The hard decision in (3.16) follows from the fact that when $p(x_i = 1|y_i) > p(x_i = -1|y_i)$, then $\frac{p(x_i=1|y_i)}{p(x_i=-1|y_i)} > 1$ and its logarithm $\alpha(x_i) > 0$. Consequently, it is safe to decide on $x_i = 1$ when $\alpha(x_i) > 0$. Similarly, $\alpha(x_i) < 0$ is equivalent to $p(x_i = 1|y_i) < p(x_i = -1|y_i)$.

Notice that the joint log-likelihood $\alpha(x_i, y_i)$ is equal to the conditioned log-likelihood $\alpha(x_i|y_i)$ since the probability $p(y_i)$ term can be canceled out. From (3.15), the probability can be easily recovered by rearrangement and substitution as

$$p(x_i = 1) = \frac{e^{\alpha(x_i)}}{1 + e^{\alpha(x_i)}}, \quad p(x_i = -1) = \frac{1}{1 + e^{\alpha(x_i)}}. \tag{3.17}$$

It is easy to show that for statistically independent random variables X_1 and X_2,

$$\alpha(x_1 \oplus x_2) = \log \frac{p(x_1 \oplus x_2 = 1)}{p(x_1 \oplus x_2 = -1)} \tag{3.18}$$

$$= \log \frac{p(x_1 = 1)p(x_2 = 1) + p(x_1 = -1)p(x_2 = -1)}{p(x_1 = -1)p(x_2 = 1) + p(x_1 = 1)p(x_2 = -1)} \tag{3.19}$$

$$= \log \frac{1 + e^{\alpha(x_1)}e^{\alpha(x_2)}}{e^{\alpha(x_1)} + e^{\alpha(x_2)}} = 2\tanh^{-1}\left(\tanh\left(\frac{\alpha(x_1)}{2}\right)\tanh\left(\frac{\alpha(x_2)}{2}\right)\right). \tag{3.20}$$

The last equality is left as an exercise (see the chapter's problems). Note that given (3.14), we have $x_1 \oplus x_2 = 1$ when $x_1 = x_2 = 1$ or $x_1 = x_2 = -1$. Similarly, $x_1 \oplus x_2 = -1$ when $x_1 = 1, x_2 = -1$ or $x_1 = -1, x_2 = 1$. The LLR of the sum has the following important approximation:

$$\alpha(x_1 \oplus x_2) \approx \text{sign}\left(\alpha(x_1)\right) \cdot \text{sign}\left(\alpha(x_2)\right) \cdot \min\left(|\alpha(x_1)|, |\alpha(x_2)|\right). \tag{3.21}$$

3.1.2 LLR Algebra

For the LLRs or the "soft"-values of binary random variables, we define a special algebra where $\boxplus$ (read as "box-plus") denotes the addition over the set of LLRs as

$$\alpha(x_1) \boxplus \alpha(x_2) \triangleq \alpha(x_1 \oplus x_2), \tag{3.22}$$

with rules

$$\alpha(x) \boxplus \infty = \alpha(x), \; \alpha(x) \boxplus -\infty = -\alpha(x), \;\; \text{and} \; \alpha(x) \boxplus 0 = 0. \tag{3.23}$$

These conditions can be interpreted as follows. The reliability of $x \boxplus$ (an element of $GF(2)$ with infinite reliability) is the same as the reliability of x. The reliability of $x \boxplus$ (an element which is totally unreliable) is negative to the reliability of x. The reliability of $x \boxplus$ (an element of completely ambiguous reliability where $\alpha(\cdot) = 0$) is completely ambiguous, which is represented by 0.

The hard decision $h(\cdot)$ over the left-hand side of (3.22) has the following property, assuming $\alpha(x_1) \cdot \alpha(x_2) \neq 0$:

$$h\left(\alpha(x_1) \boxplus \alpha(x_2)\right) = h\left(\alpha(x_1)\right) \oplus h\left(\alpha(x_2)\right). \tag{3.24}$$

3.1.3 Maximum-Likelihood (ML) Bound

Given a maximum likelihood decoder, there will still be cases where the decoder does not correct the error(s). In this section, we investigate these cases and find a block error rate performance called *ML bound* using simulations. Although we may use a particular decoder for the simulations, this bound depends on the properties of the code. The ML bound is determined by identifying instances, where the optimal ML decoder would fail. During simulations, each time a decoding error occurs, we compare the likelihood of the decoded codeword $\mu(\hat{\mathbf{x}})$ with that of the transmitted modulated codeword $\mu(\mathbf{x})$. Here, the function $\mu(\cdot)$ performs the symbol mapping. Specifically, we check whether the likelihood of the received signal given the decoded codeword, $W(\mathbf{y} \mid \mu(\hat{\mathbf{x}}))$, exceeds the likelihood of the received signal given the actual transmitted codeword, $W(\mathbf{y} \mid \mu(\mathbf{x}))$. If we have

$$W(\mathbf{y} \mid \mu(\hat{\mathbf{x}})) > W(\mathbf{y} \mid \mu(\mathbf{x})), \tag{3.25}$$

the ML decoder would also misinterpret the received signal and produce the same decoding error. By counting the number of these misinterpretations and dividing them by the total number of codewords transmitted at each SNR point, one can obtain the ML bound curve for the underlying code. As described in Section 3.2, we can use the squared Euclidean distance (SED) in (3.35) as a measure of likelihood when we use an AWGN channel. This process allows us to estimate the performance bound of an ML decoder by identifying cases, where any decoder, including the optimal one, would fail.

In Section 3.2, Example 3.1, we demonstrate how an ML decoder fails after introducing a likelihood metric.

3.1.4 Union Bound

Given a linear code C, the block error rate of C under soft-input maximum likelihood (ML) decoding over a BI-DMC is bounded by the well-known union bound. In this section, we discuss this bound, its applications, and limitations. The block error probability P_e can be upper-bounded by considering the probability that the decoder confusing the transmitted codeword $\mathbf{c}_0$ with any other codeword $\mathbf{c}_j$ in the codebook. The sum of the pairwise error probability (PEP) of mistaking $\mathbf{c}_0$ for $\mathbf{c}_j$, denoted by $P\left(\mathbf{c}_0 \rightarrow \mathbf{c}_j\right)$, gives the following union bound:

$$P_e \leq \sum_{\mathbf{c}_j \neq \mathbf{c}_0} P\left(\mathbf{c}_0 \rightarrow \mathbf{c}_j\right). \tag{3.26}$$

Now, let us assume that binary phase-shift keying (BPSK) modulation is used, with $0 \rightarrow +1$ and $1 \rightarrow -1$ mapping $\mathbf{c} \in \{0,1\}^n \rightarrow \mathbf{x} \in \{1,-1\}^n$ by $x_i = 1 - 2c_i$, and the received signal is $\mathbf{y} = \mathbf{x} + \mathbf{n}$, where $\mathbf{n}$ is AWGN with variance $\sigma_n^2 = \frac{N_0}{2}$. In this scenario, the probability of an error occurring between $\mathbf{c}_0$ and $\mathbf{c}_j$ depends on the Gaussian tail probability as

$$P\left(\mathbf{c}_0 \rightarrow \mathbf{c}_j\right) = Q\left(\sqrt{\frac{\left\|\mathbf{x}_0 - \mathbf{x}_j\right\|^2}{2\sigma_n^2}}\right), \tag{3.27}$$

where $\|\mathbf{x}_0 - \mathbf{x}_j\|^2$ is the sum of the SED between the modulated sequences $\mathbf{x}_0$ and $\mathbf{x}_j$ and $Q(x) = \frac{1}{\sqrt{2\pi}} \int_x^\infty e^{-t^2/2} dt$ is the tail probability of the Gaussian distribution. The term $\|\mathbf{x}_0 - \mathbf{x}_j\|^2$ reflects how separable the two codewords are.

For positions where the bits differ $\left(c_{0,i} \neq c_{j,i}\right)$, the difference $\left(x_{0,i} - x_{j,i}\right) = 1 - (-1) = 2$. Thus, the squared difference is $\left(x_{0,i} - x_{j,i}\right)^2 = 4$, resulting in $\|\mathbf{x}_0 - \mathbf{x}_j\|^2 = \sum_{i=1}^n \left(x_{0,i} - x_{j,i}\right)^2 = 4d_H\left(\mathbf{c}_0, \mathbf{c}_j\right) = 4w$. Substituting this and $\sigma_n^2 = N_0/2$, we get

$$P\left(\mathbf{c}_0 \rightarrow \mathbf{c}_j\right) = Q\left(\sqrt{\frac{4w}{N_0}}\right). \tag{3.28}$$

Now, the energy per symbol $\left(E_s\right)$ is given by $E_s =$ Average of $c_i^2 = 1$, because each symbol has magnitude 1, and $1^2 = 1$. The energy per bit $\left(E_b\right)$ is related to the energy per symbol as $E_b = E_s/R = 1/R$, where $R = k/n$ is the code rate (k is the number of information bits and n is the total number of bits in a codeword). Then, we have $\text{SNR} = \frac{E_s}{N_0} = \frac{E_b/R}{N_0}$ and N_0 in terms of E_b is $N_0 = \frac{E_b}{E_b/N_0} = \frac{1/R}{E_b/N_0}$. Substituting this in (3.28) gives

$$P\left(\mathbf{c}_0 \rightarrow \mathbf{c}_j\right) = Q\left(\sqrt{2w \cdot R \cdot \frac{E_b}{N_0}}\right). \tag{3.29}$$

Eventually, summing (3.29) over all codewords with different weights w gives the union bound for block error rate as

$$P_e \leq \sum_{w \geq w_{\min}} A_w Q\left(\sqrt{2w \cdot R \cdot \frac{E_b}{N_0}}\right). \tag{3.30}$$

The dominant term in the union bound often comes from the minimum distance $w_{\min} = d_{\min} = \min_{i \neq j} ||\mathbf{x}_i - \mathbf{x}_j||$, as the smallest distances contribute the most to P_e. Thus, the first term of the summation in (3.30), $A_{w_{\min}} Q\left(\sqrt{2w_{\min} \cdot R \cdot \frac{E_b}{N_0}}\right)$, gives a good approximation for the upper bound. Moreover, a larger $\frac{E_b}{N_0}$, i.e. low noise, corresponds to better separation in the Gaussian space.

To simplify the computation, the Q-function can be approximated with its exponential upper bound; i.e.

$$Q(x) \leq \frac{1}{2} \exp\left(-\frac{x^2}{2}\right).$$

We substitute $x = \sqrt{\frac{2wR \cdot E_l}{N_0}}$, which leads to an exponential approximation of the PEP as

$$P\left(\mathbf{c}_0 \rightarrow \mathbf{c}_j\right) \approx \exp\left(-R \cdot \frac{wE_b}{N_0}\right). \tag{3.31}$$

This formulation is particularly useful in the following scenarios:

- High SNR regime: In this regime, the error probability is dominated by the terms with the smallest w (specifically $w_{\min}$), and the exponential form provides a reasonable approximation for small error probabilities.
- Asymptotic analysis: The exponential form helps in identifying the error exponent and analyzing performance trends.

Nevertheless, this approximation has its own drawback. It underestimates the error probability at lower E_b/N_0 because the Q-function decays slower than the exponential approximation for moderate x values.

The exponential approximation of union bound is known as *union-Bhattacharyya bound* and expressed by

$$P_e \leq \sum_{w \geq w_{\min}} A_w \gamma^w, \tag{3.32}$$

where γ is the Bhattacharyya parameter of the underlying channel. The parameter γ for the binary erasure channel (BEC) and AWGN channel are $\gamma_{\text{BEC}} = \epsilon$ (the erasure probability), and $\gamma_{\text{AWGN}} = \exp(-R \cdot E_b/N_0)$, respectively, where E_b/N_0 is the normalized SNR per information bit and R is the code rate.

The union bound is often associated with ML decoding because it fundamentally relies on the structure and principles of ML decoding, which uses the full weight spectrum between codewords in the codebook, to derive the bound. Recall that the ML decoding rule chooses the codeword $\mathbf{c}_j$ that is closest to the received signal $\mathbf{y}$ in terms of likelihood. For AWGN channels, this translates into minimizing the Euclidean distance as

$$\hat{\mathbf{x}} = \arg\min_{\mathbf{x}_j} ||\mathbf{y} - \mathbf{x}_j||.$$

Furthermore, PEP is a fundamental concept in ML decoding because ML decoding minimizes the probability of choosing the wrong codeword given the distances between codewords.

The union bound enables us to compare the performance of two codes C_1 and C_2 over the AWGN channel, where both have the same code rate R but different $A_{C_j}(D), j = 1, 2$. The term $\gamma_{\text{AWGN}}^{w} = \exp(-w \cdot R \cdot \frac{E_b}{N_0})$ is an exponentially decaying function of E_b/N_0 and the weight w, as R is fixed. Moreover, we consider the low and medium SNR regimes. When E_b/N_0 converges linearly toward infinity, γ_{AWGN} decreases exponentially toward zero. Hence, a small decrease in the SNR has a significant impact on the bound.

3.2 Reliability-Based Universal Decoding

Besides the decoding processes based on the code structure, there is another category of decoding algorithms, called reliability-based algorithms or generic decoding algorithms (also known as universal algorithms), as they usually do not depend on the code structure. In reliability-based algorithms, the goal is to find the closest modulated codeword to the received sequence using a metric such as the likelihood function, utilizing the reliability of the soft information received for each symbol in the sequence from the channel. That is, we try to maximize the likelihood in the search toward finding the transmitted sequence. Hence, this category of decoding algorithm is called maximum likelihood (ML) or near-ML decoding.

A binary code C of length N and dimension K maps a message of K bits to a codeword $\mathbf{c}$ of N bits to be transmitted over a noisy channel. We assume that we are using BPSK modulation. The channel alters the transmitted codeword so that the receiver obtains an N-symbol vector $\mathbf{y}$. An ML decoder supposedly compares $\mathbf{y}$ with all the 2^K modulated codewords in the codebook, and selects the one closest to $\mathbf{y}$. In other words, the ML decoder finds a modulated codeword $\mu(\mathbf{c})$ such that

$$\hat{\mathbf{c}} = \arg\max_{\mathbf{c} \in C} p\,(\mathbf{y}|\mu(\mathbf{c})). \tag{3.33}$$

For AWGN channel with noise power of $\sigma_n^2 = N_0/2$, where N_0 is the noise spectral density, the conditional probability $p\,(\mathbf{y}|\mu(\mathbf{c}))$ is given by

$$p\,(\mathbf{y}|\mu(\mathbf{c})) = \frac{1}{(\sqrt{\pi N_0})^n} \exp\left(-\sum_{i=1}^{n} (r_i - \mu(c_i))^2/N_0\right). \tag{3.34}$$

Observe that maximizing $p(\mathbf{y}|\mu(\mathbf{c}))$ is equivalent to minimizing

$$d_E^2 = \sum_{i=1}^{n} (r_i - \mu(c_i))^2, \tag{3.35}$$

which is called *squared Euclidean distance* (SED). Therefore, we have

$$\hat{\mathbf{c}} = \arg\max_{\mathbf{c}\in C} p\,(\mathbf{y}|\mu(\mathbf{c})) = \arg\min_{\mathbf{c}\in C} (\mathbf{y} - \mu(\mathbf{c}))^2. \tag{3.36}$$

The process of finding $\mathbf{c}$, depending on the scheme we employ, may require checking possibly an exponentially large number of binary error sequences $\hat{\mathbf{e}}$ to select the one that satisfies

$$\mathbf{H} \cdot (\theta(\mathbf{y}) \oplus \hat{\mathbf{e}})^T = \mathbf{0}, \tag{3.37}$$

where $\theta(\mathbf{y})$ returns the hard-decision demodulation of the received vector $\mathbf{y}$ and $\mathbf{H} = [\mathbf{h}_1\, \mathbf{h}_2\, \cdots\, \mathbf{h}_{n-k}]^T$, is the parity-check matrix of code $\mathscr{C}$, where the n-element row vectors for $j \in [1, n-k]$ are denoted by $\mathbf{h}_j = [h_{j,1}\; h_{j,2}\; \cdots\; h_{j,n}]$. Note that any valid codeword $\mathbf{c} = \theta(\mathbf{y}) \oplus \hat{\mathbf{e}}$ gives $\mathbf{H} \cdot \mathbf{c}^T = \mathbf{0}$. Here, $\hat{\mathbf{e}}$ is the binary error sequence that we refer to as an error pattern.

To obtain the error patterns in ML order, one can (1) generate all possible error patterns $\hat{\mathbf{e}}$, that is, $\sum_{j=1}^{n} \binom{n}{j}$ patterns, (2) sort them according to a likelihood measure such as the SED $(\mathbf{y} - \mu\,(\theta(\mathbf{y}) + \hat{\mathbf{e}}))^2$, and then (3) check them using (3.37) one by one from the smallest distance in ascending order. However, this approach is inefficient. In Chapter 10, we will describe two universal decoding algorithms, namely, the *ordered statistics decoding* (OSD) algorithm and the *guessing random additive noise decoding* (GRAND) algorithm.

It is worth noting that an ML decoding algorithm can also fail when the closest candidate to the received signals is not the transmitted sequence. The following example demonstrates this event.

Example 3.1 Suppose the codeword

$$\mathbf{x} = [0\ 1\ 0\ 1\ 0\ 1\ 0\ 1]$$

is modulated by bipolar signaling (BPSK modulation) as

$$\mu(\mathbf{x}) = [1\ {-1}\ \ 1\ {-1}\ \ 1\ {-1}\ \ 1\ {-1}],$$

and then transmitted over an AWGN channel. We use a reliability-based decoder to decode the received sequence

$$\mathbf{y} = [0.67\ {-0.46}\ \ 0.43\ {-1.19}\ 2.12\ \ 0.17\ \ 0.72\ \ 1.76].$$

The decoder finds the following codewords in the order of SED:

$$[1\ 1\ 1\ 1\ 0\ 0\ 1\ 0],\quad d_E^2 = 7.7791, \tag{3.38}$$

$$[0\ 0\ 0\ 0\ 0\ 0\ 0\ 0],\quad d_E^2 = 9.9273, \tag{3.39}$$

$$[0\ 1\ 0\ 1\ 0\ 1\ 0\ 1],\quad d_E^2 = 11.1023, \tag{3.40}$$

In this case, the codeword with the highest likelihood is not the one transmitted. Observe that the transmitted symbol −1 was heavily impacted by noise and changed to 1.76. This increases the distance with the correct codeword, resulting in a decoding failure. We show this in detail in Example 10.3 of Chapter 10.

3.3 Recursive Decoding

For the decoding of generalized concatenated codes or recursively constructed codes such as Reed–Muller (RM) codes, as discussed in Chapter 2, a soft decoding algorithm can be employed in a recursive fashion to decompose a block code into its constituent codes. This section describes the decoding procedure for RM codes.

The RM code $\mathcal{R}(r, m)$, denoted by $\mathscr{C}$, with length $N = 2^m$, order r, dimension $K = \sum_{j=0}^{r} \binom{m}{j}$, and minimum Hamming distance $d_{\text{min}} = 2^{m-r}$ can be decomposed using the $(\mathbf{c}^{(1)} \oplus \mathbf{c}^{(2)}, \mathbf{c}^{(2)})$-construction into $\mathscr{C}^{(1)}$ of $\mathcal{R}(r, m-1)$ and $\mathscr{C}^{(2)}$ of $\mathcal{R}(r-1, m-1)$ as demonstrated in Figure 3.1.

As discussed in Chapter 2, this construction is based on the Plotkin transform or binary Walsh–Hadamard matrix of size 2 × 2 of

$$\mathbf{P}_2 = \begin{bmatrix} 1 & 0 \\ 1 & 1 \end{bmatrix},$$

such that $[\mathbf{c}^{(1)} \oplus \mathbf{c}^{(2)} \;\; \mathbf{c}^{(2)}] = [\mathbf{c}^{(1)} \;\; \mathbf{c}^{(2)}] \cdot \mathbf{P}_2$. Note that the common transform for RM codes is $\mathbf{P}'_2 = \mathbf{P}_2^T = \begin{bmatrix} 1 & 1 \\ 0 & 1 \end{bmatrix}$ however, we use its transpose to match the polar coding transform.

We can continue the decomposition of constituent codes until we arrive at the trivial codes including repetition (REP) codes $\mathcal{R}(0, g)$ for $g = 2, 3, \dots, m-r$ and single parity check (SPC) codes $\mathcal{R}(h, h+1)$ for $h = 1, 2, \dots, r$. Figure 3.2 illustrates the decomposition tree for RM codes of length $N = 2^6$ and Figure 3.3 shows the tree specifically for the code $\mathcal{R}(3, 6)$.

The SPC codes $\mathcal{R}(h, h+1)$ can be decoded using *Wagner decoder*. A Wagner code consists of a sequence of $2^{h+1} - 1$ bits and a parity. The Wagner decoder decides about the received sequence based on the LLRs by following the ML hard decisions rule in (3.16). Then the decoded sequence is checked for parity. If the parity check

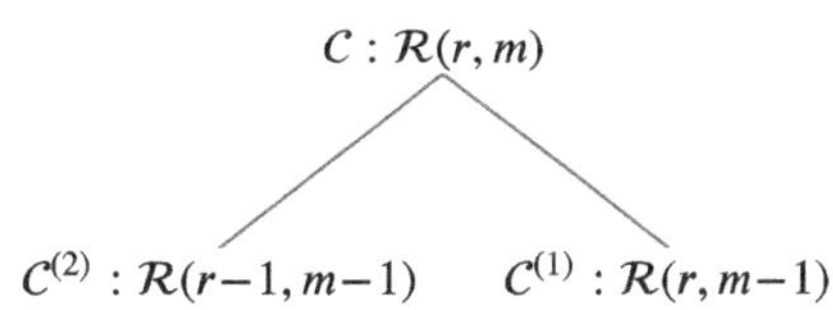

Figure 3.1 Decomposition of an RM code $\mathcal{R}(r, m)$.

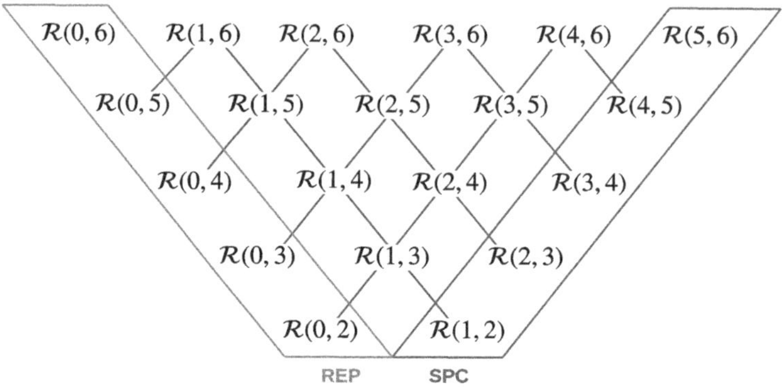

Figure 3.2 Decomposition tree for RM codes with block length up to $N = 2^6$.

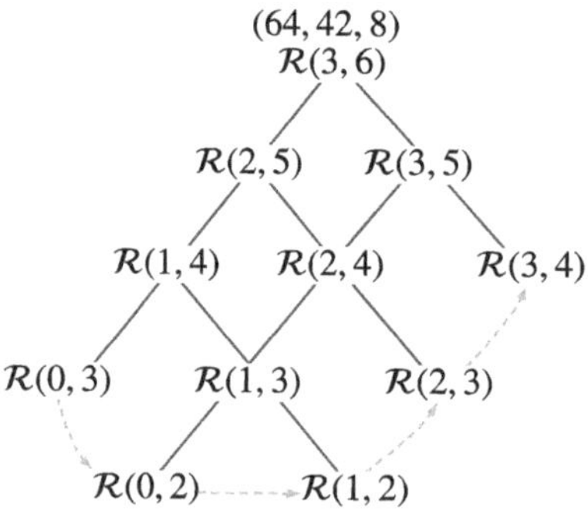

Figure 3.3 Decomposition tree for RM code $\mathcal{R}(3, 6)$. The gray dashed arrow shows the order of decoding of the trivial codes.

fails, the bit for which the LLR is the smallest is considered the bit most in doubt. Hence the decision on this bit is flipped.

The repetition (REP) codes $\mathcal{R}(0, g)$ can be decoded by *majority logic decoding*. Here, we can take advantage of soft information and decode the codeword $\hat{c}$ for which the metric $\sum_{j=0}^{2^g-1} \hat{c}_j \cdot \alpha(x_j)$ is maximized. For repetition codes, this is equivalent to the revised ML rule in (3.16) as follows:

$$\hat{c}_j = \text{sign}\left(\sum_{i=0}^{2^g-1} \alpha(x_i)\right), \quad j = 0, 1, \dots 2^g - 1. \tag{3.41}$$

The soft information for each constituent codeword $\mathbf{c}^{(1)}$ and $\mathbf{c}^{(2)}$ needs to be computed. Since the recursive decomposition is basically the inverse of recursive construction, we here develop operations on the soft information received from the channel or from the previous recursion. Assuming the soft information for single bits $x^{(1)} = c^{(1)} \oplus c^{(2)}$ and $x^{(2)} = c^{(2)}$ is available in the form of LLRs, the soft decomposition can be performed as follows:

1) Obtaining $\alpha(c^{(1)})$: Since $x^{(1)} = c^{(1)} \oplus c^{(2)}$ and $x^{(2)} = c^{(2)}$, by substitution we can recover $c^{(1)}$ by $c^{(1)} = x^{(1)} \oplus x^{(2)}$. Given $\alpha(x^{(1)}), \alpha(x^{(2)})$, we need to find $\alpha(c^{(1)}) = \alpha(x^{(1)} \oplus x^{(2)})$. To obtain $\alpha(x^{(1)} \oplus x^{(2)})$, we can use (3.20) or its approximation (3.21) as follows:

$$\alpha(c^{(1)}) = \log \frac{1 + e^{\alpha(x^{(1)})}e^{\alpha(x^{(2)})}}{e^{\alpha(x^{(1)})} + e^{\alpha(x^{(2)})}} \tag{3.42}$$

$$\approx \text{sign}\left(\alpha(x^{(1)})\right) \cdot \text{sign}\left(\alpha(x^{(2)})\right) \cdot \min\left(|\alpha(x^{(1)})|, |\alpha(x^{(2)})|\right). \tag{3.43}$$

2) Obtaining $\alpha\left(c^{(2)}\right)$: Since $x^{(1)} = c^{(1)} \oplus c^{(2)}$, after determining $\hat{c}^{(1)}$ in the preceding step, we have more information to decide on $\hat{c}^{(2)}$. Here we develop a probabilistic relationship and convert it to an LLR-based relation. This relationship depends on value decided for $\hat{c}^{(1)}$. Let us first see what this is relation for both cases of $\hat{c}^{(1)} = 1$ and $\hat{c}^{(1)} = -1$ individually. Given $\hat{c}^{(1)} = 1$, we can have $\hat{c}^{(2)} = 1$ if we have both $x^{(2)} = 1$ and $x^{(1)} = 1$. Similarly, we can have $\hat{c}^{(2)} = -1$ if we have both $x^{(2)} = -1$ and $x^{(1)} = -1$. Hence, we have

$$\begin{aligned}\alpha(c^{(2)}) &\overset{\hat{c}^{(1)}=1}{=} \log \frac{p(c^{(1)} = 1)}{p(c^{(1)} = -1)} = \log \frac{p(x^{(1)} = 1)p(x^{(2)} = 1)}{p(x^{(1)} = -1)p(x^{(2)} = -1)} \\ &= \log \frac{p(x^{(1)} = 1)}{p(x^{(1)} = -1)} + \log \frac{p(x^{(2)} = 1)}{p(x^{(2)} = -1)} = \alpha(x^{(1)}) + \alpha(x^{(2)}).\end{aligned} \tag{3.44}$$

We can come up with a similar relation when $\hat{c}^{(1)} = -1$. We can have $\hat{c}^{(2)} = -1$ if not only $x^{(2)} = -1$ but also $x^{(1)} = 1$. Similarly, we can have $\hat{c}^{(2)} = 1$ if not only $x^{(2)} = 1$ but also $x^{(1)} = -1$. Hence, we have

$$\begin{aligned}\alpha(c^{(2)}) &\overset{\hat{c}^{(1)}=-1}{=} \log \frac{p(c^{(1)} = 1)}{p(c^{(1)} = -1)} = \log \frac{p(x^{(1)} = -1)p(x^{(2)} = 1)}{p(x^{(1)} = 1)p(x^{(2)} = -1)} \\ &= -\log \frac{p(x^{(1)} = 1)}{p(x^{(1)} = -1)} + \log \frac{p(x^{(2)} = 1)}{p(x^{(2)} = -1)} = -\alpha(x^{(1)}) + \alpha(x^{(2)}).\end{aligned} \tag{3.45}$$

Combining (3.44) and (3.45), we can draw the following general relation:

$$\alpha(c^{(2)}) = \hat{c}^{(1)} \cdot \alpha(x^{(1)}) + \alpha(x^{(2)}). \tag{3.46}$$

Since the soft-decision ML decoding of the trivial codes is simple, a computationally efficient decoder for RM codes can be obtained, based on computing the soft information for constituent codes and hard-decision on trivial codes. Note that recursive decoding based on trivial codes is not ML decoding. In fact, the Wagner decoder assumes that a single error has occurred in the sequence and if more errors are present, it will propagate wrong decisions to the following stages, which cannot be recovered later in the process. This explains why even if the ML decoder for the local constituent codes is used, it does not result in ML decoding for the whole code.

3.4 Message-Passing (MP) Decoding

In this section, we introduce a decoding scheme that is traditionally used for low-density parity-check (LDPC) codes, as this scheme performs well with sparse parity-check matrices. Let us first introduce the factorization concept and the graph representation of a parity-check matrix.

3.4.1 Factorization and Factor Graph

The well-known distributive law can help us in the separation of independent variables and factorization in multivariate functions. A careful application of this simple law can significantly reduce complexity. For example, the evaluation of the expression $\sum_{i=1}^{n}\sum_{j=1}^{n} u_i v_j$ requires n^2 multiplications and $n^2 - 1$ additions (the complexity is quadratic in n). However, evaluation of the equivalent expression $(\sum_{i=1}^{n} u_i)(\sum_{j=1}^{n} v_j)$ only requires $2(n-1)$ additions and one multiplication (the complexity is linear in n), while these two are equivalent.

Consider a binary linear code $\mathscr{C}$ defined by a parity-check matrix $\mathbf{H}$. Let $\mathbb{F}_2$ be the binary field with elements in $\{0, 1\}$ and $\mathbf{x} = [x_0, x_1, \ldots, x_{N-1}]$. The *code membership function* $\mathbb{1}_{\{\cdot\}}$ is an *indicator* function that gives 1 when the condition inside the braces is fulfilled; otherwise, it returns 0. Explicitly, it is defined as

$$\mathbb{1}_{\{\mathbf{x}\in\mathscr{C}\}} = f(x_0, x_1, \ldots, x_{N-1}) = \begin{cases} 1 \text{ if } \mathbf{x}\mathbf{H}^T = \mathbf{0}, \\ 0 \text{ otherwise.} \end{cases} \tag{3.47}$$

Here,

Example 3.2 Consider the Hamming code (7,4) with parity-check matrix

$$\mathbf{H} = \begin{bmatrix} 1 & 1 & 0 & 1 & 1 & 0 & 0 \\ 1 & 0 & 1 & 1 & 0 & 1 & 0 \\ 0 & 1 & 1 & 1 & 0 & 0 & 1 \end{bmatrix},$$

we can factor f as

$$f(x_0, x_1, \ldots, x_6) = \mathbb{1}_{\{x_0+x_1+x_3+x_4=0\}}\mathbb{1}_{\{x_0+x_2+x_3+x_5=0\}}\mathbb{1}_{\{x_1+x_2+x_3+x_6=0\}}. \tag{3.48}$$

Hence, all three conditions inside the braces of the indicator functions must be fulfilled to conclude that $\mathbf{x}$ is a Hamming codeword.

Now, to associate the factorization of f (like the one in (3.48)) with a graph, we draw a circle to represent a *variable node* for each component of the codeword (variables x_0 to x_6) and a square to represent a *check node* for each factor indicator function, i.e. the parity-check equation corresponding to the rows of $\mathbf{H}$. Then, if a variable i participates in the jth parity-check constraint, i.e. if $h_{j,i} = 1$, connect

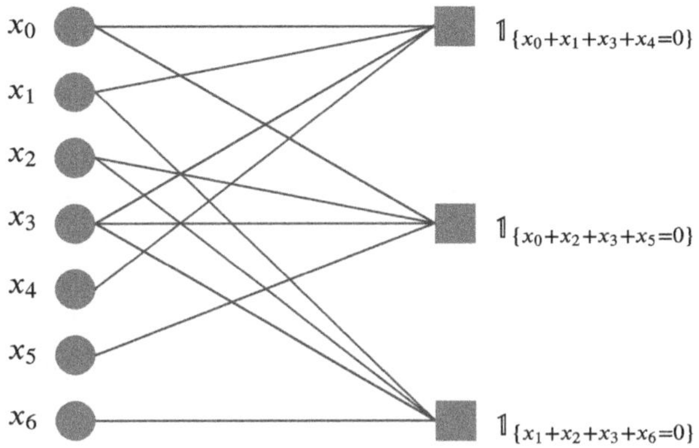

Figure 3.4 Factor graph for the code membership.

the variable node i to the check node j by an *edge*. The resulting graph visualizes the factorization of the code membership function and is called the *factor graph*. It is also a graphical representation of the parity-check matrix. The factor graph for the function (3.48) is shown in Figure 3.4. A graph such as this, consisting of two distinct sets of nodes (corresponding to the rows and columns of the $\mathbf{H}$ matrix) and having edges only between the nodes in different sets (corresponding to entry 1 in the $\mathbf{H}$ matrix), is called a *bipartite* graph. Tanner first proposed this graph for LDPC codes, and it is also known as the *Tanner* graph. Since there are many equivalent parity-check matrices representing the same code, there will be many Tanner graphs corresponding to a given $\mathscr{C}$. Although all of these Tanner graphs describe the same code, from the point of view of message-passing decoding, they are not equivalent.

3.4.2 Message Passing

Bit-wise MAP decoding aims to determine the most likely value of each transmitted bit given the received observations. The decoding process requires the computation of marginal probabilities for individual bits from a joint multivariate distribution of transmitted and received signals.

We determine the marginals of multivariate conditional probability functions to evaluate the LLRs for each bit, as given by

$$\lambda(x_i|\mathbf{y}) = \log \frac{p(x_i = 1|y_i, \{y_j, j \neq i\})}{p(x_i = 0|y_i, \{y_j, j \neq i\})}. \tag{3.49}$$

Based on Bayes' rule, the numerator becomes

$$\begin{aligned} p(x_i = 1|y_i, \{y_j, j \neq i\}) &= \frac{p(x_i = 1, y_i, \{y_j, j \neq i\})}{p(y_i, \{y_j, j \neq i\})} \\ &= \frac{p(y_i|x_i = 1, \{y_j, j \neq i\})p(x_i = 1, \{y_j, j \neq i\})}{p(y_i|\{y_j, j \neq i\})p(\{y_j, j \neq i\})} \\ &= \frac{p(y_i|x_i = 1)p(x_i = 1, \{y_j, j \neq i\})}{p(y_i|\{y_j, j \neq i\})p(\{y_j, j \neq i\})} \\ &= \frac{p(y_i|x_i = 1)p(x_i = 1|\{y_j, j \neq i\})}{p(y_i|\{y_j, j \neq i\})}. \end{aligned} \tag{3.50}$$

Note that $p(y_i|x_i = 1, \{y_j, j \neq i\}) = p(y_i|x_i = 1)$ because given x_i, y_i is independent of $\{y_j, j \neq i\}$ in memoryless channels. We derive a similar expression for the denominator of (3.49), where $x_i = 0$. Since both expressions share the same denominator $p(y_i|\{y_j, j \neq i\})$, we have

$$\begin{aligned} \lambda(x_i|\mathbf{y}) &= \log \frac{p(y_i|x_i = 1)p(x_i = 1|\{y_j, j \neq i\})}{p(y_i|x_i = 0)p(x_i = 0|\{y_j, j \neq i\})} \\ &= \underbrace{\log \frac{p(y_i|x_i = 1)}{p(y_i|x_i = 0)}}_{\text{intrinsic}} + \underbrace{\log \frac{p(x_i = 1|\{y_j, j \neq i\})}{p(x_i = 0|\{y_j, j \neq i\})}}_{\text{extrinsic}}, \end{aligned} \tag{3.51}$$

where the *intrinsic* term, the channel LLR, is determined explicitly by y_i and the *extrinsic* term is affected by all other $y_j, j \neq i$ and the structure of the code, namely the parity-check constraints. The intrinsic term for a Gaussian channel is $2\sqrt{E_s}/\sigma^2 \cdot y_i$ as shown in Section 3.1.1.

Next, we involve the code structure to express the extrinsic term in terms of the parity-check constraints. Given a parity-check matrix $\mathbf{H}$, let $\mathcal{N}_m = \{j : h_{m,j} = 1\}$ identify the set of bits participating in the mth parity-check constraint (or the set of variable nodes that are connected to the check node m in the factor graph) and $\mathcal{M}_j = \{m : h_{m,j} = 1\}$ identify the set of parity-check constraints containing the jth bit (or the set of check nodes that are connected to the variable node j in the factor graph). Given $i \in \mathcal{N}_m$, we form a parity check denoted by $p_{m,i}$, where x_i is excluded from the mth parity-check constraint. That is,

$$p_{m,i} = \sum_{j \in \mathcal{N}_m \setminus \{i\}} x_j. \tag{3.52}$$

This is in line with the extrinsic term, where we consider all other observations except for y_i, i.e. $y_j, j \neq i$. Observe that by excluding x_i from the mth parity-check constraint, we have $p_{m,i} + x_i = 0$ or $p_{m,i} = x_i$. As a result, depending on $x_i = 0$ or 1, we have $p_{m,i} = 0$ or 1, respectively, for all $m \in \mathcal{M}_i$. We can rewrite (3.51) as

$$\lambda(x_i|\mathbf{y}) = \log \frac{p(y_i|x_i = 1)}{p(y_i|x_i = 0)} + \log \frac{p(p_{m,i} = 1 \text{ for all } m \in \mathcal{M}_i|\{y_j, j \neq i\})}{p(p_{m,i} = 0 \text{ for all } m \in \mathcal{M}_i|\{y_j, j \neq i\})}. \quad (3.53)$$

If we assume that the factor graph associated with the code is cycle-free, meaning that there are no loops in the graph representation of the parity-check matrix, then the bits associated with $p_{m,i}$ are independent of the bits associated with $p_{m',i}$ for $m \neq m'$. Hence, we can write the following.

$$\lambda(x_i|\mathbf{y}) = \log \frac{p(y_i|x_i = 1)}{p(y_i|x_i = 0)} + \log \frac{\prod_{m \in \mathcal{M}_i} p(p_{m,i} = 1|\{y_j, j \neq i\})}{\prod_{m \in \mathcal{M}_i} p(p_{m,i} = 0|\{y_j, j \neq i\})}. \quad (3.54)$$

Observe that the factorization of the second term into a product over all parity-check constraints allows us to split the log of the product into a sum of log as

$$\lambda(x_i|\mathbf{y}) = \log \frac{p(y_i|x_i = 1)}{p(y_i|x_i = 0)} + \sum_{m \in \mathcal{M}_i} \log \frac{p(p_{m,i} = 1|\{y_j, j \neq i\})}{p(p_{m,i} = 0|\{y_j, j \neq i\})}. \quad (3.55)$$

Let us denote the component of the second term as the LLR of $p_{m,i}$

$$\lambda(p_{m,i}|\{y_j, j \neq i\}) = \lambda\left(\sum_{j \in \mathcal{N}_m \setminus \{i\}} x_j|\{y_j, j \neq i\}\right) = \log \frac{p(p_{m,i} = 1|\{y_j, j \neq i\})}{p(p_{m,i} = 0|\{y_j, j \neq i\})} \quad (3.56)$$

Then, we have

$$\lambda(x_i|\mathbf{y}) = \lambda(y_i|x_i) + \sum_{m \in \mathcal{M}_i} \lambda\left(\sum_{j \in \mathcal{N}_m \setminus \{i\}} x_j|\{y_j, j \neq i\}\right) \quad (3.57)$$

From LLR properties and Algebra and based on (3.20), we can use the sum-product rule to further decompose the terms and express $\lambda\left(\sum_{j \in \mathcal{N}_m \setminus \{i\}} x_j|\{y_j, j \neq i\}\right)$ as a combination of incoming LLR messages from all variable nodes connected to check node m, excluding x_i. Denote the incoming LLR message from bit x_j to check node m as $\lambda_{j \to m}$. Then, the outgoing LLR message from check node m to bit x_i is

$$\lambda_{m \to i} = \lambda\left(\sum_{j \in \mathcal{N}_m \setminus \{i\}} x_j \mid \{y_j, j \neq i\}\right) = 2 \tanh^{-1}\left(\prod_{j \in \mathcal{N}_m \setminus \{i\}} \tanh\left(\frac{\lambda_{j \to m}}{2}\right)\right). \quad (3.58)$$

Hence, the outgoing message $\lambda_{m \to i}$ depends on the incoming messages $\{\lambda_{j \to m} \mid j \in \mathcal{N}_m \setminus \{i\}\}$. The function $\tanh^{-1}$ accounts for the nonlinear nature of combining probabilities in the LLR domain. The LLR of x_i given the observations $\mathbf{y}$ in (3.57) can now be written as

$$\lambda(x_i \mid \mathbf{y}) = \lambda_i + \sum_{m \in \mathcal{M}_i} \lambda_{m \to i}, \quad (3.59)$$

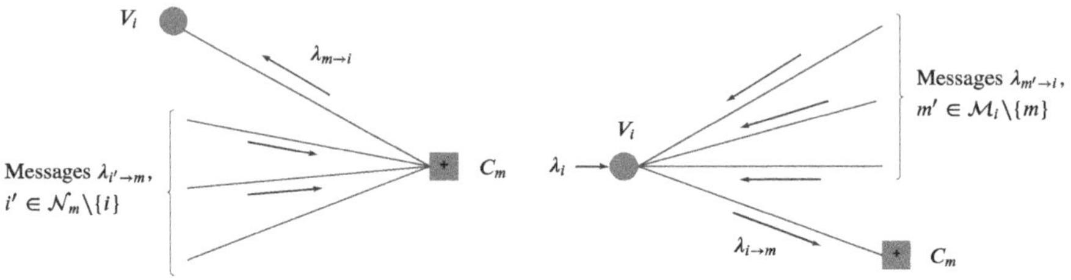

Figure 3.5 Messages from/to variable node V_i considering check node C_m.

where $\lambda_i = \lambda\left(y_i \mid x_i\right)$ is the channel LLR of variable x_i. This ensures that the message-passing algorithm maintains consistency between the incoming and outgoing messages in the factor graph.

Similarly, the message that the variable node V_i sends to the check node C_m is obtained by adding the message from all neighbors (including the channel LLR λ_i) except for the check node itself C_m, which is

$$\lambda_{i\to m} = \lambda_i + \sum_{m\in\mathcal{M}_i\backslash\{m\}} \lambda_{m\to i}. \tag{3.60}$$

Figure 3.5 illustrates the messages sent from V_i to C_m (right) and from C_m to V_i (left).

The process of passing the messages from variable nodes to check nodes and back is repeated up to a maximum number of iterations where the messages supposedly converge. Then, for each i, we decide $\hat{x}_i = 1$ if $\lambda\left(x_i \mid \mathbf{y}\right) < 0$, otherwise $\hat{x}_i = 0$. One can check if $H\hat{\mathbf{x}}^T = 0$ before the maximum number of iterations and stop the message passing algorithm earlier. We will use the concept of message passing for belief propagation decoding of polar codes in Chapter 10.

3.4.3 The Effect of Cycles on Message Passing

In a cycle-free graph, the factor graph is a tree structure, meaning there is a unique path between any two nodes. This ensures the independence of messages passed along the graph. For instance, if the LLR of x_i is expressed as

$$\lambda\left(x_i \mid \mathbf{y}\right) = \lambda\left(y_i \mid x_i\right) + \sum_{m\in\mathcal{M}_i} \lambda\left(p_{m,i} \mid \left\{y_j, j \neq i\right\}\right),$$

where each $\lambda\left(p_{m,i} \mid \left\{y_j, j \neq i\right\}\right)$ can be independently calculated based on parity-check constraints involving x_j for $j \in \mathcal{N}_m \backslash \{i\}$. The independence ensures that

$$p\left(p_{m,i} = 1 \mid \left\{y_j, j \neq i\right\}\right) = \prod_{j\in\mathcal{N}_m\backslash\{i\}} p\left(x_j \mid y_j\right).$$

Thus, the LLR are consistent, leading to correct decoding.

3.4.3.1 Scenario with Cycles

Now consider a factor graph with a loop. For simplicity, assume a loop exists between $x_i, x_j, x_k, \ldots, x_i$, meaning these variables are indirectly connected multiple times through shared parity-check constraints. The impact or cycles are as follows:

1) Double counting of information: Messages passed along the looped edges may revisit a node multiple times, causing the same information to be counted more than once. For example, suppose the LLR for x_i involves terms dependent on x_j and x_k in the loop:

$$\lambda\left(x_i \mid \mathbf{y}\right) = \lambda\left(y_i \mid x_i\right) + \sum_{m \in \mathcal{M}_i} \lambda\left(p_{m,i} \mid \left\{y_j, j \neq i\right\}\right),$$

but due to the loop, $p_{m,i}$ itself depends indirectly on x_i, violating the independence assumption:

$$p\left(p_{m,i} = 1 \mid \left\{y_j, j \neq i\right\}\right) \neq \prod_{j \in \mathcal{N}_m \setminus \{i\}} p\left(x_j \mid y_j\right).$$

The incorrect dependencies result in biased LLRs.

2) Inconsistent LLR updates: The loop causes contradictory updates to the LLRs. Suppose $p_{m,i}$ depends on both x_j and x_k, but x_j and x_k are part of a loop. Then, $\lambda\left(p_{m,i} \mid \left\{y_j, j \neq i\right\}\right)$ includes terms from x_j and x_k, but due to the loop, x_k's value is influenced by x_i, leading to circular dependencies. This inconsistency can manifest as oscillations resulting in incorrect convergence in iterative decoders such as belief propagation.

Example 3.3 Consider a parity-check matrix H with a 4-cycle between variable nodes x_1, x_2 and parity-check nodes p_1, p_2. The dependencies can be written as

$$p_1 : x_1 + x_2 + x_3 = 0, \quad p_2 : x_2 + x_3 + x_4 = 0.$$

A 4-cycle exists when there is a closed path in the graph involving 4 alternating nodes (2 variable nodes and 2 check nodes). In this example, the 4-cycle is as follows.

$$x_2 \rightarrow p_1 \rightarrow x_3 \rightarrow p_2 \rightarrow x_2,$$

This path forms a loop involving the variable nodes x_2, x_3 and the parity-check nodes p_1, p_2.

The LLR for x_2 involves contributions from both p_1 and p_2 as

$$\lambda\left(x_2 \mid \mathbf{y}\right) = \lambda\left(y_2 \mid x_2\right) + \lambda\left(p_{1,2} \mid \left\{y_j, j \neq 2\right\}\right) + \lambda\left(p_{2,2} \mid \left\{y_j, j \neq 2\right\}\right).$$

However, because of the cycle, $p_{1,2}$ depends on x_3, and $p_{2,2}$ also depends on x_3, introducing correlations that are ignored in the LLR computation as

$$p\left(p_{1,2} = 1 \mid \{y_j, j \neq 2\}\right) \quad \text{and} \quad p\left(p_{2,2} = 1 \mid \{y_j, j \neq 2\}\right)$$

are not independent. This leads to incorrect updates in the iterative decoding process.

3.5 Tree/Trellis-Based Search Decoding

In this section, we discuss a class of decoding algorithms that are based on tree traversal. Tree codes are defined by labeling the edges of a tree with symbols of some alphabet (here $\{0, 1\}$ that yield a binary tree). As the edges are randomly labeled depending on the source bits, these codes can be essentially considered distributions over labeled trees. Random tree codes were introduced by Wozencraft and used for sequential decoding. Later, Schulaman introduced a stronger notion of (deterministic) tree codes for a different purpose, which is not the subject of this book.

The binary tree is a known data structure in computer science in which each node has at most two children, and there are edges connecting these nodes. The string labeling the edges from the origin to any leaf is called a path representing a valid codeword. The edge labels belonging to parity bits connect the parent node to a single child node, as these edges depend on the label of source bits and are not free to take any value. This results in irregular trees in which $N - K$ nodes out of N nodes on each path have a single child node. Figure 3.6 illustrates an irregular tree.

A decoding algorithm can explore the paths of this tree guided by a metric to measure the distance between the received sequence and the chosen path(s). There are various tree traversal algorithms, and we describe the popular ones in the following.

3.5.1 Sequential Decoding

Recognized by John Wozencraft, sequential decoding explores the tree code in such a way that minimizes memory requirements to store the tree. This algorithm was used for decoding convolutional codes in the Pioneer 9 mission in 1968. Although it is mainly used for convolutional codes, it has found its way to other codes such as polar codes, as we will see in Chapter 10.

The fundamental idea behind sequential decoding is to explore only the most promising path(s). If a path to a node looks "bad," we can discard all the paths through this node without a significant loss in the error correction performance compared to that of a maximum likelihood decoder.

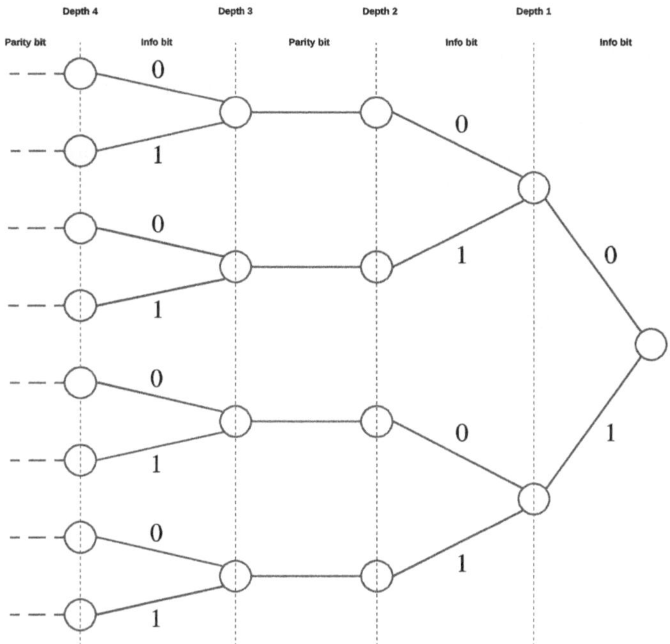

Figure 3.6 An example of code tree where 3rd and 5th bits are parity bits, represented by edges connected to a single child node.

3.5.1.1 Fano Algorithm

The *Fano algorithm*, named after Robert Fano, is a depth-first memory-efficient tree search, in which the decoder moves from a node either back to its parent node or to one of its children. The Fano decoder can visit a node only if its *path metric* μ_F is larger than or equal to a certain value called *threshold* T. The threshold takes only discrete values $0, \pm\Delta, \pm 2\Delta, \ldots$.

The complexity of sequential decoding is essentially independent of the encoder memory, since only one encoder state is examined at each step. Hence, this algorithm is a memory-efficient type of sequential decoding algorithm.

3.5.1.2 Stack Decoding

A simpler sequential decoding algorithm is the *stack algorithm*, where the algorithm maintains a stack of size/depth D of the most promising partial paths sorted in a "stack" data structure, ordered with respect to the path metric. That is, the best path is placed at the top of the stack followed by the second-best path, and so on. The algorithm extends the path with the best metric at the top of the stack. The stack decoding is a memory-intensive algorithm with a variable time complexity that, instead of backtracking as in the Fano decoding, selects to extend the best partial path in the stack at each time step. At each time step, the top node is removed from the ordered stack, and its children are placed in the stack according to their metrics. If the top node is a leaf node of the tree, the decoding terminates, and such a path is considered the decoded codeword. This algorithm requires a sorting operation and a large stack to perform well. In general, the Fano algorithm visits more nodes than the stack algorithm and requires essentially no memory for the stack of partial paths.

3.5.1.3 Path Metric

The metric used in the sequential decoding of convolutional codes is a probabilistic path metric. We consider the set $\mathcal{X} = \{\mathbf{a}^{(1)}, \mathbf{a}^{(2)}, ..., \mathbf{a}^{(M)}\}$ of M partial sequences, representing partially explored paths of different lengths, to be compared. Let $n_{\max} = \max\{n_1, n_2, ..., n_M\}$ denote the length of the longest sequence, and $\tilde{\mathbf{r}}$ the partial received sequence of length $n_{\max}$ symbols where each encoded symbol takes n bits corresponding to k information/uncoded bits, $R = k/n$. Note that the parameters n and k in stream encoding are different from the ones in block coding. Hence, the sequences $\tilde{\mathbf{r}}$ and $\mathbf{a}^{(\ell)}$ are

$$\tilde{\mathbf{r}} = (\mathbf{r}_0\mathbf{r}_1 \dots \mathbf{r}_{n_{\max}-1}) = (r_0 \dots r_{n-1} \dots r_{nn_{\max}-1})$$

$$\mathbf{a}^{(\ell)} = (\mathbf{a}_0^{(\ell)}\mathbf{a}_1^{(\ell)} \dots \mathbf{a}_{n_\ell-1}^{(\ell)}) = (a_0^{(\ell)} \dots a_{n-1}^{(\ell)} \dots a_{nn_\ell-1}^{(\ell)}).$$

Among the sequences in $\mathcal{X}$, we choose the partial sequence $\mathbf{a}^{(\ell)}$ that maximizes the a posterior probability $P(\mathbf{a}^{(\ell)}|\tilde{\mathbf{r}})$. According to Bayes' rule

$$P(\mathbf{a}^{(\ell)}|\tilde{\mathbf{r}}) = \frac{P(\mathbf{a}^{(\ell)})P(\tilde{\mathbf{r}}|\mathbf{a}^{(\ell)})}{P(\tilde{\mathbf{r}})}. \tag{3.61}$$

Assuming the channels are memoryless, since the length of the sequence $\mathbf{a}^{(\ell)}$ is $n_\ell \leq n_{\max}$, and there are no associated symbols in this sequence for $r_{n_\ell}, \dots, r_{n_{\max}-1}$, then we have

$$P(\tilde{\mathbf{r}}|\mathbf{a}^{(\ell)}) = \prod_{j=0}^{n_\ell-1} P(\mathbf{r}_j|\mathbf{a}_j^{(\ell)}) \prod_{j=n_\ell}^{n_{\max}-1} P(\mathbf{r}_j), \tag{3.62}$$

Also, we can rewrite the denominator of (3.61) as

$$P(\tilde{\mathbf{r}}) = \prod_{j=0}^{n_\ell-1} P(\mathbf{r}_j) \prod_{j=n_\ell}^{n_{\max}-1} P(\mathbf{r}_j). \tag{3.63}$$

Now, by substituting (3.62) and (3.63) in (3.61) and cancelling the common term $\prod_{j=n_\ell}^{n_{\max}-1} P(\mathbf{r}_j)$, we have

$$P(\mathbf{a}^\ell|\tilde{\mathbf{r}}) = P(\mathbf{a}^\ell) \prod_{i=0}^{nn_\ell-1} \frac{P(r_i|a_i^{(\ell)})}{P(r_i)}. \tag{3.64}$$

Suppose that each encoded bit occurs with equal probability, then each sequence $\mathbf{a}^{(\ell)}$ occurs with probability $P(\mathbf{a}^{(\ell)}) = (2^{-k})^{n_\ell} = (2^{-nR})^{n_\ell} = (2^{-R})^{nn_\ell}$. Thus, by taking the base-2 logarithm of (3.64), we have

$$\log P(\mathbf{a}^{(\ell)}|\tilde{\mathbf{r}}) = \sum_{i=0}^{nn_\ell - 1} \left(\underbrace{\log P(r_i|a_i^{(\ell)})}_{\text{ML-metric}} - \underbrace{\log P(r_i) - R}_{\text{path-length bias}} \right). \tag{3.65}$$

3.5.1.4 Computational Cutoff-Rate

The *computational cut-off rate*, denoted by R_{comp} or R_0, is a crucial parameter in digital communication. This rate is of particular significance in the context of sequential and list decoding, being restricted to be less than or equal to the channel capacity; $R_{\text{comp}} \leq C$. The following findings demonstrate why this parameter is regarded as a practical limit for communications (without a feedback channel). For sequential decoding and when $R > R_{\text{comp}}$, the average number of computations required for error-free decoding tends toward infinity, making sequential decoding impractical. The size of the candidate list in list decoding, which always includes the transmitted codeword, can decrease exponentially with the code length only for $R \leq R_{\text{comp}}$. Furthermore, the error probability gain of a list decoding over a conventional decoder—deciding only for one codeword—is significant if $R < R_{\text{comp}}$, whereas only a marginal gain can be expected for $R > R_{\text{comp}}$. Theoretically, error-free communication is possible for $R_{\text{comp}} < R < C$, but only at the cost of significantly increased decoding complexity.

Generally, R_{comp} is defined as follows:

$$R_{\text{comp}} \triangleq \max_{P(x)} \left[-\log_2 \sum_{y \in Y} \left(\sum_{x \in X} P(x) \sqrt{P(y \mid x)} \right)^2 \right], \tag{3.66}$$

where, for continuous input and output symbols x, y, the sums are replaced by integrals. For BPSK transmission, or in general antipodal signaling, over the AWGN channel, with $x \in \{1, -1\}$, R_{comp} can be expressed in closed form as:

$$R_{\text{comp}} = \log_2 \left(\frac{2}{1 + e^{-E_s/N_0}} \right) = 1 - \log_2 \left(1 + e^{-E_s/N_0} \right). \tag{3.67}$$

As shown above, increasing E_s/N_0 would increase the cutoff rate. As long as the transmission in bits per dimension, or code rate R, is less than the cutoff rate R_{comp}, the block error probability can be arbitrarily small by choosing a significantly large N as the upper bound for the block error probability is

$$P_e < 2^{-N[R_{\text{comp}} - R]}. \tag{3.68}$$

This upper bound originates from sphere-packing and random coding arguments in information theory (for further reading, we refer you to references provided in the bibliographical notes, as we will use it again in Chapter 5). When the code rate R is below R_{comp}, there exists a codebook such that the probability of decoding error is dominated by the exponentially vanishing tail of the probability distribution over the noise sphere. Let us look at the parameters of this bound in the following:

1) Exponential dependence on block length (N): The probability of error decreases exponentially with N because a larger block length provides more redundancy, allowing the decoder to distinguish codewords more reliably. 2. Rate Difference ($R_{\text{comp}} - R$): The term $R_{\text{comp}} - R$ in the exponent reflects the "gap" between the chosen code rate R and the maximum tolerable rate R_{comp}. A larger gap ensures faster error probability decay, as there is more redundancy to correct errors.
2) General validity for random coding: This bound is derived for random coding ensembles under maximum-likelihood decoding, where the performance represents the best achievable reliability for a given R and N. While practical codes may approach this bound, the bound itself provides a theoretical benchmark.

Thus, this expression quantifies how rapidly the error probability can be reduced by increasing the block length N, provided $R < R_{\text{comp}}$.

3.5.2 List Decoding

Unlike sequential decoding and sphere decoding where one path is followed at a time, in list decoding a number of paths (denoted as the *list size* L) are explored simultaneously. When reaching the leaf nodes, at depth N, a tree path on the list with the highest likelihood, the best metric, or closest to the received sequence is declared as the estimated transmitted message. We can take advantage of the detection capability of cyclic codes to detect the transmitted message as well (see Chapter 2), however, this requires concatenated coding.

3.5.3 Trellis-Based Viterbi Decoding

The Viterbi algorithm (VA) is a maximum-likelihood decoding method traditionally used to decode convolutional codes. This algorithm examines the entire state space of convolutional codes at each step (see Chapter 2). Consequently, the space complexity (storage requirement) of the Viterbi algorithm grows exponentially with the constraint length $m + 1$ of the encoder. However, by making a decision about the correct path after decoding every short-length sequence and truncating the path memory for the rest of the paths, the space complexity is constrained at the cost of insignificant degradation in error rate performance.

3.5.4 Lattice-Based Sphere Decoding

The sphere decoding is founded on a simple idea: We search over integer lattice points $\mathbf{1} - 2\mathbf{u}G$ for valid $\mathbf{u}$'s that lie in a certain sphere with radius r around a received noisy vector $\mathbf{y}$ in a N-dimensional space. In the first step of the algorithm, the sphere decoder performs a depth-first search with a very large initialized radius r_0. Radius r_0 may reduce later as the decoder explores the decoding tree looking for alternative candidates. By finding the candidate that allows the sphere to have a minimum radius, SD achieves the ML estimation for the transmitted data $\mathbf{u}$ with respect to the received vector $\mathbf{y}$.

Exercises

1 Derive the right-hand side of (3.11).

2 Derive (3.20) from (3.18).

3 In some cases, it is convenient to deal with $\{0, 1\}$-valued random variables. Let u_1 and u_2 be $\{0, 1\}$-valued variables. What is the analogous to (3.20)? What is the identity element for the addition operation in this case, analogous to the "null" element for addition over x_1 and x_2?

4 If we have $u^{(1)}$ and $u^{(2)}$ as $\{0, 1\}$-valued variables instead of $c^{(1)}, c^{(2)} \in \{1, -1\}$, show that the relation (3.46) will be

$$\alpha(u^{(2)}) = (-1)^{u^{(1)}} \cdot \alpha(x^{(1)}) + \alpha(x^{(2)}). \tag{3.69}$$

5 Using the relations

$$\tanh(x/2) = \frac{e^x - 1}{e^x + 1} \quad \text{and} \quad \log\left(\frac{1+x}{1-x}\right) = 2\tanh^{-1}x,$$

show that

$$\alpha(x_1 \oplus x_2) = 2\tanh^{-1}\left(\tanh\left(\frac{\alpha(x_1)}{2}\right) \cdot \tanh\left(\frac{\alpha(x_2)}{2}\right)\right).$$

Bibliographic Notes

Soft decision-based decoding algorithms are classified into two main categories: Code structure-based algorithms and reliability-based algorithms (a.k.a. generic/general/universal decoding algorithms) because they usually do not depend on the code structure. As originally shown by Van Tilborg et al. [1978], ML decoding is a non-deterministic polynomial (NP)-complete problem and therefore prohibitively complex for most codes. Forney [1966] proposed the generalized minimum distance (GMD) decoding algorithm, where a list of candidate codewords was produced based on the reliability of the received symbols using an algebraic decoder. Chase [1972] proposed a method in which the search was performed among a fixed number of error patterns corresponding to a particular number of least reliable bit positions with respect to the minimum distance d of the underlying code. Chase classified his algorithm into three types, according to the error pattern generation. In another effort, Snyders and Be'ery [1989] proposed to perform syndrome decoding on the received sequence and then use the syndrome information to modify and improve the original hard decision decoding.

The best-known generic decoding algorithm is perhaps the information set decoding (ISD) algorithm proposed by Prange [1962], which was later improved by Stern [1989] and Dumer [1991]. Following this approach, other generic decoding approaches were developed based on the most reliable basis (MRB), defined as the support of the most reliable independent positions (MRIPs) of the received sequence, hence forming an information set. In these approaches, each error pattern is subtracted from the hard decision of the MRIPs, and the corresponding codeword is reconstructed by encoding the corresponding information sequence. Dorsch [1974] considered error patterns restricted to the MRB in increasing a priori likelihood order. Following this approach, Fossorier and Lin [1995] proposed processing the error patterns in a deterministic order within families with increasing Hamming weight. This algorithm, which is referred to as ordered statistics decoding (OSD), is one of the most popular generic decoding algorithms today. The OSD algorithm permutes the columns of the generator matrix with respect to the reliability of the symbols for every received vector and performs elementary row operations on the independent columns extracted from the permuted generator matrix, resulting in the systematic form. Testing error patterns can have a Hamming weight of up to $l, 0 \leq l \leq k$ in l-order OSD, chosen from the most reliable k positions. The main drawback of OSD is the use of row operations to put either the generator matrix or the parity-check matrix of the code in systematic form. The complexity of row operation for an (n, k) linear block code is $O(n^3 \min\{R, 1-R\}^2)$, where R is the code rate. However, since overall complexity is an exponential function of code length, this pre-processing complexity is negligible. Moreover, having information set in a systematic form is needed only to further simplify decoding attempts. Otherwise, the error patterns can be checked without this preprocessing. The OSD algorithm evolved into the box-and-match algorithm (BMA) by Valembois and Fossorier [2004], where the matching technique was used to reduce the time complexity at the cost of space complexity. It should be noted that a similar algorithm to BMA, called the sort and match algorithm, was proposed by Dumer [1999], which has the same asymptotic complexity as BMA.

Reed–Muller codes were viewed as multiple concatenated codes that are constructed recursively, and the decoding procedures were advised by Litsyn [1988], Hemmati [1989], and Kabatyanskii [1990] in general for the semi-continuous channel and later by Schnabl and Bossert [1995] who gave specialized reliability values for the AWGN channel. Note that the idea of decoding the constituent codes of RM codes discussed in this chapter was later employed for simplified and fast decoding of polar codes, as we will discuss in Chapter 5. Among other significant work on decoding of RM codes, we can name recursive list decoding by Dumer and Shabunov [2006] which inspired the list decoding algorithm for polar codes, which is discussed in Chapter 8. Moreover, recursive projection-aggregation decoding for Reed–Muller code has been introduced by Ye and Abbe [2020] and Abbe et al. [2020].

The tree traversal algorithms are relatively old. Interested readers are referred to classical papers such as probabilistic decoding by Fano [1963], M-algorithm decoding of convolutional codes by Lin and Anderson [1986] which is similar to list decoding of polar codes, and lattice-based sphere decoding by Pohst [1981] and Viterbo and Boutros [1999].

There are also good books available as references: *Modern Coding Theory* book by Richardson and Urbanke [2008] for belief propagation decoding, *Fundamentals of Convolutional Coding* by Johannesson and Zigangirov [2015], and *Information theory and reliable communication* by Gallager [1968] for sequential decoding, path metric, cutoff-rate, and *Information Combining* by Land and Huber [2006] for information processing characteristics of coding schemes. Lastly, the LLR algebra discussed in this chapter was proposed by Hagenauer et al. [1996].

Bibliography

Emmanuel Abbe, Amir Shpilka, and Min Ye. Reed–Muller codes: Theory and algorithms. *IEEE Transactions on Information Theory*, 67(6):3251–3277, 2020.

David Chase. Class of algorithms for decoding block codes with channel measurement information. *IEEE Transactions on Information Theory*, 18(1):170–182, 1972.

B Dorsch. A decoding algorithm for binary block codes and J-ary output channels (Corresp.). *IEEE Transactions on Information Theory*, 20(3):391–394, 1974.

Ilya Dumer. On minimum distance decoding of linear codes. In *Proceedings of the 5th Joint Soviet-Swedish International Workshop Information Theory*, pages 50–52. Moscow, 1991.

Ilya Dumer. Sort-and-match algorithm for soft-decision decoding. *IEEE Transactions on Information Theory*, 45(7):2333–2338, 1999.

Ilya Dumer and Kirill Shabunov. Soft-decision decoding of Reed-Muller codes: Recursive lists. *IEEE Transactions on Information Theory*, 52(3):1260–1266, 2006.

Robert Fano. A heuristic discussion of probabilistic decoding. *IEEE Transactions on Information Theory*, 9(2):64–74, 1963.

G Forney. Generalized minimum distance decoding. *IEEE Transactions on Information Theory*, 12(2):125–131, 1966.

Marc P C Fossorier and Shu Lin. Soft-decision decoding of linear block codes based on ordered statistics. *IEEE Transactions on Information Theory*, 41(5):1379–1396, 1995.

Robert G Gallager. *Information theory and reliable communication*, volume 588. Springer, 1968.

Joachim Hagenauer, Elke Offer, and Lutz Papke. Iterative decoding of binary block and convolutional codes. *IEEE Transactions on Information Theory*, 42(2):429–445, 1996.

F Hemmati. Closest coset decoding of mod u mod u+v mod codes. *IEEE Journal on Selected Areas in Communications*, 7(6):982–988, 1989. doi: 10.1109/49.29620.

Rolf Johannesson and Kamil S h Zigangirov. *Fundamentals of convolutional coding*. John Wiley & Sons, 2015.

G A Kabatyanskii. On decoding of Reed-Muller codes in semicontinuous channels. In *Proceedings of the 2nd International Workshop Algebraic and Combinatorial Coding Theory*, pages 87–91, 1990.

Ingmar Land and Johannes Huber. Information combining. *Foundations and Trends®in Communications and Information Theory*, 3(3):227–330, 2006.

C F Lin and J B Anderson. M-algorithm decoding of channel convolutional codes. In *Princeton Conference on Information Sciences and Systems*, volume 1000, Princeton, NJ, pages 0–082, 1986.

S N Litsyn. On decoding complexity of low-rate Reed-Muller codes. In *Proceedings of the 9th All-Union Conference Coding Theory and Information Transmission*, pages 202–204, 1988.

Michael Pohst. On the computation of lattice vectors of minimal length, successive minima and reduced bases with applications. *ACM SIGSAM Bulletin*, 15(1):37–44, 1981.

Eugene Prange. The use of information sets in decoding cyclic codes. *IRE Transactions on Information Theory*, 8(5):5–9, 1962.

Tom Richardson and Ruediger Urbanke. *Modern coding theory*. Cambridge University Press, 2008.

Gottfried Schnabl and Martin Bossert. Soft-decision decoding of Reed-Muller codes as generalized multiple concatenated codes. *IEEE Transactions on Information Theory*, 41(1):304–308, 1995.

Jakov Snyders and Yair Be'ery. Maximum likelihood soft decoding of binary block codes and decoders for the golay codes. *IEEE Transactions on Information Theory*, 35(5):963–975, 1989.

Jacques Stern. A method for finding codewords of small weight. In *Coding Theory and Applications: 3rd International Colloquium Toulon*, France, November 2–4, 1988 Proceedings 3, pages 106–113. Springer, 1989.

Antoine Valembois and Marc Fossorier. Box and match techniques applied to soft-decision decoding. *IEEE Transactions on Information Theory*, 50(5):796–810, 2004.

H C A Van Tilborg, R J Mceliece, and E R Berlekamp. On the inherent intractability of certain coding problems, 1978.

Emanuele Viterbo and Joseph Boutros. A universal lattice code decoder for fading channels. *IEEE Transactions on Information Theory*, 45(5):1639–1642, 1999.

Min Ye and Emmanuel Abbe. Recursive projection-aggregation decoding of Reed-Muller codes. *IEEE Transactions on Information Theory*, 66(8):4948–4965, 2020.

4

Polar Codes

"I just wondered how things were put together."

— Claude Shannon

The idea of building synthetic channels [Arıkan, 2015] originated from concatenated schemes for convolutional codes under sequential decoding by Massey [1981] and Pinsker [1965] to boost the cutoff rate. The cutoff rate is said to be "boosted" when the sum of the cutoff rates of the synthetic channels is greater than the sum of the cutoff rates of the raw channels. The key idea to achieve this goal was to correlate the independent copies of raw channels through concatenation. In Pinsker's scheme, the identical outer convolutional transforms were employed, while the inner block code (with length N) was suggested to be chosen at random. This requires maximum-likelihood (ML) decoding with prohibitive complexity. Different from Pinsker's scheme, in multi-level coding and multi-stage decoding (MLC/MSD), originally proposed by Imai and Hirakawa [1977] as an efficient coded-modulation technique, N convolutional codes at different rates $\{R_i\}$ are used, which consequently require a chain of N outer convolutional decoders. In contrast to the aforementioned schemes, polar coding was originally designed as a low-complexity recursive channel combining and splitting operations, where the polarization effect constrains the rates R_i set to 0 or 1. This method of building synthetic channels turned out to be so effective that no outer code was employed to achieve the original objective of boosting the cutoff rate.

This chapter begins by exploring the fundamentals of polar coding, detailing the channel transformation process that drives the polarization effect. The use of the Bhattacharyya parameter in constructing polar codes is explained, providing a mathematical framework for selecting the optimal set of information and frozen bits.

The SC decoding algorithm, a cornerstone of polar coding, is introduced with an emphasis on its operation, computational complexity, and error correction

Polar Codes: From Theory to Practice, First Edition. Mohammad Rowshan and Emanuele Viterbo.

performance. The analysis highlights the strengths and limitations of SC decoding, offering insights into its practical applications and performance trade-offs.

By the end of this chapter, readers will have a solid understanding of the fundamental principles underlying polar codes and the mechanisms that make them a powerful tool for achieving reliable communication.

4.1 Introduction

In Chapter 2, we learned that one way to represent the generator matrix of Reed–Muller (RM) code $\mathcal{R}(r, m)$ is by selecting rows of the upper triangular matrix $\mathbf{K}_2^{\otimes m}$ with Hamming weights equal to or greater than 2^{m-r}, where $\mathbf{K}_2$ is

$$\mathbf{K}_2 = \begin{bmatrix} 1 & 1 \\ 0 & 1 \end{bmatrix}. \tag{4.1}$$

Recall that the Reed–Muller code $\mathcal{R}(r, m)$ accepts a particular number of information bits to be encoded, i.e. $k = \sum_{i=0}^{r} \binom{m}{i}$. A subcode of an RM code can also be considered to achieve flexibility in the choice of k. However, the question of how to optimally and explicitly choose the rows for a subcode is critical. Note that the subcode of a code is obtained by removing some codewords from the codebook. As the codewords of a code are obtained by a linear combination of the rows of its generator matrix, removing some of the rows of the generator matrix produces a subcode.

In this chapter, we change $\mathbf{K}_2$ to $\mathbf{G}_2$ as follows:

$$\mathbf{G}_2 = \begin{bmatrix} 1 & 0 \\ 1 & 1 \end{bmatrix}. \tag{4.2}$$

Erdal Arıkan in his seminal paper introduced an elegant and revolutionary approach to row selection in $\mathbf{K}_2^{\otimes m}$. This approach was based on the mutual information of the synthetic channels created based on $\mathbf{G}_2^{\otimes m}$. Polar codes are the culmination of his research into the computational cutoff rate of sequential decoding.

Consider the kernel $\mathbf{G}_2$ and the input vector $\mathbf{u} = (u_0\ u_1)$. Encoding of $\mathbf{u}$ is performed as

$$\mathbf{u}\mathbf{G}_2 = (u_0\ u_1) \begin{bmatrix} 1 & 0 \\ 1 & 1 \end{bmatrix} = \begin{bmatrix} u_0 \oplus u_1 \\ u_1 \end{bmatrix} = \begin{bmatrix} x_0 \\ x_1 \end{bmatrix} = \mathbf{x}. \tag{4.3}$$

Note that $\mathbf{G}_2$ is a lower triangular matrix with 1s on the diagonal; therefore, it is invertible. As a result, the mapping from $(u_0\ u_1)$ to $(x_0\ x_1)$ is also invertible. We leave the proof of invertibility of $\mathbf{G}_2^{\otimes m}$ as an exercise. Figure 4.1 illustrates the mapping of $(u_0\ u_1)$ to $(x_0\ x_1)$ and de-mapping (decoding) after passing them through a channel to the estimated values $(\hat{u}_0\ \hat{u}_1)$.

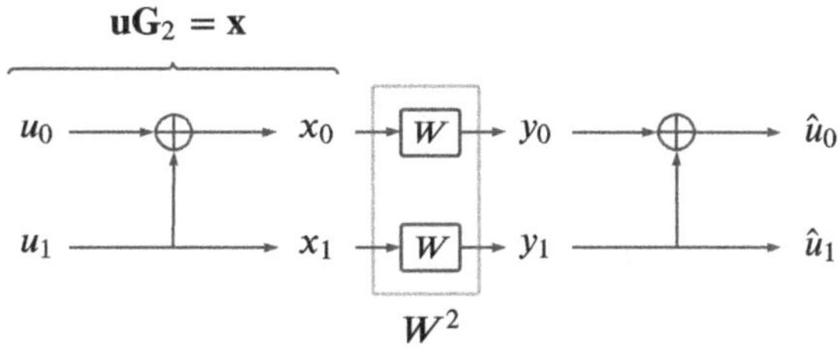

Figure 4.1 Encoding, channel transmission, and decoding based on the basic kernel $\mathbf{G}_2$.

This process can be presented as a channel transformation. First, we define a raw channel and N copies of them. A generic discrete memoryless channel (B-DMC) with input alphabet $\mathcal{X} = \{0, 1\}$, arbitrary output alphabet $\mathcal{Y}$, and arbitrary transition probabilities $P(Y = y \mid X = x) = W(y \mid x), x \in \mathcal{X}, y \in \mathcal{Y}$ is denoted by $W : \mathcal{X} \to \mathcal{Y}$. If we use N copies of this channel simultaneously, it is denoted by $W^N : \mathcal{X}^N \to \mathcal{Y}^N$ with $W^N\left(y_0^{N-1} \mid x_0^{N-1}\right) = \prod_{i=0}^{N-1} W\left(y_i \mid x_i\right)$.

4.2 Basic Channel Transform

Now, we present the basic channel transform, which is equivalent to (4.3). This linear channel transform, which is a bijection, maps the equiprobable bits $u_0^1 = (u_0\ u_1)$ to $x_0^1 = (x_0\ x_1)$ through the channel combining as shown in Figure 4.2.

This transformation creates a super-channel (also known as *vector channel*) denoted by W_2 between u_0^1 and x_0^1 with the transition probability

$$\begin{aligned} W_2\left(y_0^1 \mid u_0^1\right) &\triangleq \prod_{i=0}^{1} W\left(y_i \mid x_i\right) = \prod_{i=0}^{1} W\left(y_i \mid \left(u_0^1 \mathbf{G}_2\right)_i\right) \\ &= W\left(y_1 \mid u_0 \oplus u_1\right) W\left(y_1 \mid u_1\right). \end{aligned} \tag{4.4}$$

We use the mutual information I of two random variables and its properties in Chapter 1. Specifically, we use mutual information between the input and the output of the channels as random variables. As mentioned above, the mapping from u_0^1 to x_0^1 is invertible, hence we have

$$I\left(u_0, u_1; y_0, y_1\right) = I\left(x_0, x_1; y_0, y_1\right) = I\left(x_0; y_0\right) + I\left(x_1; y_1\right) = 2I(W), \tag{4.5}$$

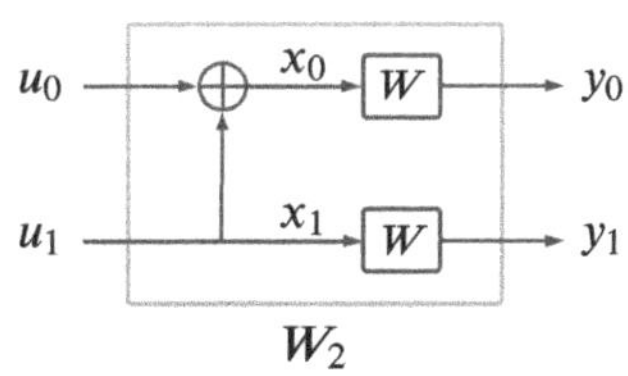

Figure 4.2 Channel combining: super channel or vector channel W_2.

where $C = I(x_0; y_0) = I(W)$ is the capacity of symmetric B-DMC. Note that x_0 is equiprobable and

$$I(W) \triangleq \sum_y W(y \mid 0) \log_2 \left[W(y \mid 0) \Big/ \left(\frac{1}{2} W(y \mid 0) + \frac{1}{2} W(y \mid 1) \right) \right] = I(x_0; y_0).$$

Therefore, the capacity of the super-channel equals the sum of the capacities of two raw channels W and we can conclude that the transformation $(x_0, x_1) = (u_0 \oplus u_1, u_1)$ preserves the total capacity of the system.

Now, we can split the term $I(u_0, u_1; y_0, y_1)$ representing the capacity of the combined channels using the chain rule of mutual information (see Chapter 1) as

$$\begin{aligned} I\left(u_0^1; y_0^1\right) &= I\left(u_0; y_0^1\right) + I\left(u_1; y_0^1 \mid u_0\right) \\ &= I\left(u_0; y_0^1\right) + I\left(u_1; y_0^1, u_0\right). \end{aligned} \tag{4.6}$$

The two terms on the right-hand side of (4.6) can be interpreted as mutual information of two synthetic channels. Therefore, this step is called channel splitting, which reflects splitting the vector channel W_2 into the following two sub-channels:

- The term $I(u_0; y_0^1)$ represents the mutual information of the channel between u_0 and the output y_0^1, where u_1 is considered noise. This channel is denoted by $W_2^{(0)}$ or simply W^- and illustrated in Figure 4.3. We can find the transition probabilities of $W_2^{(0)} : \mathcal{X} \rightarrow \mathcal{Y}^2$ by marginalizing W_2 over u_1 as

$$W_2^{(0)}\left(y_0^1 \mid u_0\right) = \frac{1}{2} \sum_{u_1} W_2\left(y_0^1 \mid u_0^1\right) = \frac{1}{2} \sum_{u_1} W\left(y_0 \mid u_0 \oplus u_1\right) W\left(y_1 \mid u_1\right). \tag{4.7}$$

- The second term $I(u_1; y_0^1, u_0)$ can be interpreted as mutual information of the channel between u_1 and y_0^1 when u_0 is available. This channel is similarly denoted by $W_2^{(1)}$ or W^+ and illustrated in Figure 4.4. We can find the transition probabilities of $W_2^{(1)} : \mathcal{X} \rightarrow \mathcal{Y}^2 \times \mathcal{X}$ by

$$W_2^{(1)}\left(y_0^1, u_0 \mid u_1\right) = \frac{1}{2} W_2\left(y_0^1 \mid u_0^1\right) = \frac{1}{2} W\left(y_0 \mid u_0 \oplus u_1\right) W\left(y_1 \mid u_1\right). \tag{4.8}$$

The two channels created above can be alternatively denoted by two distinct operations $\boxast$ and $\circledast$. For any two B-DMCs $W : \mathcal{X} \rightarrow \mathcal{Y}$, we denote

$$\begin{aligned} W \boxast W &: \mathcal{X} \longrightarrow \mathcal{Y}^2, \\ W \circledast W &: \mathcal{X} \longrightarrow \mathcal{Y}^2 \times \mathcal{X}. \end{aligned} \tag{4.9}$$

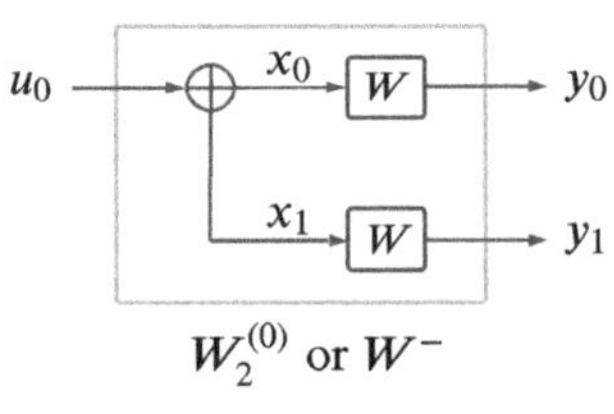

Figure 4.3 Channel splitting: representation of channel $W_2^{(0)}$.

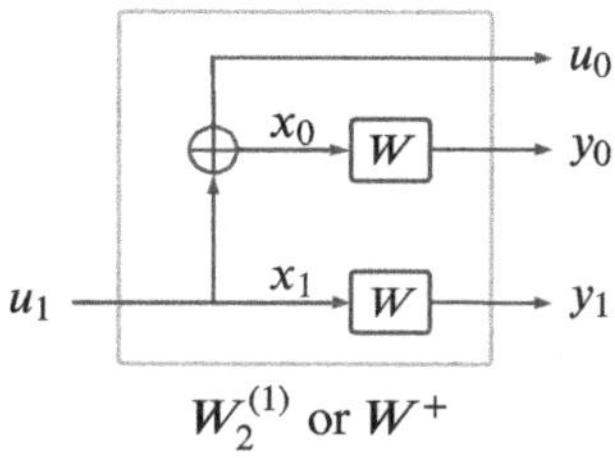

Figure 4.4 Channel splitting: representation of channel $W_2^{(1)}$.

Therefore, we have the following identical notations for these two channels:

$$W_2^{(0)} \triangleq W \boxast W, \tag{4.10}$$

$$W_2^{(1)} \triangleq W \circledast W. \tag{4.11}$$

Note that channel splitting is equivalent to the decoding process in which we first estimate u_0 based on $(y_0\ y_1)$, and then u_1 based on $(y_0\ y_1)$ and u_0. Now, let us revisit the mutual information of the two channels above. From (4.5) and (4.6), we can write

$$I\left(W_2^{(0)}\right) + I\left(W_2^{(1)}\right) = 2I(W). \tag{4.12}$$

Furthermore, using the properties of mutual information and entropy in Chapter 1, we can find the relation between the terms of equation (4.12). We first find $I\left(W_2^{(1)}\right)$ as

$$\begin{aligned} I\left(W_2^{(1)}\right) &= I\left(u_1; y_0^1, u_0\right) \\ &= H\left(u_1\right) - H\left(u_1 \mid y_0^1, u_0\right) \\ &\geq H\left(u_1\right) - H\left(u_1 \mid y_1\right) = I(W). \end{aligned} \tag{4.13}$$

Hence, we have $I\left(W_2^{(1)}\right) \geq I(W)$. Consequently, we should also have $I\left(W_2^{(0)}\right) \leq I(W)$ to obtain (4.12). Overall, we get

$$I\left(W_2^{(0)}\right) \leq I(W) \leq I\left(W_2^{(1)}\right). \tag{4.14}$$

4.3 Recursive Channel Transformation

In Section 4.1, we learned about a basic transform, a.k.a. a single-step transform, based on $\mathbf{G}_2$. This transform results in two sub-channels, one of which is less reliable than the raw channel W and the other is more reliable, as shown in (4.14). This channel transformation can then be recursively expanded to any power of two $N = 2^n$ synthetic channels for an integer $n \geq 0$. Figure 4.5 shows an overview of combining N raw channels W during encoding, which results

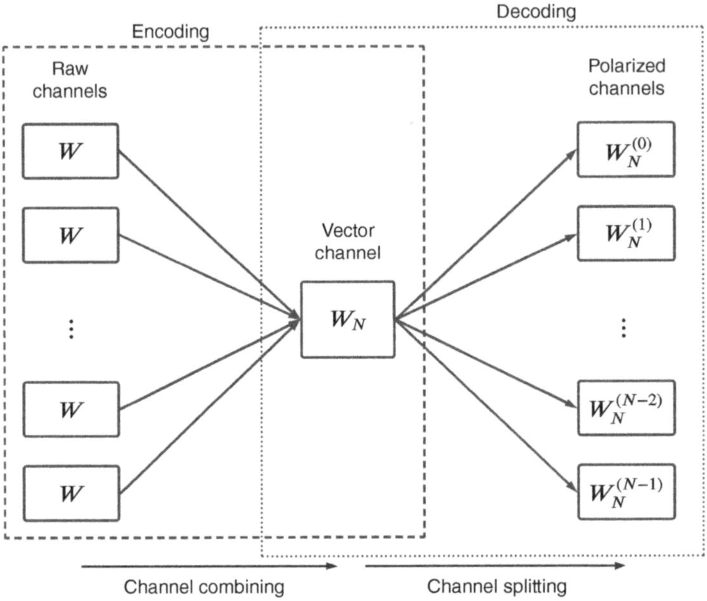

Figure 4.5 Combining *N* raw channels and splitting a vector channel.

in a vector channel W_N, and then splitting the vector channel into N polarized channels $W_N^{(i)}, 0 \leq i \leq N-1$ which occurs during decoding. We observe these two processes for $N = 2$ in Section 4.2. Now, we generalize them in this section.

For $n = 0$, we get a single raw channel $W_1 \triangleq W$. Given two independent copies of the vector channel W_2, we can combine them as shown in Figure 4.6 to form the vector channel $W_4 : \mathcal{X}^4 \rightarrow \mathcal{Y}^4$ with transition probabilities $W_4\left(y_0^3 \mid u_0^3\right) = W_2\left(y_0^1 \mid u_0 \oplus u_1, u_2 \oplus u_3\right) W_2\left(y_2^3 \mid u_1, u_3\right)$.

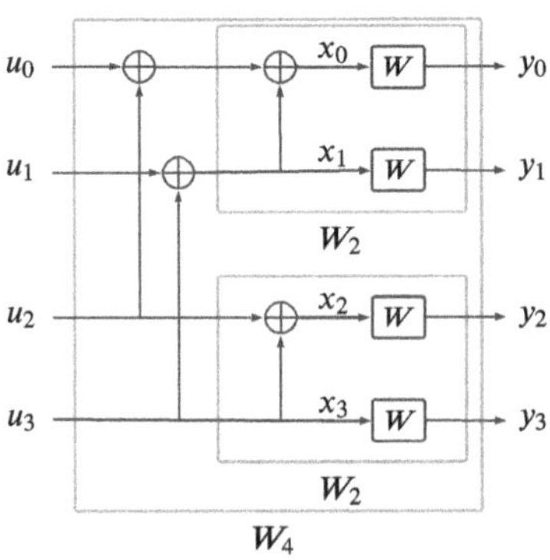

Figure 4.6 Channel transformation: four channels, W_4.

In general, at every stage $j \in [1, n]$ of recursion, two independent copies of $W_{2^{j-1}}$ are combined to produce the vector channel W_{2^j}. Extending W_2 in (4.4), the transition probabilities of vector channel $W_N : \mathcal{X}^N \to \mathcal{Y}^N$ is defined as

$$W_N\left(y_0^{N-1} \mid u_0^{N-1}\right) = W_{N/2}\left(y_0^{N/2-1} \mid u_{0,e}^{N-1} \oplus u_{0,o}^{N-1}\right) W_{N/2}\left(y_{N/2}^{N-1} \mid u_{0,o}^{N-1}\right),$$

where $u_{0,o}^{N-1} = (u_1, u_3, \dots, u_{N-1})$, or the elements of u_0^{N-1} with odd indices, and $u_{0,e}^{N-1} = (u_0, u_2, \dots, u_{N-2})$, or the elements of u_0^{N-1} with even indices. Similarly, we can define the channel $W_{N/2}$ in terms of $W_{N/4}$, and continue this recursively down to the basic transform W_2, which was discussed in Section 4.2.

Now, we find the transition probabilities of the N channels after combining and splitting operations, i.e. the probability distribution induced in the set $\mathcal{U}^N \times \mathcal{X}^N \times \mathcal{Y}^N$ due to the channel combining operation.

As $U_i, i = 0, \dots, N-1$ are independent and identically distributed (i.i.d.) Bernoulli random variables, i.e. Ber$\left(\frac{1}{2}\right)$, the probability distribution induced on the set $\mathcal{U}^N \times \mathcal{X}^N \times \mathcal{Y}^N$ due to the channel combining operation can be expressed as

$$P\left(u_0^{N-1}, x_0^{N-1}, y_0^{N-1}\right) = \frac{1}{2^N} \mathbb{1}_{\{x_0^{N-1} = u_0^{N-1} G_2^{\otimes n}\}} \prod_{i=0}^{N-1} W\left(y_i \mid x_i\right). \tag{4.15}$$

The channel between U_0^{N-1} and Y_0^{N-1} as a result of channel combining operation is defined by the transition probabilities:

$$W_N\left(y_0^{N-1} \mid x_0^{N-1}\right) = P\left(y_0^{N-1} \mid u_0^{N-1}\right) = \prod_{i=0}^{N-1} W\left(y_i \mid x_i\right). \tag{4.16}$$

The splitting operation, similar to (4.6), is defined by the chain rule as

$$I\left(U_0^{N-1}; Y_0^{N-1}\right) = \sum_{i=0}^{N-1} I\left(U_i; Y_0^{N-1}, U_0^{i-1}\right), \tag{4.17}$$

where the term $I\left(U_i; Y_0^{N-1}, U_0^{i-1}\right)$ is associated to the channel between U_i and $\left(Y_0^{N-1}, U_0^{i-1}\right)$. The transition probabilities of this channel denoted by $W_N^{(i)} : \mathcal{X} \to \mathcal{Y}^N \times \mathcal{X}^{i-1}$ are given by

$$\begin{aligned} W_N^{(i)}\left(y_0^{N-1}, u_0^{i-1} \mid u_i\right) &\triangleq P\left(y_0^{N-1}, u_0^{i-1} \mid u_i\right) = \sum_{u_{i+1}^{N-1} \in \{0,1\}} \frac{P\left(y_0^{N-1} \mid u_0^{N-1}\right) P\left(u_0^{N-1}\right)}{P\left(u_i\right)} \\ &= \frac{1}{2^{N-1}} \sum_{u_{i+1}^{N-1}} P\left(y_0^{N-1} \mid u_0^{N-1}\right) = \frac{1}{2^{N-1}} \sum_{u_{i+1}^{N-1}} W_N\left(y_0^{N-1} \mid u_0^{N-1}\right). \end{aligned} \tag{4.18}$$

Let us clarify how equation (4.18) is derived. The joint probability $P\left(y_0^{N-1}, u_0^{i-1} \mid u_i\right)$ is the probability of observing the channel outputs y_0^{N-1} and the first $i-1$ input bits u_0^{i-1}, given the ith input bit u_i. The law of total probability allows us to marginalize (sum) over all possible values of the remaining input bits u_{i+1}^{N-1}; i.e.

$$P\left(y_0^{N-1}, u_0^{i-1} \mid u_i\right) = \sum_{u_{i+1}^{N-1}} P\left(y_0^{N-1}, u_0^{i-1}, u_{i+1}^{N-1} \mid u_i\right).$$

Now, using Bayes' rule, the joint probability inside the summation can be rewritten as

$$P\left(y_0^{N-1}, u_0^{i-1}, u_{i+1}^{N-1} \mid u_i\right) = P\left(y_0^{N-1} \mid u_0^{N-1}\right) P\left(u_0^{N-1} \mid u_i\right).$$

The term $P\left(u_0^{N-1} \mid u_i\right)$ represents the probability of the entire input sequence u_0^{N-1}, conditioned on the bit u_i. This can be expanded using the chain rule of probability as

$$P\left(u_0^{N-1} \mid u_i\right) = \frac{P\left(u_0^{N-1}\right)}{P\left(u_i\right)}.$$

It is typically assumed that all input bits u_0^{N-1} are equally likely to be 0 or 1, i.e. they are uniformly distributed. Therefore, by substituting $P\left(u_0^{N-1}\right) = 1/2^N$ and $P\left(u_i\right) = 1/2$, we get

$$P\left(y_0^{N-1}, u_0^{i-1} \mid u_i\right) = \frac{1}{2^{N-1}} \sum_{u_{i+1}^{N-1}} P\left(y_0^{N-1} \mid u_0^{N-1}\right).$$

The relation between $W_N^{(i)}$ and $W_{N/2}^{(i)}$ for recursion, similar to (4.7) and (4.8), is given as follows (proofs are left as an exercise)

$$\begin{aligned} &W_N^{(2i)}\left(y_0^{N-1}, u_0^{2i-1} \mid u_{2i}\right) \\ &\quad = \frac{1}{2} \sum_{u_{2i+1}} W_{N/2}^{(i)}\left(y_0^{N/2-1}, u_{0,o}^{2i-1} \mid u_{2i+1}\right) W_{N/2}^{(i)}\left(y_{N/2}^{N-1}, u_{0,e}^{2i-1} \oplus u_{0,o}^{2i-1} \mid u_{2i} \oplus u_{2i+1}\right), \end{aligned} \tag{4.19}$$

$$\begin{aligned} &W_N^{(2i+1)}\left(y_0^{N-1}, u_0^{2i} \mid u_{2i+1}\right) \\ &\quad = \frac{1}{2} W_{N/2}^{(i)}\left(y_0^{N/2-1}, u_{0,o}^{2i-1} \mid u_{2i+1}\right) W_{N/2}^{(i)}\left(y_{N/2}^{N-1}, u_{0,e}^{2i-1} \oplus u_{0,o}^{2i-1} \mid u_{2i} \oplus u_{2i+1}\right). \end{aligned} \tag{4.20}$$

Furthermore, the relations (4.10) and (4.11) can be generalized and a combination of two identical channels $W_{N/2}^{(i)}$ recursively with two distinct operations $\boxast$ and $\circledast$ can be defined as

$$W_N^{(2i)} = W_{N/2}^{(i)} \boxast W_{N/2}^{(i)}. \tag{4.21}$$

$$W_N^{(2i+1)} = W_{N/2}^{(i)} \circledast W_{N/2}^{(i)}. \tag{4.22}$$

Now, given the n-bit binary representation of the channel index i as $\left(b_0, \ldots, b_{n-1}\right)$, by defining $W_N^{(i)} \triangleq W^{(b_0, \ldots, b_{n-1})} = \left(\left(W^{b_0}\right)^{b_1} \ldots\right)^{b_{n-1}}$, we can obtain

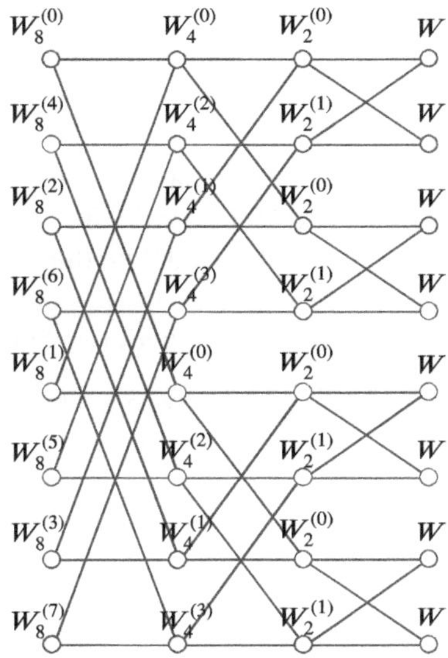

Figure 4.7 Transformation of $N = 8$ raw channels.

$W_N^{(i)}$ recursively using the following operations on the channels:

$$W^{(b_0,\ldots,b_{k-1},b_k)} = \begin{cases} W^{(b_1,\ldots,b_{k-1})} \boxast W^{(b_0,\ldots,b_{k-1})}, & \text{if } b_k = 0, \\ W^{(b_0,\ldots,b_{k-1})} \circledast W^{(b_1,\ldots,b_{k-1})}, & \text{if } b_k = 1. \end{cases} \tag{4.23}$$

for $k \in [1, n-1]$, i.e. we consider k least significant bits of the n bits, and $b_k \in \{0, 1\}$, where

$$W^{(b_0)} = \begin{cases} W \boxast W, & \text{if } b_0 = 0, \\ W \circledast W, & \text{if } b_0 = 1. \end{cases}$$

Figure 4.7 demonstrates the transformation of eight raw channels W to sub-channels $\{W_8^{(i)} : i \in [0, 7]\}$ from right to left.

4.4 Channel Polarization

We discussed the recursive transformation of N copies of channel W to N distinct subchannels $W_N^{(i)}, i = 0, \ldots, N-1$ in Section 4.3. The recursive channel transformation also transforms the mutual information $I\left(W_N^{(i)}\right)$ and the Bhattacharyya parameter $Z\left(W_N^{(i)}\right)$ of the sub-channels. The relationships will remain similar to (4.14) and (4.12) as follows:

$$I\left(W_N^{(2i)}\right) \le I\left(W_{N/2}^{(i)}\right) \le I\left(W_N^{(2i+1)}\right)$$
$$I\left(W_N^{(2i)}\right) + I\left(W_N^{(2i+1)}\right) = 2I\left(W_{N/2}^{(i)}\right), \tag{4.24}$$

$$Z\left(W_N^{(2i)}\right) \leq 2Z\left(W_{N/2}^{(i)}\right)^2 - Z\left(W_{N/2}^{(i)}\right)$$
$$Z\left(W_N^{(2i+1)}\right) = Z\left(W_{N/2}^{(i)}\right)^2. \tag{4.25}$$

To analyze the behavior of sub-channels, we represent them through a random process. Let $\{B_n : n \geq 1\}$ be a sequence of i.i.d. symmetric Bernoulli random variables. We do not define the probability space for this random variable, as our aim is to provide an overview of the process, rather than proving it. $\{W_n, n \geq 0\}$ denote a tree process with two branches defined as

$$W_{n+1} = \begin{cases} W_n \boxast W_n & \text{if} B_n = 0, \\ W_n \circledast W_n & \text{if} B_n = 1. \end{cases}$$

The output space of the random variable W_n is the set of channels $\left\{W_{2^n}^{(i)}\right\}_{i=0}^{2^n-1}$. Moreover, W_n is uniformly distributed over the channels $\left\{W_{2^n}^{(i)}\right\}_{i=0}^{2^n-1}$. To show channel polarization, it is sufficient to track the mutual information $\left\{I\left(W_N^{(i)}\right)\right\}$ and the Bhattacharyya parameters $\left\{Z\left(W_N^{(i)}\right)\right\}$. Let $\{I_n : n \geq 0\} := \{I(W_n) : n \geq 0\}$ and $\{Z_n : n \geq 0\} := \{Z(W_n) : n \geq 0\}$ be two processes. Then, one can show that

1) the sequence $\{I_n\}$ converges almost certainly to a random variable I_∞ where

$$I_\infty = \begin{cases} 1 \text{ with probability} I(W), \\ 0 \text{ with probability} 1 - I(W). \end{cases}$$

2) the sequence $\{Z_n\}$ converges almost certainly to a random variable Z_∞ where

$$Z_\infty = \begin{cases} 1 \text{ with probability} 1 - I(W) \\ 0 \text{ with probability} I(W). \end{cases}$$

Then, we have for any small $\delta > 0$

$$\lim_{n\to\infty} \frac{\left|\left\{i : I\left(W_{2^n}^{(i)}\right) \in (1-\delta, 1]\right\}\right|}{2^n} = \lim_{n\to\infty} \Pr\left[I_n \in (1-\delta, 1]\right] = I(W), \tag{4.26}$$

and

$$\lim_{n\to\infty} \frac{\left|\left\{i : I\left(W_{2^n}^{(i)}\right) \in [0, \delta)\right\}\right|}{2^n} = \lim_{n\to\infty} \Pr\left[I_n \in [0, \delta)\right] = 1 - I(W). \tag{4.27}$$

Observe that as n and consequently $N = 2^n$ get larger, almost all the channels $W_N^{(i)}, i \in [0, N-1]$ get polarized to either perfect channels (mutual information close to 1) or unreliable channels (mutual information close to 0). Thus, we will refer to the channels with mutual information close to 1 as "good channels" and the remaining channels as "bad channels."

In other words, the fraction of channels with $I\left(W_{2^n}^{(i)}\right) \in (\delta, 1-\delta)$ decreases. That is,

$$\lim_{n\to\infty} \frac{\left|\left\{i : I\left(W_{2^n}^{(i)}\right) \in (\delta, 1-\delta)\right\}\right|}{2^n} = 0.$$

Also, given (4.12), by induction on

$$\sum_{i\in[0,2^n-1]} I\left(W_{2^n}^{i}\right) = 2^n I(W),$$

we can conclude that the fraction of almost perfect channels approaches the symmetric capacity.

Figure 4.8 illustrates the polarization effect for the binary erasure channel (BEC) W where the erasure probability is 0.5. In the case of BEC W, the rate $I(W_N^{(i)})$ can be computed using the following recursive relations for $0 \leq i \leq N/2-1$:

$$\begin{aligned} I\left(W_N^{(2i)}\right) &= I\left(W_{N/2}^{(i)}\right)^2, \\ I\left(W_N^{(2i+1)}\right) &= 2I\left(W_{N/2}^{(i)}\right) - I\left(W_{N/2}^{(i)}\right)^2. \end{aligned} \quad (4.28)$$

Observe that the second term of $I\left(W_N^{(2i+1)}\right)$ is equivalent to $I\left(W_N^{(2i)}\right)$. This preserves the total of $2I\left(W_{N/2}^{(i)}\right)$.

Example 4.1 Suppose that we have a raw BEC W with $I(W) = \varepsilon = 0.5$, the erasure probability of the BEC. By recursive channel transformation, we combine two copies of this channel and form a vector channel consisting of W^- and W^+ as discussed in Section 4.2. Then, we have

$$\begin{aligned} I\left(W^-\right) &= I(W)^2 = 0.5^2 = 0.25, \\ I\left(W^+\right) &= 2I(W) - I(W)^2 = 2 \times 0.5 - 0.25 = 0.75. \end{aligned} \quad (4.29)$$

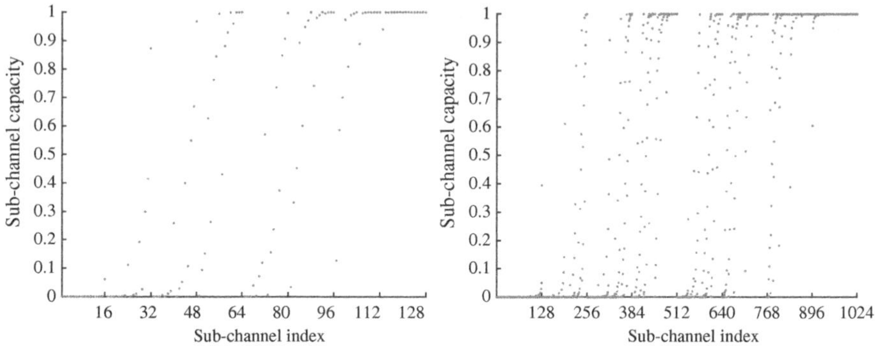

Figure 4.8 The capacity of sub-channels due to the polarization effect for $N = 128, 1024$ over BEC with $\varepsilon = 0.5$.

Now, following the recursive channel transformation in Section 4.3, we form four subchannels after two stages of transformation. This results in four sub-channels of W^{--}, W^{-+}, W^{+-}, and W^{++} with mutual information:

$$
\begin{aligned}
&I\left(W^{--}\right) = I(W^{-})^2 = 0.25^2 = 0.0625, \\
&I\left(W^{-+}\right) = 2I\left(W^{-}\right) - I(W^{-})^2 = 2 \times 0.25 - 0.0625 = 0.4375, \\
&I\left(W^{+-}\right) = I\left(W^{+}\right)^2 = 0.75^2 = 0.5625, \\
&I\left(W^{++}\right) = 2I\left(W^{+}\right) - I\left(W^{+}\right)^2 = 2 \times 0.75 - 0.5625 = 0.9375.
\end{aligned} \tag{4.30}
$$

Figure below illustrates the channel polarizations for four channels in this example.

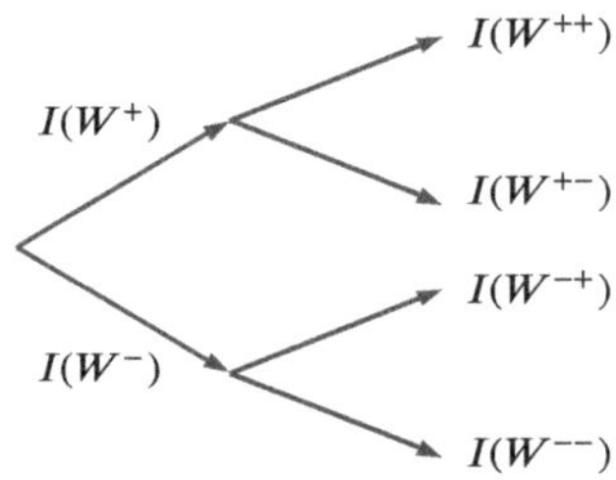

Following the example above, let us now look at larger N. Figure 4.8 visualizes the sub-channel capacities $I\left(W_N^{(i)}\right)$ for $N = 128$ and 1024 for sub-channel indices $i \in [0, N-1]$. Figures 4.9 and 4.10 show the sub-channel capacities in increasing values (the indexes are permuted accordingly) for $N = 128$ and 1024. The sorted capacity plots highlight the good, the bad, and the intermediate sub-channels. As expected from the theory, if we further increase N, the portion of the sub-channels with a capacity away from 0 or 1 diminishes.

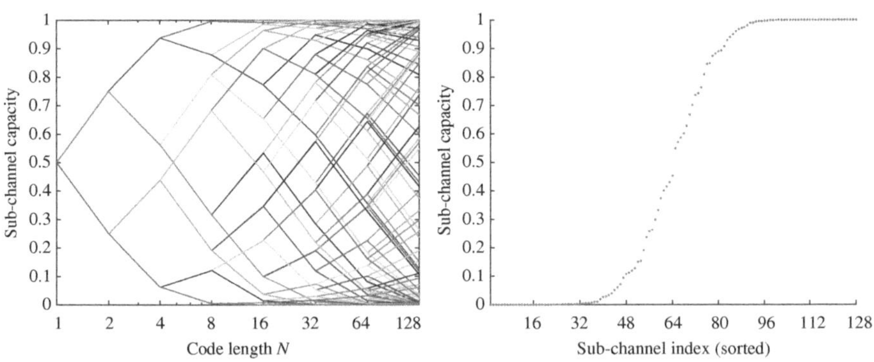

Figure 4.9 The polarization effect for $N = 2^7 = 128$ over BEC with $\varepsilon = 0.5$.

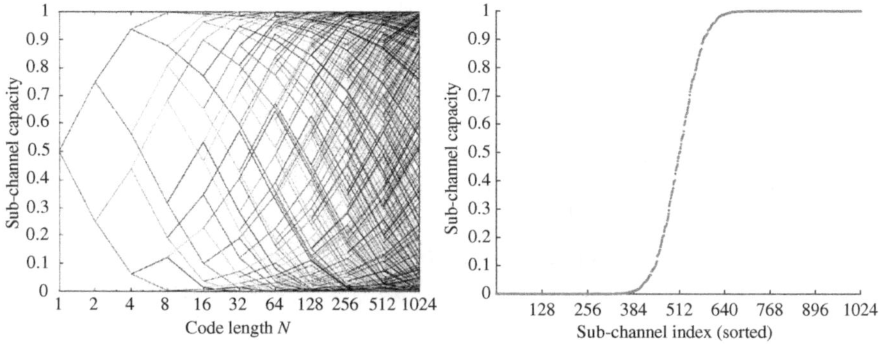

Figure 4.10 The polarization effect for $N = 2^{10} = 1024$ over BEC with $\varepsilon = 0.5$.

4.5 Code Construction Based on $Z\left(W_N^{(i)}\right)$

In Section 4.4, we observed how sub-channels are polarized in a recursive transformation. We indicated this transformation using two metrics: mutual information and the Bhattacharyya parameter. Therefore, we can say that the construction of polar codes is explicitly defined. That is, one can recursively evaluate $I\left(W_N^{(i)}\right)$ for $i \in [0, N-1]$ and choose the K most reliable sub-channels with the highest $I\left(W_N^{(i)}\right)$ values. We collect the indices of these sub-channels in set $\mathcal{A}$ called *information set*.

Unfortunately, the exact evaluation of $I(W_N^{(i)})$ for an additive white Gaussian noise channel is intractable, as it depends on the calculation of the channel parameters when the output alphabets grow exponentially in code length. If W is a BEC, then the sub-channels $W_N^{(i)}$ for $i \in [0, N-1]$ are all BECs. Therefore, we just need to track their erasure probabilities ϵ_i, which are exactly equal to the Bhattacharyya parameters. For a BEC W, $Z(W) = \varepsilon$ and the channel capacity is $I(W) = 1 - Z(W)$.

For other channels, the Bhattacharyya parameter $Z(W)$ only defines the upper bound of the probability of error when transmitting one bit, recalling Chapter 1, Section 1.3.5:

$$Z(W) = \sum \sqrt{W(y|0)W(y|1)}. \tag{4.31}$$

The Bhattacharyya parameter for the synthesized channel $W_N^{(i)}$ is defined as

$$Z\left(W_N^{(i)}\right) = \sum_{y_0^{N-1}, u_0^{i-1}} \sqrt{W_N^{(i)}\left(y_0^{N-1}, u_0^{i-1} \mid 0\right) W_N^{(i)}\left(y_0^{N-1}, u_0^{i-1} \mid 1\right)}, \tag{4.32}$$

which generalizes the concept of Bhattacharyya parameter and is derived from the raw channel W through the recursions as described in the following.

The recursions in (4.23) for BECs can be simplified to

$$Z\left(W^{(b_0,\ldots,b_{k-1},b_k)}\right) = \begin{cases} 2Z\left(W^{(b_0,\ldots,b_{k-1})}\right) - Z\left(W^{(b_0,\ldots,b_{k-1})}\right)^2 & \text{if } b_k = 0, \\ Z\left(W^{(b_0,\ldots,b_{k-1})}\right)^2 & \text{if } b_k = 1, \end{cases}$$

where $b_0 \ldots b_{n-1}$ is the n-bit binary expansion of i, $k \in [1, n-1]$, i.e. we consider K least significant bits of the n bits, and $Z\left(W_N^{(i)}\right) = Z\left(W^{(b_0,\ldots,b_{n-1})}\right)$.

To begin the recursion, we set $Z(W_1^{(0)}) = \varepsilon$. Then, the recursions to find the Bhattacharyya parameters $Z\left(W_N^{(i)}\right)$ for N sub-channels, where $N = 2^n$, $n \geq 0$, and $0 \leq i \leq N/2 - 1$, are executed using the following relations:

$$Z\left(W_N^{(2i)}\right) = 2 \cdot Z\left(W_{\frac{N}{2}}^{(i)}\right) - Z\left(W_{\frac{N}{2}}^{(i)}\right)^2 \tag{4.33}$$

$$Z\left(W_N^{(2i+1)}\right) = Z\left(W_{\frac{N}{2}}^{(i)}\right)^2. \tag{4.34}$$

The capacities of the binary erasure subchannels $W_N^{(i)}$ are $I(W_N^{(i)}) = 1 - Z(W_N^{(i)})$ and can also be considered as the reliabilities of the subchannels.

Example 4.2 Construct the polar code (8, 4) using the Bhattacharyya parameter over BEC with $\varepsilon = 0.5$. The recursion of the Bhattachariyaa parameters is executed using (4.33) and (4.34) and results in the tree shown in Figure 4.11. After ranking the sub-channels, four subchannels with the highest reliabilities are selected

$Z(W_1^i)$	$Z(W_2^i)$	$Z(W_4^i)$	$Z(W_8^i)$	$I(W_8^i)$	Rank		
0.5	0.75	0.9375	0.9961	0.0039	8	frozen	$u_0 = 0$
			0.8789	0.1211	7	frozen	$u_1 = 0$
		0.5625	0.8086	0.1914	6	frozen	$u_2 = 0$
			0.3164	0.6836	4		$u_3 \in \{0, 1\}$
	0.25	0.4375	0.6836	0.3164	5	frozen	$u_4 = 0$
			0.1914	0.8086	3		$u_5 \in \{0, 1\}$
		0.0625	0.1211	0.8789	2		$u_6 \in \{0, 1\}$
			0.0039	0.9961	1		$u_7 \in \{0, 1\}$

Figure 4.11 The polar code (8, 4) constructed using Bhattacharyya parameter over BEC with $\varepsilon = 0.5$.

to transmit information bits $u_i, i \in \mathcal{I} = \{3, 5, 6, 7\}$ and the rest transmit a known value, by default $u_i = 0, i \in \mathcal{I}^c = \{0, 1, 2, 4\}$.

Unlike BECs, the capacities of additive white Gaussian noise (AWGN) channels are not easy to calculate. We will introduce other methods for AWGN channels in Chapter 6. The listing in Algorithm 4.1 gives a procedure to construct a polar code using Arıkan's Bhattacharyya parameter method. In this algorithm, the Bhattacharyya parameter of the raw channel or the initial parameter value $Z_0 = Z\left(W_1^{(1)}\right)$ for the BEC channels is set to the erasure probability ϵ.

For the AWGN channel, we can obtain this value using a design parameter called *design-SNR*, denoted by SNR_d. Given the design-SNR SNR_d in dB, according to (1.19) in Chapter 1, we can find the initial Bhattacharyya parameter using

$$Z_0 = \exp\left(-10^{\text{SNR}_d/10}\right), \tag{4.35}$$

where the term $10^{\text{SNR}_d/10}$ is used for converting design-SNR from dB to linear form. Note that if we consider the Bhattacharyya parameter as an approximate measure of the error probability, the value obtained using this algorithm gives an upper bound for the probability, and hence this method is sometimes called the Bhattacharyya bound method. If we compare them using a more accurate probability estimate, it is observed that the bound is tight for the more reliable sub-channels and loose for the less reliable ones. Thus, this method cannot accurately select the reliable sub-channels for set $\mathcal{A}$.

Algorithm 4.1: Bhattacharyya Parameter Method

input : N, K, Initial Bhattacharyya parameter $Z_0 = Z\left(W_1^{(0)}\right)$
output: The indices of the most reliable sub-channels $\mathcal{A}$

$\mathbf{Z} \leftarrow [Z_0 \cdots Z_0]$ `// Initialization with an N-length vector`
for $\ell \leftarrow 1$ **to** $\log_2 N$ **do**
  $S \leftarrow 2^\ell$ `// The number of tree branches at layer ℓ`
  for $j \leftarrow 0$ **to** $\frac{S}{2} - 1$ **do**
    $z \leftarrow \mathbf{Z}[j]$
    $\mathbf{Z}[j] \leftarrow 2z - z^2$ `// Upper branch/sub-channel`
    $\mathbf{Z}[\frac{S}{2} + j] \leftarrow z^2$ `// Lower branch/sub-channel`
$\mathcal{A} \leftarrow$ Find the K indices of the smallest elements in vector $\mathbf{Z}$
return $\mathcal{A}$

Clearly, the computation of each $Z\left(W_N^{(i)}\right)$ can be accomplished using $O(\log N)$ arithmetic operations. Hence, the total complexity is $O(N \log N)$, under the assumption that infinite precision arithmetic can be implemented at unit cost.

4.6 $\mathbf{G}_N$-Coset Codes

Let us first consider a broader class of linear block codes containing polar codes which are based on the linear transform $\mathbf{G}_N$. We call this transform the *polar transform*. The block length is a power of two ($N = 2^n$) and the encoding ($\mathbf{u} \mapsto \mathrm{x}$) is performed as shown in Section 4.1 as

$$\mathbf{x} = \mathbf{u}\mathbf{G}_N, \tag{4.36}$$

which relates the input-output of n-stage polar transformation.

Let $\mathcal{A}$ be an arbitrary subset of $\{0, 1, \dots, N-1\}$ and define the "complementary" set $\mathcal{A}^c \triangleq \{0, 1, \dots, N-1\} \backslash \mathcal{A}$. Then, (4.36) can be rewritten as:

$$\mathrm{x} = \mathbf{u}_{\mathcal{A}}\mathbf{G}_N[\mathcal{A}] + \mathbf{u}_{\mathcal{A}^c}\mathbf{G}_N\left[\mathcal{A}^c\right]$$

where $\mathbf{G}_N[\mathcal{A}]$ is the sub-matrix formed by keeping the rows of $\mathbf{G}_N$ whose indices are in $\mathcal{A}$.

By fixing $\mathcal{A}$ and $\mathbf{u}_{\mathcal{A}^c}$, we obtain a mapping from the binary vectors $\mathbf{u}_{\mathcal{A}}$ of length $K \triangleq |\mathcal{A}|$ to codewords of length N. This mapping is called G_N-coset code.

Clearly, the rate of this code is $R = \frac{K}{N}$. Recall from Section 4.5 that we refer to $\mathcal{A}$ as an information set, and the bits in $\mathbf{u}_{\mathcal{A}^c} \in \mathbb{F}_2^{N-K}$ are called *frozen bits*. A particular $\mathbf{G}_N$-coset code is identified by a set of parameters $\left(N, K, \mathcal{A}, \mathbf{u}_{\mathcal{A}^c}\right)$. The frozen bits can be omitted from this definition when they are taken to be all zeros.

4.7 Successive Cancellation (SC) Decoding

This section describes the successive cancelation decoding algorithm, closely related to the recursive multi-stage decoding described in Chapter 3. Given a $\mathbf{G}_N$-coset code with parameters $\left(N, K, \mathcal{A}, \mathbf{u}_{\mathcal{A}^c}\right)$, a binary message vector $\mathbf{u}_{\mathcal{A}}$ can be encoded according to (4.36), resulting in a codeword $\mathbf{c}$. Assuming a bipolar mapping $\{0, 1\} \rightarrow \{1, -1\}$ is performed, the symbols are transmitted through a channel with a transition probability $W^N(\mathbf{y} \mid \mathbf{x})$. The receiver observes the corresponding vector $\mathbf{y}$. Knowing the code parameters, the decoder should estimate the transmitted message vector $\mathbf{u}_{\mathcal{A}}$, denoted by $\hat{\mathbf{u}}_{\mathcal{A}}$, knowing $\mathbf{u}_{\mathcal{A}^c}$.

The decoding is performed using the channel splitting, represented by (4.18).

Having received the channel output y_0^{N-1}, the decision on the estimate $\hat{u}_0^{N-1}$ of u_0^{N-1} is made bit by bit in ascending order of the indices. At the ith stage, if

$i \in \mathcal{A}^c$, the frozen bit is known, and thus $\hat{u}_i = u_i$. If $i \in \mathcal{A}$, we use a maximum likelihood decision rule as

$$\hat{u}_i \leftarrow \arg\max_{u_i \in \{0,1\}} W_n^{(i)}\left(y_0^{N-1}, \hat{u}_0^{i-1} \mid u_i\right). \tag{4.37}$$

As detailed in Chapter 3, the efficient way to make this decision is based on the log-likelihood ratio (LLR). Hence, the SC decoder computes the *decision LLR* as

$$L_N^{(i)}\left(y_0^{N-1}, \hat{u}_0^{i}\right) \triangleq \log \frac{W_N^{(i)}\left(y_0^{N-1}, \hat{u}_0^{i-1} \mid \hat{u}_i = 0\right)}{W_N^{(i)}\left(y_0^{N-1}, \hat{u}_0^{i-1} \mid \hat{u}_i = 1\right)}, \tag{4.38}$$

and generates the decision $\hat{u}_i$ as

$$\hat{u}_i = h\left[L_N^{(i)}\left(y_0^{N-1}, \hat{u}_0^{i}\right)\right] \triangleq \begin{cases} \frac{1}{2}\left[1 - \operatorname{sign}\left(L_N^{(i)}\left(y_0^{N-1}, \hat{u}_0^{i}\right)\right)\right], & \text{if } L_N^{(i)}\left(y_0^{N-1}, \hat{u}_0^{i}\right) \neq 0 \\ \text{Bernoulli}\left(\frac{1}{2}\right). & \text{if } L_N^{(i)}\left(y_0^{N-1}, \hat{u}_0^{i}\right) = 0 \end{cases}. \tag{4.39}$$

Note that an incorrect estimation of any $u_i, i \in \mathcal{A}$ results in the propagation of this error and therefore the decoding failure. This decoding is sub-optimal as the maximum likelihood rule is locally valid (i.e. at the ith sub-channel), not for the entire sequence similar to ML decoding approach discussed in Chapter 3.

Algorithm 4.2: SC Decoding

input : N, y_0^{N-1}, Information set $\mathcal{A}$, $u_{\mathcal{A}^c}$
output: Estimated information bits $\hat{u}_{\mathcal{A}}$
for $i \leftarrow 0$ **to** $N-1$ **do**
  if $i \notin \mathcal{A}$ **then**
    $\hat{u}_i \leftarrow u_i$
  else
    $\hat{u}_i \leftarrow h\left[L_N^{(i)}\left(y_0^{N-1}, \hat{u}_0^{i}\right)\right]$ `// cf.` (4.39)
return $\hat{u}_{\mathcal{A}}$

Now, the question is how to find $L_N^{(i)}\left(y_0^{N-1}, \hat{u}_0^{i}\right)$ making use of the recursive channel splitting process. The LLRs can be recursively computed via the update rule

$$L_N^{(2i)}\left(y_0^{N-1}, \hat{u}_0^{2i-1}\right) = f_-\left[L_{N/2}^{(i)}\left(y_{0,e}^{N-1}, \hat{u}_{0,e}^{2i-1} \oplus \hat{u}_{0,o}^{2i-1}\right), L_{N/2}^{(i)}\left(y_{0,o}^{N-1}, \hat{u}_{0,o}^{2i-1}\right)\right]$$

$$L_N^{(2i+1)}\left(y_0^{N-1}, \hat{u}_0^{2i}\right) = f_+\left[L_{N/2}^{(i)}\left(y_{0,e}^{N-1}, \hat{u}_{0,e}^{2i-1} \oplus \hat{u}_{0,o}^{2i-1}\right), L_{N/2}^{(i)}\left(y_{0,o}^{N-1}, \hat{u}_{0,o}^{2i-1}\right), \hat{u}_{2i}\right] \tag{4.40}$$

where

$$f_-(\alpha, \beta) \triangleq \log\left(\frac{e^{\alpha+\beta}+1}{e^{\alpha}+e^{\beta}}\right), \tag{4.41}$$

$$f_+(\alpha, \beta, u) \triangleq (-1)^u\alpha + \beta. \tag{4.42}$$

For details on the LLR algebra, see Chapter 3. The recursion ends at code length 1 with $L_1^{(1)}(y_i) = \log\left(W(y_i \mid \hat{c}_i = 0) / W(y_i \mid \hat{c}_i = 1)\right)$ and $\hat{c}_0^{N-1} = \hat{u}_0^{N-1}\mathbf{G}_N$. For a more computationally efficient recursion for odd indices, $f_-(\alpha, \beta)$ can be approximated by

$$f_-(\alpha, \beta) \approx \tilde{f}_-(\alpha, \beta) \triangleq \text{sign}(\alpha)\text{sign}(\beta)\min\{|\alpha|, |\beta|\}. \tag{4.43}$$

The replacement of f_- with $\tilde{f}_-$ is referred to as *min-sum SC decoding.*

Example 4.3 Let us illustrate the step-by-step successive cancellation decoding of the polar code $(8, 4, \{3, 5, 6, 7\})$.

We assume that the message data $\mathbf{d} = [1,\ 1,\ 0,\ 1]$ is sent. First, the message is rate-profiled into $\mathbf{u} = [0,\ 0,\ 0,\ 1,\ 0,\ 1,\ 0,\ 1]$ and encoded as $\mathbf{c} = [1, 1, 0, 0, 0, 0, 1, 1]$. With binary phase-shift keying (BPSK) modulation, the codeword becomes $\mathbf{x} = [-1, -1,\ 1,\ 1,\ 1,\ 1, -1, -1]$.

Let $\mathbf{x}$ be transmitted over the AWGN channel with $E_b/N_0 = 2\,\text{dB}$, resulting in the received sequence $\mathbf{y} = [-0.4167, -0.6357,\ 1.6604, -0.5944,\ 0.6982,\ 1.6727, -1.7881, -1.4540]$. The output values of the AWGN channel are obtained as explained in Chapter 1, Section 1.2.3.

As shown in Chapter 3, Section 3.1.1, the LLRs of the received symbols y_i are evaluated as $L_1^{(1)}(y_i) = \alpha(y_i) = \frac{2}{\sigma^2}y_i$, where $\sigma^2 = \frac{1}{2R \cdot 10^{\frac{\text{SNR [dB]}}{10}}} = 0.6310$:

$$\alpha(\mathbf{y}) = [-1.321, -2.015,\ 5.263, -1.884,\ 2.213,\ 5.302, -5.668, -4.609].$$

Then, we calculate the decision LLRs for all $i \in [0, 7]$ recursively as illustrated in Figures 4.12 and 4.13.

4.8 Performance of Polar Codes Under SC Decoding

For any BI-DMC W with symmetric capacity $I(W)$ and any rate $R < I(W)$, the block error probability of polar codes under successive cancellation decoding satisfies

$$P_e(N, R) \leq 2^{-N^{\beta}}, \tag{4.44}$$

for any $\beta < \frac{1}{2}$ when the block-length N is large enough. This exponential decay bound describes the scaling law or the asymptotic behavior of the block error probability of polar codes under successive cancelation decoding as the block

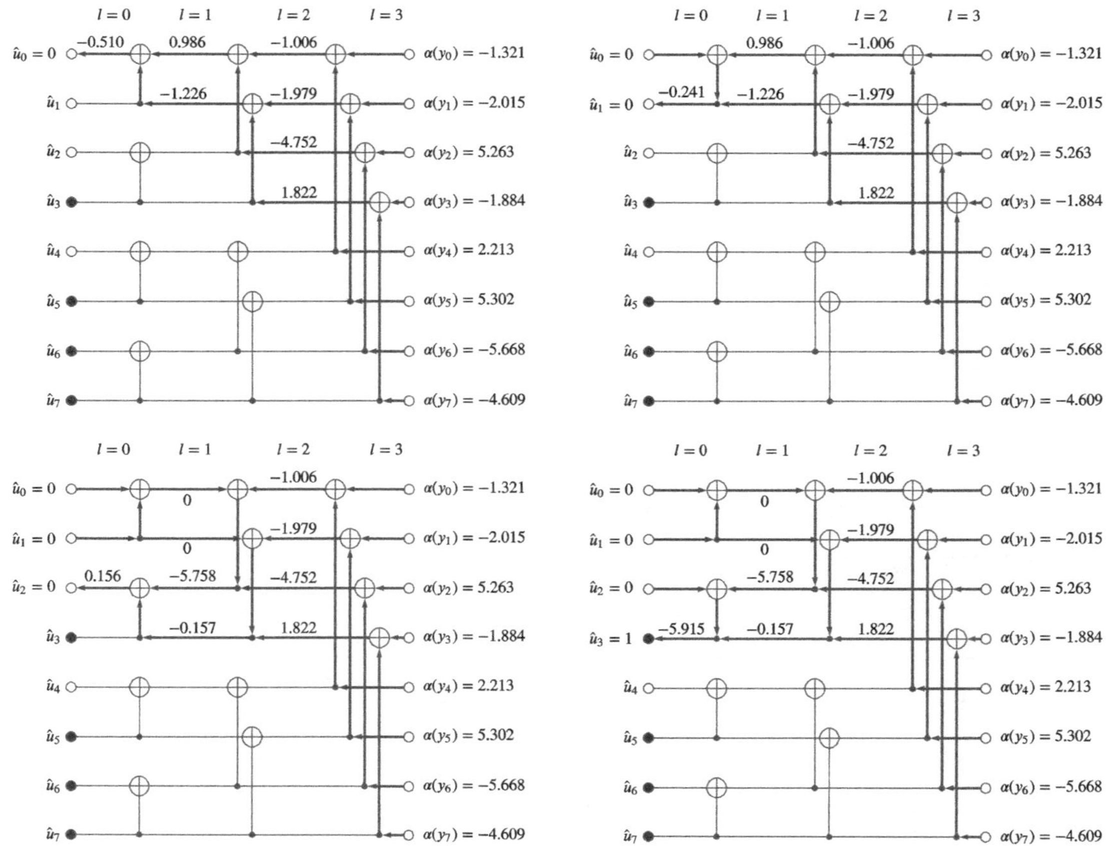

Figure 4.12 Decoding procedure of polar code (8,4) from bit 0 to bit 3 in Example 4.3.

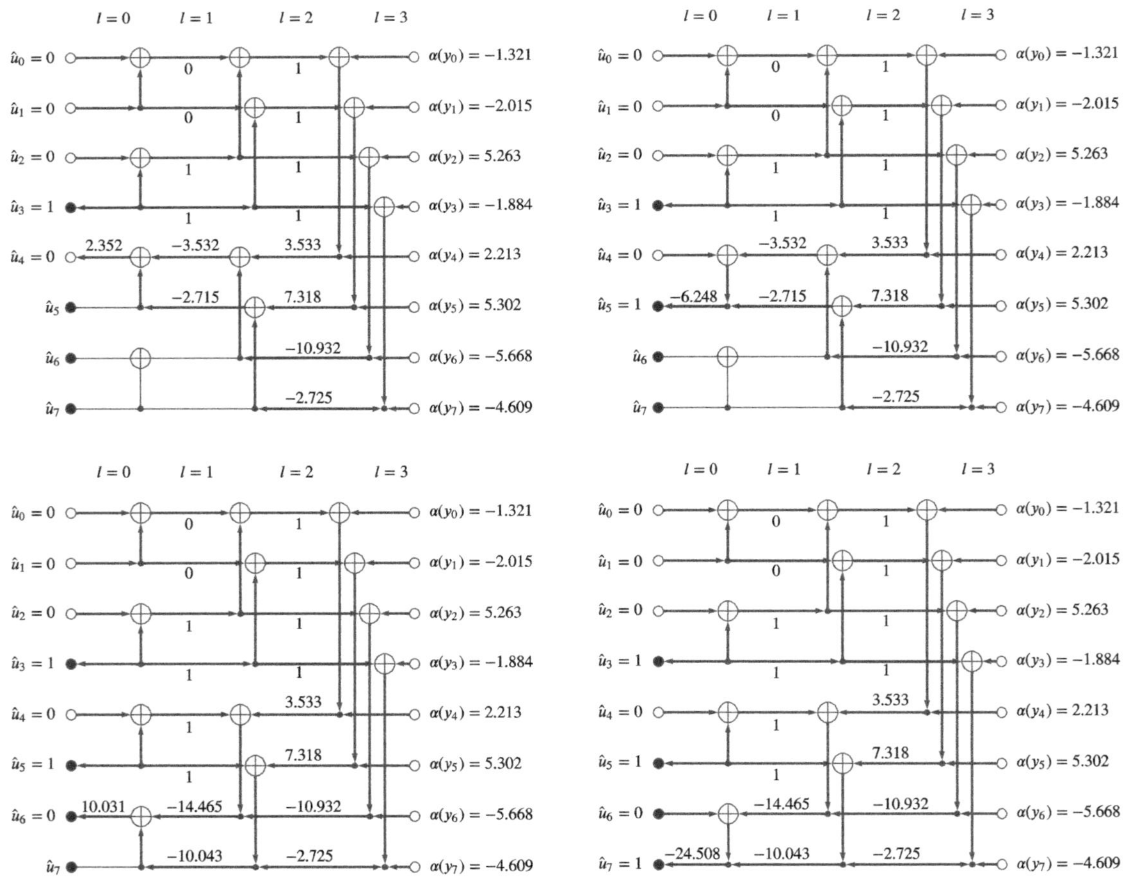

Figure 4.13 Decoding procedure of polar code (8,4) from bit 4 to bit 7 in Example 4.3.

length N grows toward infinity. The probability decreases exponentially in N^{β}, where β depends on the rate R and the gap to the channel capacity $I(W)$. We discuss this further in Section 5.9 of Chapter 5 and examine it for larger kernels. In the following, we establish a non-asymptotic bound directly involving the Bhattacharyya parameters of the bit-channels, which holds for any block length N.

Given the set of input–output pairs $(\mathbf{u}, \mathbf{y})$, the *block error event* $\mathcal{E}$ is defined as

$$\mathcal{E} = \left\{(\mathbf{u}, \mathbf{y}) \in \mathcal{X}^N \times \mathcal{Y}^N : \hat{\mathbf{u}}_{\mathcal{A}} \neq \mathbf{u}_{\mathcal{A}}\right\}.$$

The event $\mathcal{B}_i$ denotes the occurrence of the first decision error at the ith bit, which is defined as

$$\mathcal{B}_i = \left\{(\mathbf{u}, \mathbf{y}) \in \mathcal{X}^N \times \mathcal{Y}^N : u_0^{i-1} = \hat{u}_0^{i-1}, u_i \neq \hat{u}_i\left(\mathbf{y}, u_0^{i-1}\right)\right\},$$

where $\hat{u}_i$ is obtained using the ML rule in (4.39). Since the $\mathcal{B}_i$ events are mutually exclusive, we have

$$\mathcal{E} = \bigcup_{i \in \mathcal{A}} \mathcal{B}_i. \tag{4.45}$$

Observe that

$$\begin{aligned} \mathcal{B}_i &\subset \left\{(\mathbf{u}, \mathbf{y}) \in \mathcal{X}^N \times \mathcal{Y}^N : u_i \neq \hat{u}_i\left(y_0^{N-1}, u_0^{i-1}\right)\right\} \\ &\subset \left\{(\mathbf{u}, \mathbf{y}) \in \mathcal{X}^N \times \mathcal{Y}^N : W_N^{(i)}\left(y_0^{N-1}, u_0^{i-1} \mid u_i\right) \leqslant W_N^{(i)}\left(y_0^{N-1}, u_0^{i-1} \mid \hat{u}_i\right)\right\} \triangleq \mathcal{E}_i, \end{aligned} \tag{4.46}$$

where $\mathcal{E}_i$ is the event that a single bit sent through channel $W_N^{(i)}$ using ML decision function (4.39) is decoded wrongly by favoring $\hat{u}_i$ in case of a tie. Note that the event $\mathcal{B}_i$ requires that $u_0^{i-1} = \hat{u}_0^{i-1}$ for the vector $\mathbf{u}$, whereas the set $(\mathbf{u}, \mathbf{y}) \in \mathcal{X}^N \times \mathcal{Y}^N$: $u_i \neq \hat{u}_i\left(y_0^{N-1}, u_0^{i-1}\right)$ does not impose this condition.

The second inclusion (subset relation) in (4.46) is due to the lack of knowledge about the decision when the ties are broken by coin flip, i.e. Bernoulli (1/2), in the decision function (4.39). A tie event occurs when $W_N^{(i)}\left(y_0^{N-1}, u_0^{i-1} \mid u_i\right) = W_N^{(i)}\left(y_0^{N-1}, u_0^{i-1} \mid \hat{u}_i\right)$.

The probability of any event $\mathcal{E}_i$ is the sum of the probabilities of all input–output pairs $(\mathbf{u}, \mathrm{y})$ that belong to the event. This is expressed as

$$\mathbb{P}\left[\mathcal{E}_i\right] = \sum_{(\mathbf{u},\mathbf{y}) \in \mathcal{E}_i} \mathbb{P}[(\mathbf{u}, \mathbf{y})].$$

The joint probability $\mathbb{P}[(\mathbf{u}, \mathbf{y})]$ is given by

$$\mathbb{P}[(\mathbf{u}, \mathbf{y})] = \mathbb{P}[\mathbf{u}] \cdot W_N(\mathbf{y} \mid \mathbf{u}). \tag{4.47}$$

where $\mathbb{P}[\mathbf{u}]$ is the probability of selecting the input vector $\mathbf{u}$. Since we assume uniformly distributed input bits, we have $\mathbb{P}[\mathbf{u}] = \frac{1}{2^N}$. To account for whether $(\mathbf{u}, \mathbf{y})$ is part of the error event $\mathcal{E}_i$, an indicator function $\mathbf{1}_{\mathcal{E}_i}(\mathbf{u}, \mathbf{y})$ is introduced as

$$\mathbf{1}_{\mathcal{E}_i}(\mathbf{u}, \mathbf{y}) = \begin{cases} 1, & \text{if } (\mathbf{u}, \mathbf{y}) \in \mathcal{E}_i \\ 0, & \text{otherwise} \end{cases} \tag{4.48}$$

Thus, using (4.47) and (4.48), the probability of $\mathcal{E}_i$ can be expressed as

$$\mathbb{P}\left[\mathcal{E}_i\right] = \sum_{(\mathbf{u},\mathbf{y})\in\mathcal{X}^N\times\mathcal{Y}^N} \mathbb{P}[(\mathbf{u}, \mathbf{y})]\mathbf{1}_{\mathcal{E}_i}(\mathbf{u}, \mathbf{y}) = \sum_{(\mathbf{u},\mathbf{y})\in\mathcal{X}^N\times\mathcal{Y}^N} \frac{1}{2^N} W_N(\mathbf{y} \mid \mathbf{u})\mathbf{1}_{\mathcal{E}_i}(\mathbf{u}, \mathbf{y}).$$

The probability of $\mathcal{E}_i$ is upper bounded by the Bhattacharyya parameter of the channel as

$$\begin{aligned} \mathbb{P}\left[\mathcal{E}_i\right] &= \sum_{(\mathbf{u},\mathbf{y})\in\mathcal{X}^N\times\mathcal{Y}^N} \frac{1}{2^N} W_N(\mathbf{y} \mid \mathbf{u})\mathbf{1}_{\mathcal{E}_i}(\mathbf{u}, \mathbf{y}) \\ &\leqslant \sum_{(\mathbf{u},\mathbf{y})\in\mathcal{X}^N\times\mathcal{Y}^N} \frac{1}{2^N} W_N(\mathbf{y} \mid \mathbf{u}) \sqrt{\frac{W_N^{(i)}\left(y_0^{N-1}, u_0^{i-1} \mid \hat{u}_i\right)}{W_N^{(i)}\left(y_0^{N-1}, u_0^{i-1} \mid u_i\right)}} \leqslant Z\left(W_N^{(i)}\right). \end{aligned} \tag{4.49}$$

Directly summing the indicator function is combinatorially complex. Replacing it with the likelihood-based term $\sqrt{\frac{W_N^{(i)}(y_0^{N-1}, u_0^{i-1} \mid \hat{i}_i)}{W_N^{(0)}(y_0^{N-1}, u_0^{i-1} \mid u_i)}}$ gives a simpler, looser bound that is easier to analyze mathematically. This term is a local likelihood comparison for specific y_0^{N-1} and u_0^{i-1}, which is summed over all possible combinations. The ratio inside the square root reflects how "confusable" the likelihoods are for u_i and $\hat{u}_i$. This is related to the Bhattacharyya parameter $Z\left(W_N^{(i)}\right)$ defined in (4.32) which provides a measure of the error by summing over all possible y_0^{N-1}, u_0^{i-1} combinations. Note that when $W_N^{(i)}\left(y_0^{N-1}, u_0^{i-1} \mid u_i\right) = W_N^{(i)}\left(y_0^{N-1}, u_0^{i-1} \mid \hat{u}_i\right)$, i.e. the event $\mathcal{E}_i$ only includes the erasures, the last inequality in (4.49) reduces to equality.

We conclude our discussion with the following: for each given $(N, K, \mathcal{A})$, there exists at least one frozen vector $\mathbf{u}_{\mathcal{A}^c}$ such that the block error probability of the $\mathbf{G}_N$-coset code defined by parameters $\left(N, K, \mathcal{A}, \mathbf{u}_{\mathcal{A}}\right)$ satisfies

$$P_e\left(N, K, \mathcal{A}, \mathbf{u}_{\mathcal{A}^c}\right) \leqslant \sum_{i\in\mathcal{A}} Z\left(W_N^{(i)}\right).$$

This is commonly referred to as the union bound, as it sums over the individual probabilities of errors (upper bounds of $\mathbb{P}\left[\mathcal{E}_i\right]$) for the bit indices in $\mathcal{A}$.

This guarantees that an explicit construction of $\mathbf{G}_N$-coset codes is achievable. Specifically, for symmetric binary discrete memoryless channels (B-DMCs), this provides a systematic method for constructing polar codes by appropriately selecting the frozen vector $\mathbf{u}_{\mathcal{A}^c}$ based on the indices $\mathcal{A}$ of the best channels; i.e. $\mathcal{A}^c = [0, N-1]\backslash\mathcal{A}$. These indices are chosen to minimize the upper bound on the block error probability.

Exercises

1 Construct the polar code $(16, 8)$ using the Bhattacharyya parameter over BEC with $\varepsilon = 0.2$.

2 Analyze the error probability for the polar code $(16, 8)$ under SC decoding. Calculate the probability that an information bit in one of the less reliable channels is incorrectly decoded due to channel erasures.

3 Construct the polar code $(8, 6)$ based on the results of Example 4.2.

4 In SC decoding, explain why frozen bits are fixed and do not require decoding. For the $(8, 6)$ polar code, illustrate the decoding process by showing the path taken by the decoder and how the information bits are estimated in each step. Assume the received sequence in Example 4.3.

5 Demonstrate the asymptotic behavior of polar codes as $N \to \infty$. Explain how channel polarization impacts the reliability of each channel under SC decoding and show that as N grows, the fraction of reliable channels approaches the symmetric capacity of the channel.

Bibliographical Notes

Polar codes were invented by Arıkan [2009] in his seminal work, where he showed that they achieve the capacity of any binary-input discrete memoryless channel in the limit of large block lengths $N \to \infty$ under $O(N \log N)$ encoding and decoding complexity via the successive cancellation decoder. In particular, Arıkan proved that the transform $\mathbf{G}_2^{\otimes n}$ is polarizing in the limit of $n \to \infty$, in the sense that for any fixed $\gamma > 0$, the fraction of indices for which $H\left(\mathbf{W}_i \mid \mathbf{W}_{<i}\right) \in (\gamma, 1-\gamma)$ is vanishing

for large n. Shortly after, Arıkan and Telatar [2009] demonstrated that it is possible to choose $\gamma < 2^{-\Omega(2^{\beta n})}$ for any $\beta < 1/2$. This result contributed to an enhanced decoding error probability of $2^{-N^{-\beta}}$. The decrease in the parameter γ with respect to n was denoted as the "rate" of (limiting) polarization.

Polar codes are widely described as the first capacity-achieving codes with explicit construction based on the polarization phenomenon. Moreover, polarization has proven to be a versatile technique for asymptotically resolving various fundamental problems. These include, but are not limited to, problems such as the lossless and lossy source coding problem—Korada et al. [2010], coding for broadcast - Goela et al. [2014], multiple access—Abbe and Telatar [2012], and wiretap channels—Mahdavifar and Vardy [2011].

The first decoding algorithm for polar codes, known as successive cancellation decoding, was proposed by Arıkan [2009]. This algorithm has a hard output. Later, Fayyaz and Barry [2014] proposed using soft-output, which improves the performance of the decoder. This algorithm is called soft cancellation (SCAN) decoding.

Bibliography

Emmanuel Abbe and Emre Telatar. Polar codes for the m-user multiple access channel. *IEEE Transactions on Information Theory*, 58(8):5437–5448, 2012.

Erdal Arıkan. Channel polarization: A method for constructing capacity-achieving codes for symmetric binary-input memoryless channels. *IEEE Transactions on Information Theory*, 55(7):3051–3073, 2009.

Erdal Arıkan. On the origin of polar coding. *IEEE Journal on Selected Areas in Communications*, 34(2):209–223, 2015.

Erdal Arıkan and Emre Telatar. On the rate of channel polarization. In *2009 IEEE International Symposium on Information Theory*, pages 1493–1495. IEEE, 2009.

Ubaid U Fayyaz and John R Barry. Low-complexity soft-output decoding of polar codes. *IEEE Journal on Selected Areas in Communications*, 32(5):958–966, 2014.

Naveen Goela, Emmanuel Abbe, and Michael Gastpar. Polar codes for broadcast channels. *IEEE Transactions on Information Theory*, 61(2):758–782, 2014.

Hideki Imai and Shuji Hirakawa. A new multilevel coding method using error-correcting codes. *IEEE Transactions on Information Theory*, 23(3):371–377, 1977.

Satish Babu Korada, Eren Şaşoğlu, and Rüdiger Urbanke. Polar codes: Characterization of exponent, bounds, and constructions. *IEEE Transactions on Information Theory*, 56(12):6253–6264, 2010.

Hessam Mahdavifar and Alexander Vardy. Achieving the secrecy capacity of wiretap channels using polar codes. *IEEE Transactions on Information Theory*, 57(10):6428–6443, 2011.

J Massey. Capacity, cutoff rate, and coding for a direct-detection optical channel. *IEEE Transactions on Communications*, 29(11):1615–1621, 1981.

Mark Semenovich Pinsker. On the complexity of decoding. *Problemy Peredachi Informatsii*, 1(1):113–116, 1965.

5

Properties of Polar Codes

"All truths are easy to understand once they are discovered; the point is to discover them."

— Galileo Galilei

Beyond their ability to achieve channel capacity, polar codes exhibit a rich set of structural and mathematical properties that influence their design, analysis, and performance across a wide range of applications.

This chapter delves into the key properties of polar codes, offering insights into their mathematical foundations and practical implications. We begin by examining the polar transform and its connection to linear algebra, highlighting the role of permutation matrices and the construction of generator and parity-check matrices, which are essential for encoding and decoding processes. We also explore the concept of systematic polar codes.

The relationship between polar codes and Reed–Muller codes is another focal point, providing a deeper understanding of how these two classes of codes intersect and diverge. This discussion naturally leads to the exploration of the partial order of sub-channels, a critical tool for selecting information and frozen bits.

The chapter also addresses the error correction performance of polar codes through an analysis of their minimum distance and the construction and enumeration of minimum weight codewords. These properties are fundamental to evaluating the performance of polar codes under near ML decoding scenarios.

We further extend the discussion to polar codes with large kernels and mixed/multi-kernel constructions, which aim to enhance polarization and improve performance. The exponent of the polarizing matrix is introduced as a measure of the efficiency of these alternative kernels in polarizing sub-channels.

The practical aspects of decoding are explored in depth, with a particular focus on the properties of SC decoding. Topics such as updating intermediate log-likelihood ratios (LLRs), managing partial sums, and implementing partial

Polar Codes: From Theory to Practice, First Edition. Mohammad Rowshan and Emanuele Viterbo.

rewind mechanisms are covered, along with the role of polar constituent codes in decoding. Special attention is given to decoding strategies involving constituent nodes, including rate-1, rate-0, repetition, and parity-check nodes, which simplify SC decoding for certain code structures.

Through this comprehensive exploration of polar code properties, readers will gain a deeper appreciation of the theoretical underpinnings and practical design choices that make polar codes one of the most versatile and powerful tools in modern coding theory.

Before delving into details, let us review the notations. As we discussed in Chapter 4, for polar encoding, the information bits $\mathbf{d} = [d_0 \cdots d_{K-1}]$ are carefully inserted into the input vector $\mathbf{u} = [u_0 \cdots u_{N-1}]$ at the $i \in \mathcal{A}$ positions, while $u_i = 0$ for $i \in \mathcal{A}^c$, that is, the bad subchannels are used for the transmission of known values (by default 0). Therefore, we can encode an information sequence $\mathbf{d} = \mathbf{u}_\mathcal{A}$ using $\mathbf{dG}_\mathcal{A} = \mathbf{uG}_N = \mathbf{x}$. In this chapter, we delve deeper into the properties of the polar transform, the generator matrix, the permutation matrix, and more.

We use the notation supp(i) for the support of the binary expansion of i, that is, the set of coordinates of 1s in $\text{bin}(i) = i_{n-1} \cdots i_1 i_0$. In addition, we use $\overline{\text{supp}(i)}$ for the complementary set $[0, n-1] \backslash \text{supp}(i)$. For example, with $n = 5$ and $i = 19$, $\text{bin}(i) = 10011$, $\text{supp}(i) = \{0, 1, 4\}$, and $\overline{\text{supp}(i)} = \{2, 3\}$.

5.1 Polar Transform

The n-fold Kronecker product $\mathbf{G}_N = \mathbf{G}_2^{\otimes n}$ of the basic transform, also known as Arıkan kernel,

$$\mathbf{G}_2 = \begin{bmatrix} 1 & 0 \\ 1 & 1 \end{bmatrix}$$

reveals interesting properties. Clearly, the resulting matrix is of lower triangular form. In this section, we discuss other characteristics of this transform. We omit the proofs, but the reader can refer to the bibliographical notes for further reading.

Let $g_{i,j}$ be the (i,j)th entry of $\mathbf{G}_N$ with indices i, j in $[0, N-1]$. Then the relation between this entry and its row and column indices is

$$g_{i,j} = \begin{cases} 1, & \text{if } \text{supp}(j) \subseteq \text{supp}(i), \\ 0, & \text{if otherwise}. \end{cases} \tag{5.1}$$

Example 5.1 Find the row $\mathbf{g}_5$ of $\mathbf{G}_2^{\otimes 3}$ without taking the Kronecker power of $\mathbf{G}_2$.

We know that $\text{bin}(5) = 101$ and hence $\text{supp}(5) = \{0, 2\}$. To find the eight elements of this row with indices $j \in [0, n-1]$, we check condition $\text{supp}(j) \subseteq \text{supp}(i)$. Only column indices j with $1 \in \text{supp}(j)$ will have 0 entries because in this case

$\text{supp}(j) \nsubseteq \{0, 2\}$. These coordinates are $j = \{2, 3, 6, 7\}$. Hence, the i = 5th row is

$$\mathbf{g}_5 = [1\ 1\ 0\ 0\ 1\ 1\ 0\ 0].$$

Now, we obtain the weight of a single row and of the sum of two rows based on their indices. Given row index $i \in [0, 2^n - 1]$, the weight of ith row $\mathbf{g}_i$ is

$$\text{w}(\mathbf{g}_i) = 2^{|\text{supp}(i)|}. \tag{5.2}$$

This can be seen by applying (5.1) and the fact that $\text{w}(\mathbf{g}_i) = |\text{supp}(\mathbf{g}_i)| = |\{c \in [0, N-1] : \text{supp}(c) \subseteq \text{supp}(i)\}| = 2^{|\text{supp}(i)|}$. The sum of ith and jth rows has weight

$$\text{w}(\mathbf{g}_i + \mathbf{g}_j) = 2^{|\text{supp}(i)|} + 2^{|\text{supp}(j)|} - 2 \times 2^{|\text{supp}(i) \cap \text{supp}(j)|}. \tag{5.3}$$

This comes from the principle of inclusion and exclusion and from the fact that $\text{supp}(\mathbf{g}_i) \cap \text{supp}(\mathbf{g}_j) = \{c \in [0, N-1] : \text{supp}(c) \subseteq \text{supp}(i) \cap \text{supp}(j)\} = 2^{|\text{supp}(i) \cap \text{supp}(j)|}$. Recall that the number of all subsets of any set S is $2^{|S|}$.

Example 5.2 Give the weight of $\mathbf{g}_{11}$ and $\mathbf{g}_{11} + \mathbf{g}_5$ in the transform $\mathbf{G}_2^{\otimes 4}$.

Knowing the support of 11 and 5 are $\text{supp}(11) = \{0, 1, 3\}$ and $\text{supp}(5) = \{0, 2\}$, we have

$$\begin{aligned} \text{w}(\mathbf{g}_{11}) &= 2^{|\text{supp}(11)|} = 2^3 = 8, \\ \text{w}(\mathbf{g}_5 + \mathbf{g}_{11}) &= 2^{|\text{supp}(5)|} + 2^{|\text{supp}(11)|} - 2 \times 2^{|\text{supp}(5) \cap \text{supp}(11)|} \\ &= 2^3 + 2^2 - 2 \times 2^1 = 8. \end{aligned}$$

In general, the relation between the weight of ith row and the sum of this row with any subset $\mathcal{J} \subseteq [i+1, 2^n - 1]$ of rows with indices larger than i in the transform $\mathbf{G}_2^{\otimes n}$ is

$$\text{w}\left(\mathbf{g}_i + \sum_{j \in \mathcal{J} \subseteq [i+1, 2^n - 1]} \mathbf{g}_j\right) \geq \text{w}(\mathbf{g}_i). \tag{5.4}$$

This property is due to the lower triangular structure of $\mathbf{G}_2^{\otimes n}$.

5.2 Permutation Matrix

The *bit-reversal permutation matrix* $\mathbf{B}_N$ is obtained recursively using

$$\mathbf{B}_N = \mathbf{R}_N \left(I_2 \otimes \mathbf{B}_{\frac{N}{2}}\right), \tag{5.5}$$

where $\mathbf{B}_2 = \mathbf{I}_2$ and $\mathbf{R}_N$ is the *reverse shuffle matrix* that shuffles an N-tuple as

$$\left(s_0, s_1, s_2, \ldots, s_{N-1}\right) \mathbf{R}_N = \left(s_0, s_2, \ldots, s_{N-2}, s_1, s_3, \ldots, s_{N-1}\right). \tag{5.6}$$

Observe that a matrix $\mathbf{R}_N$ can be divided into two submatrices of size $N \times \frac{N}{2}$ as $\mathbf{R}_N = [\mathbf{R}^l_{N/2}\ \ \mathbf{R}^r_{N/2}]$, where the columns of $\mathbf{R}^l_{N/2}$ contain a single 1 in even rows and the columns of $\mathbf{R}^r_{N/2}$ contain a single 1 in odd rows, assuming that the rows are indexed from 0 to $N-1$. The matrix below shows the matrix $\mathbf{R}_8$:

$$\mathbf{R}_8 = \begin{bmatrix} 1&0&0&0&0&0&0&0\\ 0&0&0&0&1&0&0&0\\ 0&1&0&0&0&0&0&0\\ 0&0&0&0&0&1&0&0\\ 0&0&1&0&0&0&0&0\\ 0&0&0&0&0&0&1&0\\ 0&0&0&1&0&0&0&0\\ 0&0&0&0&0&0&0&1 \end{bmatrix}.$$

Example 5.3 Find $\mathbf{B}_8$.

Let us first find $\mathbf{B}_4$ using $\mathbf{B}_2 = \mathbf{I}_2$ and $\mathbf{R}_4$:

$$\mathbf{B}_4 = \mathbf{R}_4\left(\mathbf{I}_2 \otimes \mathbf{B}_2\right) = \mathbf{R}_4 \cdot \mathbf{I}_4 = \begin{bmatrix} 1&0&0&0\\ 0&0&1&0\\ 0&1&0&0\\ 0&0&0&1 \end{bmatrix}.$$

Then, we find $\mathbf{B}_8$ using $\mathbf{B}_4$ and $\mathbf{R}_8$:

$$\mathbf{B}_8 = \mathbf{R}_8\left(\mathbf{I}_2 \otimes \mathbf{B}_4\right) = \begin{bmatrix} 1&0&0&0&0&0&0&0\\ 0&0&0&0&1&0&0&0\\ 0&0&1&0&0&0&0&0\\ 0&0&0&0&0&0&1&0\\ 0&1&0&0&0&0&0&0\\ 0&0&0&0&0&1&0&0\\ 0&0&0&1&0&0&0&0\\ 0&0&0&0&0&0&0&1 \end{bmatrix}.$$

The matrix $\mathbf{B}_N$ is symmetric, that is, the (i, j)th entry is equal to the (j, i)th entry for all i and j. Hence, we have

$$\mathbf{B}_N^T = \mathbf{B}_N. \tag{5.7}$$

The inverse of a permutation matrix is the matrix that reverses the permutation performed by the original matrix. Then we have

$$\mathbf{B}_N^{-1} = \mathbf{B}_N^T = \mathbf{B}_N. \tag{5.8}$$

On the other hand, since $\mathbf{B}^T = \mathbf{B}^{-1}$, $\mathbf{G}_2^{\otimes n}$ is invariant under bit-reversal, that is, $\mathbf{G}_2^{\otimes n} = \mathbf{B}_N \mathbf{G}_2^{\otimes n} \mathbf{B}_N^T$. Hence, $\mathbf{B}_N$ and $\mathbf{G}_2^{\otimes n}$ commute, that is, permuting the rows or the columns of $\mathbf{G}_2^{\otimes n}$ yields the same matrix ($\mathbf{B}_N\mathbf{G}_2^{\otimes n} = \mathbf{G}_2^{\otimes n}\mathbf{B}_N$), which is no longer lower triangular.

Note that for encoding $\mathbf{u}$ via a row or column permuted $\mathbf{G}_2^{\otimes n}$ matrix, that is, $\mathbf{x} = \mathbf{u}\mathbf{B}_N\mathbf{G}_2^{\otimes n} = \mathbf{u}\mathbf{G}_2^{\otimes n}\mathbf{B}_N$, we can alternatively permute $\mathbf{u}$ instead of the matrix $\mathbf{G}_2^{\otimes n}$ since $\mathbf{x} = \mathbf{u}\left(\mathbf{B}_N\mathbf{G}_2^{\otimes n}\right) = \left(\mathbf{u}\mathbf{B}_N\right)\mathbf{G}_2^{\otimes n}$. This enables us to take advantage of the fast recursive encoding operation that comes with the lower triangular matrix structure.

Example 5.4 Given the information sequence $\mathbf{u} = [00010101]$. Encode $\mathbf{u}$ using both permuted $\mathbf{u}$ and permuted $\mathbf{G}_2^{\otimes n}$.

$$\mathbf{x} = \mathbf{u}\left(\mathbf{B}_N\mathbf{G}_2^{\otimes n}\right) = [00010101]\begin{bmatrix} 1&0&0&0&0&0&0&0\\ 1&0&0&0&1&0&0&0\\ 1&0&1&0&0&0&0&0\\ 1&0&1&0&1&0&1&0\\ 1&1&0&0&0&0&0&0\\ 1&1&0&0&1&1&0&0\\ 1&1&1&1&0&0&0&0\\ 1&1&1&1&1&1&1&1 \end{bmatrix} = [10011001].$$

$$\mathbf{x} = \left(\mathbf{u}\mathbf{B}_N\right)\mathbf{G}_2^{\otimes n} = [00000111]\begin{bmatrix} 1&0&0&0&0&0&0&0\\ 1&1&0&0&0&0&0&0\\ 1&0&1&0&0&0&0&0\\ 1&1&1&1&0&0&0&0\\ 1&0&0&0&1&0&0&0\\ 1&1&0&0&1&1&0&0\\ 1&0&1&0&1&0&1&0\\ 1&1&1&1&1&1&1&1 \end{bmatrix} = [10011001].$$

5.3 Generator and Parity Check Matrices

The generator matrix $\mathbf{G}$ of a (K, N) polar code is a $K \times N$ submatrix of the polar transform $\mathbf{B}_N\mathbf{G}_2^{\otimes n} = [\mathbf{g}_1 \cdots \mathbf{g}_N]^T$ consisting of rows $\mathbf{g}_i$ with indices $i \in \mathcal{A}$, where $\mathbf{B}_N$ is a bit-reversal permutation matrix.

As $\mathbf{G}_2^{\otimes n}$ is a lower-triangular matrix with 1s on the diagonal, it is invertible and its inverse is itself. Given $\mathbf{x} = \mathbf{u}\mathbf{G}_2^{\otimes n}$, we have $\mathbf{u} = \mathbf{x}\mathbf{G}_2^{\otimes n}$. Furthermore, since $\mathbf{G}_2^{-1} = \mathbf{G}_2$ and $\mathbf{B}_N$ commutes $\mathbf{G}_2^{\otimes n}$, we have

$$(\mathbf{B}_N\mathbf{G}_2^{\otimes n})^{-1} = (\mathbf{G}_2^{\otimes n})^{-1}\mathbf{B}_N^{-1} = (\mathbf{G}_2^{-1})^{\otimes n}\mathbf{B}_N = \mathbf{B}_N\mathbf{G}_2^{\otimes n}. \tag{5.9}$$

Then, given $\mathbf{x} = \mathbf{u}\mathbf{B}_N\mathbf{G}_2^{\otimes n}$, we have $\mathbf{u} = \mathbf{x}\mathbf{B}_N\mathbf{G}_2^{\otimes n}$.

Next, we characterize the parity check matrix $\mathbf{H}$ of a polar code.

To impose parity check constraints $\mathbf{x} \cdot \mathbf{H}^T = \mathbf{0}$ on $\mathbf{x}$, it suffices to select the columns of $\mathbf{B}_N \mathbf{G}_2^{\otimes n}$ with indices $j \in \mathcal{A}^c$. Therefore, the parity check matrix $\mathbf{H}$ of polar codes is a submatrix of $\mathbf{B}_N \mathbf{G}_2^{\otimes n}$ consisting of columns $j \in \mathcal{A}^c$.

Example 5.5 Given the (8,4) polar code with $\mathcal{A} = \{0, 1, 2, 4\}$. Find the generator matrix and parity check matrix for this code.

Let us first find $\mathbf{B}_8 \mathbf{G}_2^{\otimes 3}$:

$$\mathbf{B}_8 \mathbf{G}_2^{\otimes 3} = \begin{bmatrix} 1&0&0&0&0&0&0&0 \\ 1&0&0&0&1&0&0&0 \\ 1&0&1&0&0&0&0&0 \\ 1&0&1&0&1&0&1&0 \\ 1&1&0&0&0&0&0&0 \\ 1&1&0&0&1&1&0&0 \\ 1&1&1&1&0&0&0&0 \\ 1&1&1&1&1&1&1&1 \end{bmatrix}.$$

Then, the generator and the parity check matrices are:

$$\mathbf{G} = \begin{bmatrix} 1&1&1&1&0&0&0&0 \\ 1&1&0&0&1&1&0&0 \\ 1&0&1&0&1&0&1&0 \\ 1&1&1&1&1&1&1&1 \end{bmatrix},$$

$$\mathbf{H} = \begin{bmatrix} 1&0&0&0 \\ 1&1&0&0 \\ 1&0&1&0 \\ 1&1&1&0 \\ 1&0&0&1 \\ 1&1&0&1 \\ 1&0&1&1 \\ 1&1&1&1 \end{bmatrix}.$$

5.4 Systematic Polar Codes

Let us denote the elements of the row $\mathbf{g}_i$ of the matrix $\mathbf{G} = \mathbf{G}_2^{\otimes n}$ by $\{g_{i,j}\}$, $j \in [0, N-1]$. Then, the submatrix $\mathbf{G}_\mathcal{A}$ includes the rows $\mathbf{g}_i, i \in \mathcal{A}$, while the submatrix $\mathbf{G}_{\mathcal{A}\mathcal{B}}$ consists of $g_{i,j}, i \in \mathcal{A}, j \in \mathcal{B}$. Also, let $\mathbf{u}_\mathcal{A}$ denote the subvector of $\mathbf{u}$ consisting of $u_i, i \in \mathcal{A}$.

A systematic code allows the original message to be recovered directly from the received codeword. That is, assuming that the encoded bits are $\mathbf{x} = [\mathbf{x}_{\mathcal{A}^c}, \mathbf{x}_\mathcal{A}]$, the systematic coding results in $\mathbf{x}_{\mathcal{A}} = \mathbf{u}_{\mathcal{A}}$. This can be demonstrated as

$$\mathbf{x}_{\mathcal{A}} = \mathbf{u}_{\mathcal{A}}\left(\mathbf{G}_{\mathcal{A}\mathcal{A}}\right)^{-1}\mathbf{G}_{\mathcal{A}\mathcal{A}} = \mathbf{u}_{\mathcal{A}}\mathbf{I}_K = \mathbf{u}_{\mathcal{A}}. \tag{5.10}$$

Note that the matrix $\mathbf{G}_{\mathcal{A}\mathcal{A}}$ has a size $K \times K$ and a lower triangular form where all entries on the diagonal are 1, hence, it is invertible.

On the other hand, to find the parity bits $\mathbf{x}_{\mathcal{A}^c}$, we use the following operation:

$$\mathbf{x}_{\mathcal{A}^c} = \mathbf{u}_{\mathcal{A}}\left(\mathbf{G}_{\mathcal{A}\mathcal{A}}\right)^{-1}\mathbf{G}_{\mathcal{A}\mathcal{A}^c} = \mathbf{u}_{\mathcal{A}}\mathbf{P}, \tag{5.11}$$

where $\mathbf{P} = \left(\mathbf{G}_{\mathcal{A}\mathcal{A}}\right)^{-1}\mathbf{G}_{\mathcal{A}\mathcal{A}^c}$.

Note that this process is equivalent to applying the non-systematic encoding (without applying the bit-reversal permutation) twice in a row, where a reset of the values of the frozen indices is performed in between. That is,

$$\mathbf{x} = \mathbf{u}\mathbf{G} \rightarrow \mathbf{x}' : \text{for all } \ i \in \mathcal{A}^c, \ \ \text{set} \ \ x_i = 0 \ \ \text{in} \ \ \mathbf{x} \rightarrow \mathbf{x}'' = \mathbf{x}'\mathbf{G}.$$

Example 5.6 Given the (8,4) polar code with $\mathcal{A} = \{0, 1, 2, 4\}$. Give the systematically encoded information sequence $\mathbf{u} = [00010101]$.

First, we find $\mathbf{u}_{\mathcal{A}}$, which is $\mathbf{u}_{\mathcal{A}} = \mathbf{d} = [1101]$, $\mathbf{G}_{\mathcal{A}\mathcal{A}}$, and $\mathbf{G}_{\mathcal{A}\mathcal{A}^c}$ as below:

$$\mathbf{G}_{\mathcal{A}\mathcal{A}} = \begin{bmatrix} 1 & 0 & 0 & 0 \\ 0 & 1 & 0 & 0 \\ 0 & 0 & 1 & 0 \\ 1 & 1 & 1 & 1 \end{bmatrix},$$

$$\mathbf{G}_{\mathcal{A}\mathcal{A}^c} = \begin{bmatrix} 1 & 1 & 1 & 0 \\ 1 & 1 & 0 & 1 \\ 1 & 0 & 1 & 1 \\ 1 & 1 & 1 & 1 \end{bmatrix}.$$

Then, we can find the parities as follows:

$$\mathbf{x}_{\mathcal{A}} = \mathbf{u}_{\mathcal{A}}\left(\mathbf{G}_{\mathcal{A}\mathcal{A}}\right)^{-1}\mathbf{G}_{\mathcal{A}\mathcal{A}^c} = [1101]\begin{bmatrix} 1 & 0 & 0 & 0 \\ 0 & 1 & 0 & 0 \\ 0 & 0 & 1 & 0 \\ 1 & 1 & 1 & 1 \end{bmatrix}\begin{bmatrix} 1 & 1 & 1 & 0 \\ 1 & 1 & 0 & 1 \\ 1 & 0 & 1 & 1 \\ 1 & 1 & 1 & 1 \end{bmatrix} = [0100].$$

Finally, we have the codeword $\mathbf{x} = [\mathbf{x}_{\mathcal{A}^c}, \mathbf{x}_{\mathcal{A}}] = [\underline{0100}1101]$, where the underlined digits are the parity bits. We can reorder the coordinates to match the indices in $\mathcal{A}$ by placing the information bits at positions $\mathcal{A}$. In this case, we get $\mathbf{x}'' = [\underline{010}1\underline{0}101]$.

Alternatively, this can be performed in two-stage encoding as follows. First, we encode as

$$\mathbf{x} = \mathbf{u}\mathbf{G} = [11000011],$$

Then, we reset the coordinates $i \in \mathcal{A}^c$ as $x_i = 0$ to get $\mathbf{x}' = [00000011]$ and repeat the encoding:

$$\mathbf{x}'' = \mathbf{x}'\mathbf{G} = [00000011]\mathbf{G} = [\underline{010}1\underline{0}101].$$

As can be seen, the result of the two approaches is identical.

5.5 Reed–Muller Codes Versus Polar Codes

A binary Reed–Muller (RM) code of length $N = 2^n$, order r, dimension $K = \sum_{i=0}^{r} \binom{n}{i}$, and minimum distance $d = 2^{n-r}$ can be represented based on the transformation matrix $\mathbf{G}_N = \mathbf{G}_2^{\otimes n}$, where the generator matrix is a $K \times N$ submatrix of $\mathbf{G}_N = [\mathbf{g}_1 \cdots \mathbf{g}_N]^T$ consisting of rows $\mathbf{g}_i$ with indices $i \in \{j \in [N] \mid w(\mathbf{g}_j) \geq 2^{n-r}\}$. We refer to this approach of selecting rows based on their weights as *RM rule*. The differences between RM codes and polar codes are mainly in the choice of rows of the transformation matrix $\mathbf{G}_N$, which offers complete flexibility in the code dimension K.

From a decoding perspective, successive cancellation decoding (see Section 4.7) in Chapter 4 closely resembles recursive decoding of RM codes [Arıkan, 2010].

The generator matrix of RM code of length 2^n and order r, denoted $\mathrm{RM}(r, n)$, is formed by the set of all rows $\mathbf{g}_i, i \in [0, N-1]$, of weight $\mathrm{w}(\mathbf{g}_i) \geq 2^{n-r}$, which is the minimum-weight $w_{\min}$ of the code. Therefore, the information set $\mathcal{A}_{RM}$ of $\mathrm{RM}(r, n)$ is created as follows.

$$\mathcal{A}_{RM} = \{i \in [0, 2^n - 1] : \mathrm{w}(\mathbf{g}_i) \geq 2^{n-r}\}.$$

The dimension of $\mathrm{RM}(r, n)$ is $K = |\mathcal{A}_{RM}| = \sum_{\ell=0}^{r} \binom{n}{\ell}$. The concept of order r in the $\mathrm{RM}(r, n)$ code comes from the wedge products of N-tuple $\mathbf{v}_{i+1} = \mathbf{g}_{N-2^i-1}, i \in [0, n-1]$ up to degree r, where $\mathbf{g}_j$ is the jth rows of $\boldsymbol{G}_N$. By default, $\mathbf{v}_0 = \mathbf{g}_{N-1} = [1\ 1\ \ldots\ 1]$. For instance, when $n = 3$ we obtain

$$\begin{aligned}
\mathbf{v}_0 &= \mathbf{g}_7 = [1\ 1\ 1\ 1\ 1\ 1\ 1\ 1],\\
\mathbf{v}_1 &= \mathbf{g}_6 = [1\ 0\ 1\ 0\ 1\ 0\ 1\ 0],\\
\mathbf{v}_2 &= \mathbf{g}_5 = [1\ 1\ 0\ 0\ 1\ 1\ 0\ 0],\\
\mathbf{v}_3 &= \mathbf{g}_3 = [1\ 1\ 1\ 1\ 0\ 0\ 0\ 0].
\end{aligned}$$

The vectors $\mathbf{v}_i, i \in [0, 3]$ in the example above form the generator matrix of RM(1,3). As an example for order 2, the generator matrix for $\mathrm{RM}(2, 3)$ is given by

$$\mathbf{G}_{\mathrm{RM}(2,3)} = \begin{bmatrix} \mathbf{v}_2 \wedge \mathbf{v}_3 \\ \mathbf{v}_1 \wedge \mathbf{v}_3 \\ \mathbf{v}_3 \\ \mathbf{v}_1 \wedge \mathbf{v}_2 \\ \mathbf{v}_2 \\ \mathbf{v}_1 \\ \mathbf{v}_0 \end{bmatrix} = \begin{bmatrix} 1&1&0&0&0&0&0&0 \\ 1&0&1&0&0&0&0&0 \\ 1&1&1&1&0&0&0&0 \\ 1&0&0&0&1&0&0&0 \\ 1&1&0&0&1&1&0&0 \\ 1&0&1&0&1&0&1&0 \\ 1&1&1&1&1&1&1&1 \end{bmatrix}, \tag{5.12}$$

which has rows with minimum Hamming weight $2^{n-r} = 2^{3-2} = 2$. One can observe that the generator matrix of RM(n, n) is $\boldsymbol{G}_N$. In the above example, only $\mathbf{g}_0$ of $\mathbf{G}_8$ which has weight 1 is not included in (5.12).

Now, let us demonstrate that the information sets $\mathcal{A}$ of Reed–Muller codes and polar codes satisfy the Partial Order Property. For the RM(r, n), we have

$$\begin{aligned} \mathcal{A}_{RM} &= \{i \in [0, 2^n - 1] : \mathrm{w}(\mathbf{g}_i) \geq 2^{n-r}\} \\ &= \{i \in [0, 2^n - 1] : |\mathrm{supp}(i)| \geq n - r\}, \end{aligned} \tag{5.13}$$

where the second equality is due to $\mathrm{w}(\mathbf{g}_i) = 2^{|\mathrm{supp}(i)|}$. Clearly, if $i^c \in \mathcal{A}^c$ then $|\mathrm{supp}(i^c)| < n - r \leq |\mathrm{supp}(i)|$.

For a polar code, its information set $\mathcal{A}$ must satisfy the condition that the sub-channel $W_N^{(i)}$ is more reliable than the sub-channel $W_N^{(i^c)}$ for every $i \in \mathcal{A}$ and $i^c \in \mathcal{A}^c$. Note that the selected rows in the polar codes of length $N = 2^n$ and minimum row weight $w_{\min} = \min \mathrm{w}(\mathbf{g}_i), i \in \mathcal{A}$ are a subset of the rows in the generator matrix for RM(r, n) where $2^{n-r} = w_{\min}$. As a result, any polar code is a subcode of some RM code with a common minimum distance, which implies $\mathcal{A} \subseteq \mathcal{A}_{RM}$.

5.6 Partial Order of Sub-channels

In mathematics, a partially ordered set, often abbreviated as POSET, formalizes the intuitive concept of ordering the elements of a set. A poset comprises a set along with a binary relation that denotes that one element precedes the other in the ordering for certain pairs of elements in the set. The relation itself is termed a *partial order*. The term "partial" in "partial order" signifies that not every pair of elements is comparable. In other words, there may exist pairs of elements for which neither element precedes the other in the poset. Partial orders thus extend the concept of total orders, where every pair is comparable.

Polar codes and Reed–Muller codes can be defined based on partial orders using binary representations of the sub-channel indices. Partial orders abstract away the cumbersome notion of sub-channel reliability. We first define in Definition 5.2

these partial orders, combined as a single partial order, and then the so-called *Partial Order Property* that the information set $\mathcal{A}$ of these codes needs to satisfy in Definition 5.3. Let us first look at the concept of *stochastic degradation* indicating the sub-channel that precedes the other in the reliability ordering.

Definition 5.1 ***(Stochastic Degradation)*** Given two BMS channels

$$W_1 : \mathcal{X} \to \mathcal{Y}_1,$$
$$W_2 : \mathcal{X} \to \mathcal{Y}_2,$$

with the following respective transition probabilities:

$$p_{Y_1|X}\left(y_1 \mid x\right),$$
$$p_{Y_2|X}\left(y_2 \mid x\right),$$

where $x \in \mathcal{X}, y_1 \in \mathcal{Y}_1$, and $y_2 \in \mathcal{Y}_2$. It is said that W_1 is *stochastically degraded* with respect to W_2 and denoted by $W_1 \preceq W_2$ if there exists a memoryless channel with transition probabilities $p_{Y_1|Y_2}(y_1 \mid y_2)$ such that for all $x \in \mathcal{X}$ and $y_1 \in \mathcal{Y}_1$, we have

$$p_{Y_1|X}\left(y_1 \mid x\right) = \sum_{y_2 \in \mathcal{Y}_2} p_{Y_1|Y_2}\left(y_1 \mid y_2\right) p_{Y_2|X}\left(y_2 \mid x\right).$$

As the name implies, the reliability measures (see Chapter 4) of the stochastically degraded channel is worse. That is,

$$I\left(W_1\right) \leq I\left(W_2\right),$$
$$Z\left(W_1\right) \geq Z\left(W_2\right).$$

The notation $\preceq$ is called "precedes" or "preorder," where $W_i \preceq W_j$ indicates that the channel W_i is considered to be less reliable than or equally reliable as the channel W_j. For simplicity, we also use the index of the channels to show the partial orders throughout this book, that is, we simply write $i \preceq j$ for $W_i \preceq W_j$.

Recall that $\text{supp}(i) \triangleq \text{supp}(\text{bin}(i)) \subseteq [0, n-1]$ is the set of indices, where i has a "1" in its binary representation.

Definition 5.2 ***(Partial Order)*** Given $i, j \in [0, 2^n - 1]$, we denote $i \preceq j$ or $j \succeq i$ if they satisfy one of the following conditions:

- $\text{supp}(i) \subseteq \text{supp}(j)$. Observe that j can be obtained by flipping at least one 0 in $\text{bin}(i)$.
- $\text{supp}(j) = (\text{supp}(i) \setminus \{a\}) \cup \{b\}$ for some $a \in \text{supp}(i)$, $b \notin \text{supp}(i)$ and $a < b$ Equivalently, $\text{bin}(j)$ is obtained from $\text{bin}(i)$ by swapping a "1" in $\text{bin}(i)$ and a "0" at a more significant bit position, i.e., a *left-swap* of a "1" with a "0."
- there exists $k \in [0, 2^n - 1]$ satisfying $i \preceq k$ and $k \preceq j$.

Note that $i \preceq j$ implies that $i \leq j$ but not vice versa.

It turns out that the relative reliability of some pairs of bit channels with indices in $[0, N-1]$ can be determined using the partial order defined in Definition 5.2 as follows: If $j \succeq i$ then the sub-channel $W_N^{(j)}$ is more reliable than the sub-channel $W_N^{(i)}$.

Example 5.7 Take $n = 4$ and the following row indices:

$$i = 10 = (1010)_2 \text{ with } \mathrm{supp}(i) = \{1, 3\},$$

$$j = 12 = (1100)_2 \text{ with } \mathrm{supp}(j) = \{2, 3\},$$

$$k = 14 = (1110)_2 \text{ with } \mathrm{supp}(k) = \{1, 2, 3\}.$$

Since $\mathrm{supp}(i) \subseteq \mathrm{supp}(k)$ and $\mathrm{supp}(j) \subseteq \mathrm{supp}(k)$ (the first case in Definition 5.2), we conclude that $i \preceq k$ and $j \preceq k$, respectively. Furthermore, $\mathrm{supp}(j) = (\mathrm{supp}(i) \setminus \{1\}) \cup \{2\}$ (the second case in Definition 5.2), then $i \preceq j$ and the position of j in this order is between i and k (the third case in Definition 5.2). Observing the binary representation, we get j from i by swapping the bit "1" at position 1 with the "0" on the left at position 2.

Note that the third case in Definition 5.2 can generalize a single left-swap in the second case to multiple left-swaps.

Example 5.8 Consider $a = 9 = (1001)_2$ with $\mathrm{supp}(a) = \{0, 3\}$ and $b = 15 = (1111)_2$ with $\mathrm{supp}(b) = \{0, 1, 2, 3\}$ along with i, j, and k from Example 5.7. Observe that the supports of i, j, and k are all subsets of $\mathrm{supp}(b)$. Furthermore, comparing $9 = (1001)_2$ with $10 = (1010)_2$ and $11 = (1011)_2$ reveals that 10 and 11 can be obtained from a single left-swap operation on 9 and a single flip operation (inverting 0 to 1), respectively. That is,

$$9 = (10\underline{01})_2 \xrightarrow{\text{left-swap}} 10 = (10\underline{10})_2$$

$$9 = (10\underline{0}1)_2 \xrightarrow{\text{flip}} 11 = (10\underline{1}1)_2.$$

Definition 5.3 ***(Partial Order Property)*** A set $\mathcal{A} \subseteq [0, 2^n - 1]$ is said to satisfy the *Partial Order Property* if $i \preceq i^c$ for every $i \in \mathcal{A}$ and $i^c \in \mathcal{A}^c$. In other words, none of the indices in $\mathcal{A}$ is smaller than (i.e., precedes) or equal to another index in $\mathcal{A}^c$ according to the partial order defined in Definition 5.2. Equivalently, for every $i \in \mathcal{A}$ and $j \in [0, N-1]$, if $j \succeq i$ then $j \in \mathcal{A}$.

5.7 Minimum Distance of Polar Codes

As discussed in (2.8), the minimum distance d of a linear code C equals the minimum weight $w_{\min}$ of the code. Here, we find the minimum weight of a polar code. To this end, let us partition the codewords as below.

Definition 5.4 Given a set $\mathcal{A} \subseteq [0, N-1]$, we partition all codewords of a polar code $C(\mathcal{A})$ (excluding the all-zero codeword $\mathbf{0}$) into cosets $C_i(\mathcal{A})$ defined as

$$C_i(\mathcal{A}) \triangleq \left\{ \mathbf{g}_i + \sum_{h\in\mathcal{H}_i} \mathbf{g}_h \quad \text{for all } \mathcal{H}_i \subseteq \mathcal{A}\backslash[0,i] \right\} \subseteq C(\mathcal{A}). \tag{5.14}$$

In other words, coset $C_i(\mathcal{A})$ is the subcode $C(\mathcal{A}\backslash[0,i])$ generated by $\{\mathbf{g}_h : h \in \mathcal{A}\backslash[0,i]\}$ and the coset leader $\mathbf{g}_i$ (also called *leading row*), where $\mathbf{g}_i$ is the ith row of the polar transform $\mathbf{G}_2^{\otimes n}$. Observe that the subcode $C(\mathcal{A}\backslash[0,i])$ used to construct coset$C_i(\mathcal{A})$ is disjoint from the subcode $C(\mathcal{A}\backslash[0,j])$ used for $C_j(\mathcal{A})$ if $i \neq j$, because the sets $\mathcal{A}\backslash[0,i]$ and $\mathcal{A}\backslash[0,j]$ are distinct for $i \neq j$. Hence, all codewords are

$$C(\mathcal{A}) = \{\mathbf{0}\} \cup \bigcup_{i\in\mathcal{A}} C_i(\mathcal{A}). \tag{5.15}$$

Note that each coset $C_i(\mathcal{A})$ is uniquely associated with the coset leader $\mathbf{g}_i$, which is unique to each $i \in \mathcal{A}$ because $\mathbf{g}_i$ is a row of the polar transform matrix $\mathbf{G}_2^{\otimes n}$, and the rows of this matrix are linearly independent. Now, if two cosets $C_i(\mathcal{A})$ and $C_j(\mathcal{A})$ (with $i \neq j$) overlap, there would exist two distinct leaders $\mathbf{g}_i$ and $\mathbf{g}_j$ that lead to the same codeword. This would contradict the linear independence of the rows of $\mathbf{G}_2^{\otimes n}$.

Now, the question is: what is the minimum weight of the codewords in each coset $C_i(\mathcal{A})$. To answer this, we use the relation in (5.4). The weight of any codeword in the coset $C_i(\mathcal{A})$ cannot be less than the weight of the coset leader and follows the relationship below.

$$\mathrm{w}(\mathbf{g}_i + \sum_{j\in\mathcal{H}_i} \mathbf{g}_h) \geq \mathrm{w}(\mathbf{g}_i), \tag{5.16}$$

where $i \in \mathcal{A}$ and for any $\mathcal{H}_i \subseteq [i+1, N-1]$. In other words, due to the lower-triangular structure of the matrix $\mathbf{G}_2^{\otimes n}$, the minimum weight of all codewords in the coset $C_i(\mathcal{A})$ equals the weight of the coset leader $\mathrm{w}(\mathbf{g}_i)$. Hence, the minimum distance $\mathrm{d}(C)$ of a polar code C is equal to the minimum weight of the rows of the generator matrix; that is,

$$\mathrm{d}_{\min} = \min\{\mathrm{w}(\mathbf{g}_i), i \in \mathcal{A}\}. \tag{5.17}$$

Example 5.9 Given the (8,4) polar code with $\mathcal{A} = \{0, 1, 2, 4\}$. Find d_{min}.
Based on $\mathbf{G}_2^{\otimes n}$ and $\mathcal{A}$, we have

$$d_{min} = \min \{w(\mathbf{g}_i), i \in \mathcal{A}\} = \min \{4, 8\} = 4.$$

You may check the weights of the rows of the generator matrix of this code obtained in Example 5.5.

5.8 Minimum Weight Codewords

As discussed in Section 2.2, the weight distribution of a code and, in particular, the number of codewords with small weights contribute the most to the error performance of a code under (near) ML decoding. As mentioned in Section 3.1.4, the reason is that the number of minimum weight codewords and the minimum distance are the elements of the dominant terms in the computation of the upper bound for the block error rate of a code.

In this section, we describe how the codewords with minimum weight w_{min} are formed by characterizing the $\boldsymbol{G}_2^{\otimes n}$-row decomposition of these codewords. Then, we discuss how these codewords are counted using a closed-form formula. In Chapter 7, we provide an alternative representation of these codewords and the formula using the notion of monomials.

5.8.1 Construction of Minimum Weight Codewords

Let us first define an important set of rows of the polar transform $\mathbf{G}_2^{\otimes n}$ that play a key role in the formation of minimum weight codewords.

Definition 5.5 For each index $i \in [0, N-1]$, we define

$$\mathcal{K}_i \triangleq \{j \in [i+1, N-1] \, : \, w(\mathbf{g}_j) \geq w(\mathbf{g}_i + \mathbf{g}_j) = w(\mathbf{g}_i)\}$$

as the set of indices $j \in [i+1, N-1]$ such that $\mathbf{g}_j$ is at distance $w(\mathbf{g}_i)$ from $\mathbf{g}_i$ and has weight at least $w(\mathbf{g}_i)$.

We can show that the set $\mathcal{K}_i$ satisfies the following property:

$$\begin{aligned}\mathcal{K}_i = \{j \in [i+1, N-1] \, : \, &|\text{supp}(j)\backslash\text{supp}(i)| = 1, |\text{supp}(j)| \\ &= |\text{supp}(i)| \;\text{ or }\; |\text{supp}(i)| + 1\}.\end{aligned} \tag{5.18}$$

Hence, according to this property, given a row index i, the row indices in the set $\mathcal{K}_i$ have the same weight as row i or twice its weight. That is, $w(\mathbf{g}_j) = w(\mathbf{g}_i)$ or $w(\mathbf{g}_j) = 2w(\mathbf{g}_i)$.

This leads us to a simple approach to finding the elements of the set $\mathcal{K}_i$ by applying the *addition* and *left-swap* operations on bin(i) as follows:

- Addition operation: If we flip every "0" in bin(i) one at a time, we get all $j \in \mathcal{K}_i$ that have weight $\mathrm{w}(\mathbf{g}_j) = 2\mathrm{w}(\mathbf{g}_i)$. In this case, we have $\mathrm{supp}(j) \supseteq \mathrm{supp}(i)$. Note that in total, we can apply this operation $n - |\mathrm{supp}(i)|$ times. Hence, since $|\mathrm{supp}(i)| = \mathrm{w}(\mathrm{bin}(i)) = \log_2(d_{\min})$ for every $i \in \mathcal{A}_{\min}$, the total application of addition operations will be $n - \log_2(d_{\min}) = |\overline{\mathrm{supp}(i)}|$, where $\overline{\mathrm{supp}(i)} \triangleq [0, n-1] \backslash \mathrm{supp}(i)$.
- Left-swap operation: If we swap every "1" in bin(i) with every "0" on its left, one at a time, we get all $j \in \mathcal{K}_i$, which have weight $\mathrm{w}(\mathbf{g}_j) = \mathrm{w}(\mathbf{g}_i)$. In fact, supp($j$) is obtained from supp($i$) by replacing an index $k \in \mathrm{supp}(i)$ with another index $h > k$ (because $j > i$). Note that in total, we can apply this operation $\sum_{k \in \mathrm{supp}(i)} \sum_{\ell > k} \bar{i}_\ell$ times, where $\bar{i}_\ell = i_\ell \oplus 1$ and $\mathrm{bin}(i) = i_{n-1} \ldots i_1 i_0$.

Note that j obtained in these two cases follows $j \succeq i$, by Definition 5.2. Since $\mathcal{A}$ satisfies the Partial Order Property and $j \in \mathcal{A}$, therefore, $\mathcal{K}_i \subseteq \mathcal{A}$.

The size of $\mathcal{K}_i$ is therefore

$$|\mathcal{K}_i| = |\overline{\mathrm{supp}(i)}| + \sum_{k \in \mathrm{supp}(i)} \sum_{\ell > k} \bar{i}_\ell. \tag{5.19}$$

Example 5.10 Suppose $i = (13)_{10} = (01101)_2$ is the index of a row in the polar transform $\mathbf{G}_2^{\otimes 5}$. Find the set $\mathcal{K}_i$.

First, we find all $j > i$ in $\mathcal{A}$ with weight $2\mathrm{w}(\mathbf{g}_i)$ by addition operation. These rows are

$$\{(01111)_2, (11101)_2\} = \{15, 29\} \subset \mathcal{K}_i.$$

Observe that the size of this subset can be found even without listing the elements simply by counting the number of zero-value positions in bin(i) as

$$n - |\mathrm{supp}(i)| = 5 - \mathrm{w}(01101) = 2.$$

In the next step, we find all $j > i$ in $\mathcal{A}$ with weight $\mathrm{w}(\mathbf{g}_i)$ by a left-swap operation over bin(i). These rows are

$$\{(01110)_2, (11100)_2, (11001)_2, (10101)_2\} = \{14, 28, 25, 21\} \subset \mathcal{K}_i.$$

The size of this subset can also be found without listing the elements as

$$\left(5 - \mathrm{w}(0110\underline{1})\right) + \left(3 - \mathrm{w}(01\underline{1})\right) + \left(2 - \mathrm{w}(0\underline{1})\right) = 4.$$

Here, we counted the number of "0" bits in bin(i) at positions larger than k, where on the left of each of the "1" bits in bin(i) underlined. Hence,

$$|\mathcal{K}_i| = 2 + 4 = 6,$$

and the complete set is

$$\mathcal{K}_i = \{14, 15, 21, 25, 28, 29\}.$$

We learned in Section 5.7 that the minimum weight codewords are all located in the cosets C_i where $\mathrm{w}(\mathbf{g}_i) = w_{\min}$. Hence, we define the set that contains all such rows as:

$$\mathcal{A}_{\min} = \{i \in \mathcal{A} : \mathrm{w}(\mathbf{g}_i) = w_{\min}\}. \tag{5.20}$$

Then, within each coset $C_i(\mathcal{A}), i \in \mathcal{A}_{\min}$, every minimum weight codeword is constructed by only adding specific rows of $\mathbf{G}_2^{\otimes n}$ as

$$\mathrm{w}\left(\mathbf{g}_i + \underbrace{\sum_{j\in\mathcal{J}} \mathbf{g}_j}_{\text{core rows}} + \underbrace{\sum_{m\in\mathcal{M}(\mathcal{J})} \mathbf{g}_m}_{\text{balancing rows}}\right) = w_{\min}, \tag{5.21}$$

where we can choose an arbitrary subset $\mathcal{J} \subseteq \mathcal{K}_i$ and the set $\mathcal{M}(\mathcal{J}) \subseteq \mathcal{A}\backslash\mathcal{K}_i$ exits for any such set $\mathcal{J}$ as will be defined next. Since the rows indexed by elements in $\mathcal{K}_i$ allow one to form all minimum weight codewords in $C_i(\mathcal{A})$ by choosing all the subsets $\mathcal{J} \subseteq \mathcal{K}_i$, we call them *core rows*. Furthermore, the rows in $\mathcal{M}(\mathcal{J})$ are called *balancing rows* as their inclusion brings the weight of the sum down to $w_{\min}$ when the sum of the coset leader and a subset of core rows has a weight exceeding $w_{\min}$. We will illustrate this with an example.

5.8.1.1 $\mathcal{M}$-Construction

Suppose that $\mathcal{A} \subseteq [0, N-1]$ satisfies the Partial Order Property and $i \in \mathcal{A}$ satisfies $\mathrm{w}(\mathbf{g}_i) = w_{\min}$. For any $\emptyset \neq \mathcal{J} \subseteq \mathcal{K}_i$, our aim is to construct a set $\mathcal{M}(\mathcal{J}) \subseteq \mathcal{A}\backslash(\mathcal{K}_i \cup [0, i])$ satisfying (5.21). First, let the set of subsets of $\mathcal{J}$ be

$$\mathfrak{J}(\mathcal{J}) \triangleq \left\{\mathcal{J}' \subseteq \mathcal{J} : |\mathcal{J}'| \geq 2 \text{ and for every } j \in \mathcal{J}', \text{supp}(j)\backslash\text{supp}(i) \text{ is distinct}\right\}. \tag{5.22}$$

Recall from (5.18) that $|\text{supp}(j)\backslash\text{supp}(i)| = 1$. Next, for every such $\mathcal{J}' \in \mathfrak{J}(\mathcal{J})$, let $m_{\mathcal{J}'} \in [0, 2^n - 1]$ such that

$$\text{supp}(m_{\mathcal{J}'}) \triangleq \bigcup_{j\in\mathcal{J}'} (\text{supp}(j)\backslash\text{supp}(i)) \cup \left(\text{supp}(i) \cap \left(\bigcap_{j\in\mathcal{J}'} \text{supp}(j)\right)\right), \tag{5.23}$$

The set $\mathcal{M}(\mathcal{J})$ consists of all such $m_{\mathcal{J}'}$ indices with odd multiplicities. More specifically,

$$\mathcal{M}(\mathcal{J}) \triangleq \left\{h \in [0, N-1] : \left|\mathfrak{J}_h(\mathcal{J})\right| \text{ is odd}\right\}, \tag{5.24}$$

where

$$\mathfrak{J}_h(\mathcal{J}) \triangleq \left\{\mathcal{J}' \in \mathfrak{J}(\mathcal{J}) : m_{\mathcal{J}'} = h\right\}. \tag{5.25}$$

Remark 5.1 An equivalent way to define $\mathfrak{J}(\mathcal{J})$ and $\text{supp}(m_{\mathcal{J}'})$ in the $\mathcal{M}$-Construction is as follows. First, let

$$\mathcal{R} = \bigcup_{j \in \mathcal{J}} (\text{supp}(j) \backslash \text{supp}(i)) \subseteq \overline{\text{supp}(i)} \triangleq [0, n-1] \backslash \text{supp}(i).$$

Then, we can verify that

$$\mathfrak{J}(\mathcal{J}) = \left\{ \mathcal{J}' \subseteq \mathcal{J} : |\mathcal{J}'| \geq 2, \left| \{ j \in \mathcal{J}' : j_k = 1 \} \right| \leq 1, \forall k \in \mathcal{R} \right\},$$

and

$$\text{supp}(m_{\mathcal{J}'}) = \left\{ k \in \mathcal{R} : \exists j \in \mathcal{J}', j_k = 1 \right\} \cup \left(\text{supp}(i) \cap \left(\bigcap_{j \in \mathcal{J}'} \text{supp}(j) \right) \right).$$

Remark 5.2 Note that in the $\mathcal{M}$-Construction, for $\mathcal{J} \subseteq \mathcal{K}_i$, $|\mathcal{J}| \leq 1$, we have $\mathcal{M}(\mathcal{J}) = \emptyset$ because there are no $\mathcal{J}' \subseteq \mathcal{J}$ with $|\mathcal{J}'| \geq 2$ and hence $\mathfrak{J}(\mathcal{J}) = \emptyset$. This is consistent with our goal to form codewords of minimum weight: if $\mathcal{J} = \emptyset$, then $\mathbf{c} = \mathbf{g}_i$ itself has weight $w_{\min}$; if $\mathcal{J} = \{j\}$, then $\mathbf{c} = \mathbf{g}_i + \mathbf{g}_j$ also has weight $w_{\min}$ due to the definition of $\mathcal{K}_i$.

Figure 5.1 demonstrates the $\mathcal{M}$-construction, in particular, how to find $m_{\mathcal{J}'}$ for every $\mathcal{J}'$. Figure 5.2 shows the Venn diagram associated with various sets defined in relation to the formation of minimum weight codewords of polar codes.

We provide below a few examples to demonstrate the $\mathcal{M}$-Construction.

Example 5.11 Let $n = 4$, $N = 2^n = 16$, and $i = 3 = (0011)_2$. Then, $\text{supp}(i) = \{0, 1\}$, $\overline{\text{supp}(i)} = \{0, 1, 2, 3\} \backslash \text{supp}(i) = \{2, 3\}$, and

$$\begin{aligned} \mathcal{K}_i &= \{5, 6, 7, 9, 10, 11\} \\ &= \left\{ (0101)_2, (0110)_2, (0111)_2, (1001)_2, (1010)_2, (1011)_2 \right\}. \end{aligned}$$

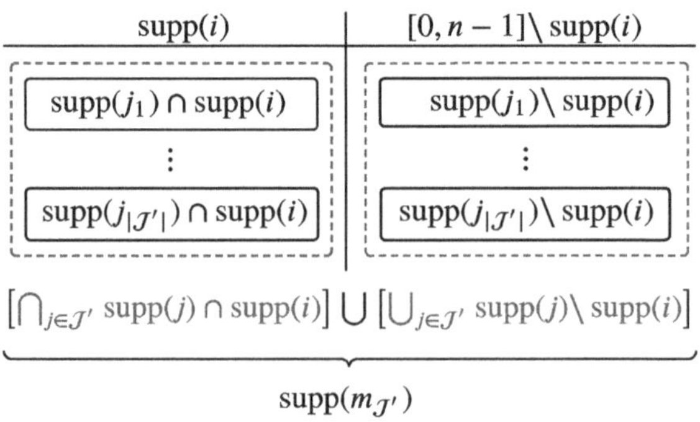

Figure 5.1 Finding the elements of the set $\mathcal{M}$ corresponding to every $\mathcal{J}' \subseteq \mathcal{J} \subseteq \mathcal{K}_i$ such that $|\mathcal{J}'| > 1$ and $|\{j \in \mathcal{J}' : j_k = 1\}| \leq 1$ for any $k \in [0, n-1] \backslash \text{supp}(i)$.

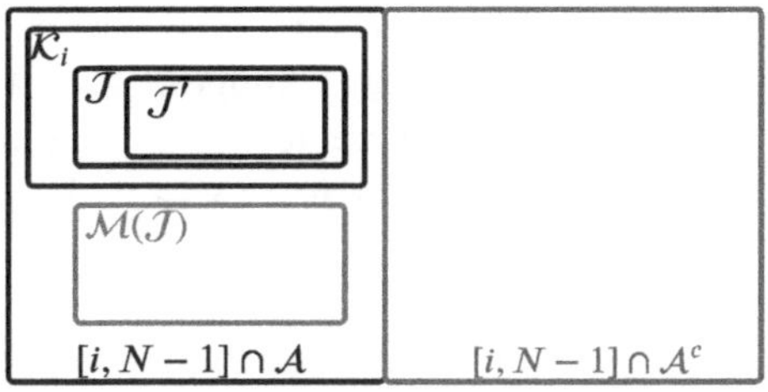

Figure 5.2 The Venn diagram illustrating the relation between different sets in relation to the set $\mathcal{K}_i$.

Take

$$\begin{aligned}\mathcal{J} &= \{5, 6, 7, 9, 10\} \\ &= \left\{(0101)_2, (0110)_2, (0111)_2, (1001)_2, (1010)_2\right\} \subset \mathcal{K}_i.\end{aligned}$$

We have supp(5) = $\{0, 2\}$, supp(6) = $\{1, 2\}$, supp(7) = $\{0, 1, 2\}$, supp(9) = $\{0, 3\}$, supp(10) = $\{1, 3\}$. Therefore, supp(5)\supp(3) = $\{2\}$, supp(6)\supp(3) = $\{2\}$, supp(7)\supp(3) = $\{2\}$, supp(9)\supp(3) = $\{3\}$, supp(10)\supp(3) = $\{3\}$. As $\mathfrak{J}(\mathcal{J})$ consists of the subsets $\mathcal{J}' \subseteq \mathcal{J}$, $|\mathcal{J}'| \geq 2$, such that supp(j)\supp(i), $j \in \mathcal{J}'$, are all distinct, we have

$$\mathfrak{J}(\mathcal{J}) = \{\{5, 9\}, \{5, 10\}, \{6, 9\}, \{6, 10\}, \{7, 9\}, \{7, 10\}\}.$$

From (5.23), we obtain supp($m_{\mathcal{J}'}$) for all $\mathcal{J}' \in \mathfrak{J}(\mathcal{J})$ as follows.

$$\begin{aligned}
\text{supp}(m_{\{5,9\}}) &= ((\text{supp}(5)\backslash\text{supp}(3)) \cup (\text{supp}(9)\backslash\text{supp}(3)) \\
&\quad \cup (\text{supp}(3) \cap \text{supp}(5) \cap \text{supp}(9)) = \{0, 2, 3\}. \\
\text{supp}(m_{\{5,10\}}) &= ((\text{supp}(5)\backslash\text{supp}(3)) \cup (\text{supp}(10)\backslash\text{supp}(3))) \\
&\quad \cup (\text{supp}(3) \cap \text{supp}(5) \cap \text{supp}(10)) = \{2, 3\}. \\
\text{supp}(m_{\{6,9\}}) &= ((\text{supp}(6)\backslash\text{supp}(3)) \cup (\text{supp}(9)\backslash\text{supp}(3))) \\
&\quad \cup (\text{supp}(3) \cap \text{supp}(6) \cap \text{supp}(9)) = \{2, 3\}. \\
\text{supp}(m_{\{6,10\}}) &= ((\text{supp}(6)\backslash\text{supp}(3)) \cup (\text{supp}(10\backslash\text{supp}(3))) \\
&\quad \cup (\text{supp}(3 \cap \text{supp}(6 \cap \text{supp}(10)) = \{1, 2, 3\}. \\
\text{supp}(m_{\{7,9\}}) &= ((\text{supp}(7\backslash\text{supp}(3)) \cup (\text{supp}(9)\backslash\text{supp}(3))) \\
&\quad \cup (\text{supp}(3) \cap \text{supp}(7) \cap \text{supp}(9)) = \{0, 2, 3\}. \\
\text{supp}(m_{\{7,10\}}) &= ((\text{supp}(7\backslash\text{supp}(3) \cup (\text{supp}(10)\backslash\text{supp}(3))) \\
&\quad \cup (\text{supp}(3) \cap \text{supp}(7) \cap \text{supp}(10)) = \{1, 2, 3\}.
\end{aligned}$$

These supports correspond to $m_{\mathcal{J}'} = 13, 12, 12, 14, 13, 14$. Therefore, according to (5.25), $|\mathfrak{J}_h(\mathcal{J})| = 2$ for $h = 12, 13, 14$. As the cardinalities of $\mathfrak{J}_h(\mathcal{J})$ are even for all $h = 12, 13, 14$, according to (5.24), $\mathcal{M}(\mathcal{J}) = \emptyset$.

Example 5.12 We assume the same parameters $n = 4$ and $i = 3$ as in Example 5.11 but pick $\mathcal{J} = \{5, 6, 9, 10\}$. Then $\mathfrak{J}(\mathcal{J}) = \{\{5, 9\}, \{5, 10\}, \{6, 9\},$

$\{6, 10\}\}$. As already computed in Example 5.11, we have $\text{supp}(m_{\{5,9\}}) = \{0, 2, 3\}$, $\text{supp}(m_{\{5,10\}}) = \{2, 3\}$, $\text{supp}(m_{\{6,9\}}) = \{2, 3\}$, and $\text{supp}(m_{\{6,10\}}) = \{1, 2, 3\}$. These sets correspond to $m_{\mathcal{J}'} = 13, 12, 12, 14$. Therefore, according to (5.25), $|\mathfrak{J}_h(\mathcal{J})| = 1$ (odd) for $h = 13, 14$ and $|\mathfrak{J}_h(\mathcal{J})| = 2$ (even) for $h = 12$. By (5.24), $\mathcal{M}(\mathcal{J}) = \{13, 14\}$.

Example 5.13 Consider the polar code (8,7) with $d_{\min} = w_{\min} = 2$ and $\mathcal{A} = [1, 7]$ and take the coset

$$C_1(\mathcal{A}) \triangleq \left\{ \mathbf{g}_1 + \sum_{h \in \mathcal{H}_1} \mathbf{g}_h \,:\, \mathcal{H}_1 \subseteq \mathcal{A} \backslash [0, 1] \right\} \subseteq C(\mathcal{A}), \tag{5.26}$$

where the leading row has index $i = (001)_2$ and its support is $\text{supp}(i) = \{0\}$. Considering (5.18) for the properties of $\mathcal{K}_i$ on $(001)_2$ (or alternatively applying the addition and left-swap operations), we find the indices of the core rows as

$$\mathcal{K}_i = \{(010)_2, (011)_2, (100)_2, (101)_2\} = \{2, 3, 4, 5\}.$$

Now, as Figure 5.3 illustrates, the linear combination of $\mathbf{g}_i, i = 1$ with the subset $\mathcal{J}' = \mathcal{J} = \{3, 4\} = \{(011)_2, (100)_2\} \subset \mathcal{K}_i$ does not give a codeword with weight $w(\mathbf{g}_i) = 2$. That is,

$$w\left(\mathbf{g}_1 + \sum_{j \in \{3,4\}} \mathbf{g}_j\right) = 4 \neq w_{\min}.$$

However, if we add a balancing term $\sum_{m \in \mathcal{M}(\mathcal{J})} \mathbf{g}_m$ by finding set $\mathcal{M}(\mathcal{J})$ as a function of $\mathcal{J}$, we have

$$w\left(\mathbf{g}_1 + \sum_{j \in \{3,4\}} \mathbf{g}_j + \sum_{m \in \{6\}} \mathbf{g}_m\right) = 2 = 2^{|\text{supp}(i)|} = w_{\min}.$$

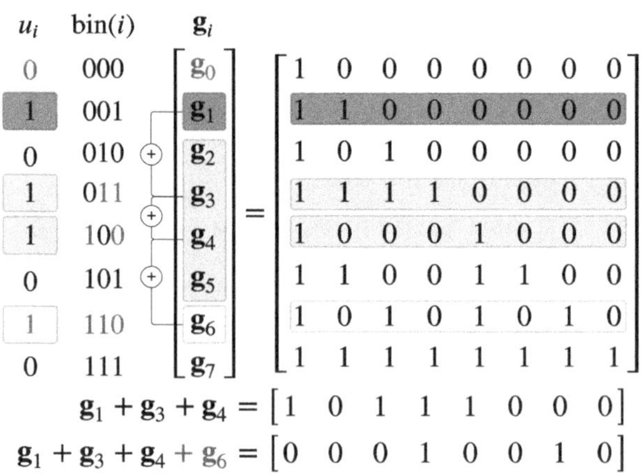

Figure 5.3 An example showing how a min-weight codeword is formed by adding a balancing row $\mathbf{g}_m$ to the subset $\mathcal{J} = \{3, 4\}$ of core rows. Core rows are shown in light gray boxes.

In this case, as $|\mathcal{J}| = 2$, we need only one balancing row $\mathbf{g}_m, m = 6$, that is, $|\mathcal{M}(\mathcal{J})| = |\{6\}| = 1$. According to (5.23), this balancing row is obtained by

$$\text{supp}(m_{\mathcal{J}'}) = \{1,2\} \cup \emptyset, \quad \text{or} \quad m_{\mathcal{J}'} = (110)_2.$$

Observe that we may not need balancing rows for some $\mathcal{J} \subseteq \mathcal{K}_i$, that is, we will have $\mathcal{M}(\mathcal{J}) = \emptyset$. For example, when $\mathcal{J} = \{2,3\}$, there are no distinct elements in the support of every $j \in \mathcal{J}' = \mathcal{J}$ outside of the support of i. In other words, we have $\left|\bigsqcup_{j\in\mathcal{J}} (\text{supp}(j)\backslash\text{supp}(i))\right| = |\{1\}| = 1 < |\mathcal{J}'|$, here $\bigsqcup$ denotes *disjoint union*, while to form an $\text{supp}(m_{\mathcal{J}'})$, we needed it to be equal to $|\mathcal{J}'| = 2$. As a result, we have

$$w\left(\mathbf{g}_1 + \sum_{j\in\{2,3\}} \mathbf{g}_j\right) = 2 = 2^{|\text{supp}(i)|} = w_{\min}.$$

Since larger weight codewords with weight less than $2w_{\min}$ are the result of the summation of two or more $w_{\min}$-weight codewords under special circumstances, the construction discussed in this section can also be used for larger weight codewords. For further reading, see the bibliographical notes.

5.8.2 Enumeration of Minimum Weight Codewords

We learned that every subset $\mathcal{J} \subseteq \mathcal{K}_i$, where $i \in \mathcal{A}_{\min}$, forms a minimum weight codeword. Hence, to enumerate all minimum weight codewords, we simply need to count the number of all possible subsets of $\mathcal{K}_i$, including the empty set and $\mathcal{K}_i$ itself, for every coset $C_i(\mathcal{A})$. Let us denote this number by $A_{i,w_{\min}}$. This is given by the binomial coefficient as follows:

$$A_{i,w_{\min}}(\mathcal{A}) = \binom{|\mathcal{K}_i|}{0} + \binom{|\mathcal{K}_i|}{1} + \cdots + \binom{|\mathcal{K}_i|}{|\mathcal{K}_i|} = 2^{|\mathcal{K}_i|}. \tag{5.27}$$

Therefore, the total number of minimum weight codewords of the polar code $C(\mathcal{A})$ is

$$A_{w_{\min}}(\mathcal{A}) = \sum_{i\in\mathcal{A}_{\min}} A_{i,w_{\min}}(\mathcal{A}) = \sum_{i\in\mathcal{A}_{\min}} 2^{|\mathcal{K}_i|}. \tag{5.28}$$

Note that $\binom{|\mathcal{K}_i|}{0} = 1$ in (5.27) refers to the row $\mathbf{g}_i$ alone, which is also a minimum weight codeword.

Example 5.14 Let us take the (64, 32) polar code with $d_{\min} = w_{\min} = 8$ and

$$\mathcal{A}_{\min} = \{26, 28, 38, 41, 42, 44, 49, 50, 52, 56\}.$$

The table below shows the contribution of the cosets C_i for $i \in \mathcal{A}_{\min}$ of the code to $A_{w_{\min}}(\mathcal{A})$.

i	26	28	38	41	42	44	49	50	52	56	Total
$A_{i,8}(\mathcal{A})$	128	64	128	128	64	32	64	32	16	8	$A_8(\mathcal{A}) = 664$

Observe that all $A_{i,8}(\mathcal{A})$ are powers of two and smaller i results in larger $|\mathcal{K}_i|$. Note that a smaller i will allow more swap operations as there are usually more zeros in the most significant bit positions, while all $i \in \mathcal{A}_{\min}$ allow the same number of addition operations. For example, compare $26 = (011010)_2$ with $56 = (111000)_2$.

5.9 Exponent of Polarizing Matrix

For the kernel matrix $\boldsymbol{G}_2$, block-length $N = 2^n$, and rate $R < I(W)$, the probability of block error $P_e(N, R)$ for polar coding and successive cancellation decoding can be bounded as $P_e(N, R) \leq 2^{-N^\beta}$, or $P_e(N, R) = O(2^{-N^\beta})$, for any $\beta < \frac{1}{2}$. The parameter $\frac{1}{2}$ is called the *exponent* of $\boldsymbol{G}_2$, denoted by $\mathrm{E}(\boldsymbol{G}_2) = \frac{1}{2}$ and is a performance measure. This implies that when n is sufficiently large, there exists a set $\mathcal{A}$ of size $N \cdot R$ such that $\sum_{i \in \mathcal{A}} Z\left(W_{2^n}^{(i)}\right) \leq 2^{-N^\beta}$. To quantify the exponent of a kernel matrix, let us define the notion of *partial distances* as follows.

Definition 5.6 The partial distance $D_i, i = 0, \ldots, \ell - 1$ of an $\ell \times \ell$ matrix $\boldsymbol{G} = \left[\mathbf{g}_0, \ldots, \mathbf{g}_{\ell-1}\right]^T$ as

$$D_i \triangleq d_H\left(\mathbf{g}_i, \langle \mathbf{g}_{i+1}, \ldots, \mathbf{g}_{\ell-1} \rangle\right), \quad i = 0, \ldots, \ell - 2$$
$$D_{\ell-1} \triangleq d_H\left(\mathbf{g}_{\ell-1}, 0\right),$$

where d_H and $\langle \cdot \rangle$ denote the minimum Hamming distance and row span, then the *rate of polarization* $\mathrm{E}(\boldsymbol{G})$ is

$$\mathrm{E}(\boldsymbol{G}) = \frac{1}{\ell} \sum_{i=0}^{\ell-1} \log_\ell D_i.$$

Example 5.15 Given the matrix below,

$$\boldsymbol{G} = \begin{bmatrix} 1 & 0 & 0 \\ 1 & 0 & 1 \\ 1 & 1 & 1 \end{bmatrix}$$

find the partial distances and the rate of polarization.

The partial distances are as follows:

$$D_0 = \min\{\mathrm{w}(\mathbf{g}_1 + \mathbf{g}_2) = 1, \mathrm{w}(\mathbf{g}_1 + \mathbf{g}_3) = 2, \mathrm{w}(\mathbf{g}_1 + \mathbf{g}_2 + \mathbf{g}_3) = 2\} = 1,$$

$$D_1 = \mathrm{w}(\mathbf{g}_2 + \mathbf{g}_3) = 1,$$

$$D_2 = \mathrm{w}(\mathbf{g}_3) = 3.$$

Then, we can find $\mathrm{E}(\boldsymbol{G})$ as

$$\mathrm{E}(\boldsymbol{G}) = \frac{1}{3}\left(\log_3 1 + \log_3 1 + \log_3 3\right) = \frac{1}{3} < \mathrm{E}(\boldsymbol{G}_2).$$

The exponent of an $\ell \times \ell$ kernel matrix $\mathbf{G}$ indicates how fast the error probability P_e decays in the code length N assuming that the rate R is fixed. Therefore, this exponent can be used to asymptotically characterize the performance of polar codes based on underlying kernel matrices. In the following example, we show an example kernel matrix with a larger exponent than $\mathrm{E}(\boldsymbol{G}_2) = \frac{1}{2} = 0.5$.

Example 5.16 Find the partial distances and the exponent of the following 16×16 matrix, which have been obtained from the 31×31 BCH matrix by careful removing of removing rows and columns.

$$\boldsymbol{G} = \begin{bmatrix}
1&0&0&1&1&1&0&0&0&0&1&1&1&1&0&1\\
0&1&0&0&1&0&0&1&0&1&1&1&0&0&1&1\\
0&0&1&1&1&1&1&0&0&1&1&0&1&1&1&0\\
0&1&0&1&0&1&1&0&1&0&1&0&0&0&0&0\\
1&1&1&1&0&0&0&0&0&0&1&0&1&1&0&1\\
0&0&1&0&0&1&0&1&0&1&0&0&0&1&1&0\\
0&0&1&0&0&0&0&0&0&1&1&1&0&0&0&0\\
0&1&0&1&1&1&0&0&1&0&1&1&0&0&1&0\\
1&1&1&0&0&1&1&0&1&0&0&1&0&1&0&0\\
1&0&1&0&1&0&1&1&1&0&1&1&0&1&0&1\\
1&1&1&0&0&0&0&0&0&0&0&1&1&0&1&0\\
1&0&0&1&1&0&0&0&0&1&0&1&1&0&1&1\\
1&1&1&1&1&0&1&0&0&0&0&1&0&1&0&0\\
1&0&1&0&1&1&1&1&0&1&0&0&0&0&0&1\\
1&0&1&0&0&0&0&1&0&1&1&1&1&1&0&0\\
1&1&1&1&1&1&1&1&1&1&1&1&1&1&1&1
\end{bmatrix}.$$

The partial distances of this matrix are

$$\mathbf{D} = \left(D_0, D_1, \ldots, D_{15}\right) = (1, 2, 2, 2, 2, 4, 4, 4, 4, 6, 6, 8, 8, 8, 8, 16).$$

Hence, the exponent of this matrix can be found accordingly as

$$\mathrm{E}(\boldsymbol{G}) = \frac{1}{16}\sum_{i=0}^{15} \log_{16} D_i = 0.5183 > \mathrm{E}(\boldsymbol{G}_2).$$

The upper and lower bounds for the achievable exponents of the kernels show that there are no matrices of a size smaller than 15×15 with exponents $\mathrm{E}(\boldsymbol{G})$ exceeding $\frac{1}{2}$.

In general, the partial distances of a polarizing matrix constructed from the Kronecker product $\boldsymbol{T}_1 \otimes \boldsymbol{T}_2 \otimes \cdots \otimes \boldsymbol{T}_n$ can be expressed as a product of those of its component matrices $\boldsymbol{T}_i, i = 1, \ldots, n$ with size $\ell_i \times \ell_i$. As a result, the exponent of the polarizing matrix is a weighted sum of the exponents of its component matrices. The exponent of the polarizing matrix $\boldsymbol{T}_1 \otimes \boldsymbol{T}_2 \otimes \cdots \otimes \boldsymbol{T}_N$ is given by

$$\mathrm{E}\left(\boldsymbol{T}_1 \otimes \boldsymbol{T}_2 \otimes \cdots \otimes \boldsymbol{T}_n\right) = \sum_{i=1}^{n} \frac{\mathrm{E}\left(\boldsymbol{T}_i\right)}{\log_{l_i} l_1 l_2 \ldots l_n}. \tag{5.29}$$

5.10 Polar Codes with Large and Mixed Kernels

5.10.1 Large Kernel Polar Codes

Arıkan's pioneering work was based on the linear binary *kernel matrix* $\boldsymbol{F}_2 = \boldsymbol{G}_2$ of dimension two and exponent $\frac{1}{2}$. To improve the exponent and, consequently, the performance of polar codes under SC decoding, as we discussed in Section 5.9, large non-singular kernel matrices $\boldsymbol{G}$ can be used. The polarization effect over BI-DMC still holds for such an $\ell \times \ell$ kernel matrix $\boldsymbol{G}$ provided that none of its column permutations is an upper triangle matrix. The block error probability is found to be $O\left(2^{-\ell^{n\beta}}\right)$ if β is less than the exponent of the kernel matrix.

The SC decoding algorithm for large kernel polar codes has the complexity of order $O(2^{\ell} N \log N)$ operations, where $O(2^{\ell})$ is the complexity of kernel processing based on trivial implementation. Although this complexity can be slightly reduced by more efficient algorithms (see the Bibliographical notes for further reading), direct calculation remains prohibitive for kernels with relatively large sizes, such as $\ell = 16, 32$.

5.10.2 Mixed/Multi-Kernel Polar Codes

As we learned, the code construction based on the n-fold Kronecker product of $\boldsymbol{F}_2 = \boldsymbol{G}_2$ restricts the length N of polar codes to be powers of two, i.e., $N = 2^n$. In the case of larger kernels such as the one in Examples 5.15 and 5.16, the code length becomes a power of integers, that is, $N = \ell^n$, where l is the dimension of the kernel. One can employ puncturing, shortening, and repetition techniques to adjust the block length. However, these techniques come with disadvantages. In addition to performance degradation, as the punctured and shortened codes are decoded based on their mother polar codes, the decoding latency is proportional to the actual code length. Furthermore, the coordinates of dummy bits representing shortened or punctured bits affect the polarization of the codes, and improper selection of these coordinates may result in catastrophic error correction performance.

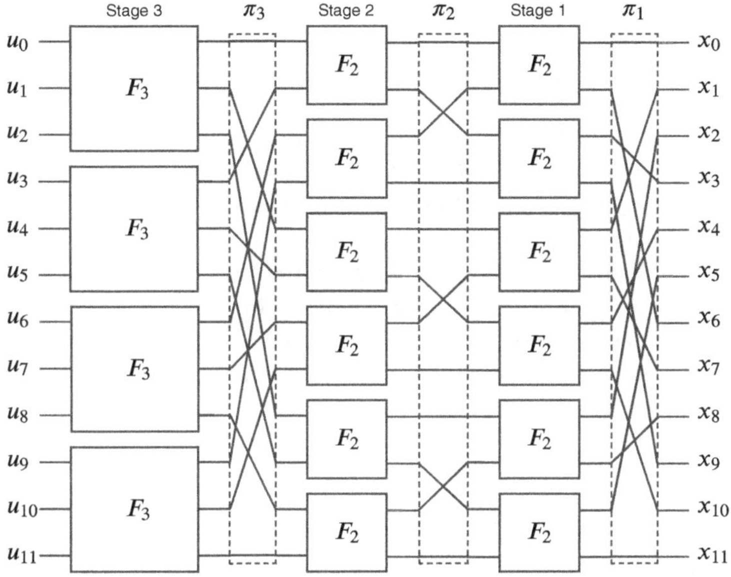

Figure 5.4 Factor graph of the multi-kernel polar code with the transformation matrix $\boldsymbol{T}_{12} = \boldsymbol{T}_2 \times \boldsymbol{T}_2 \times \boldsymbol{T}_3$.

The multi-kernel polar codes use a mix of different kernel sizes over the binary alphabet, while the polarization effect is preserved. Encoding of multi-kernel polar codes follows the general structure of polar codes, and decoding can be performed by any SC-based decoding.

An (N, K) multi-kernel polar code is defined by an $N \times N$ transformation matrix $\boldsymbol{T}_N$ as

$$\boldsymbol{T}_N = \boldsymbol{T}_{p_1} \otimes \boldsymbol{T}_{p_2} \otimes \cdots \otimes \boldsymbol{T}_{p_s}, \tag{5.30}$$

where the code length is $N = p_1 p_2 \ldots p_s$ and the binary polarizing matrices $\boldsymbol{T}_{p_i}$ of size $p_i \times p_i$ for $i = 1, \ldots, s$, called kernels of dimension p_i. Note that the Kronecker product in (5.30) is not commutative. Figure 5.4 illustrates the Tanner graph of a polar code with length $N = 12 = 2 \times 2 \times 3$ composed of two kernels of dimension 2 and one kernel of dimension 3. Observe that when all kernels are composed of kernel $\boldsymbol{T}_2 = \boldsymbol{G}_2$, we will have the polar codes discussed earlier in this chapter and Chapter 4.

5.11 Properties of SC Decoding

Decoding of polar codes and their variants requires passing the channel LLRs through the factor graph. The evolved information at the output of the factor graph is used to make hard decision or to calculate a metric in SC-based decoders. The evolved LLR, a.k.a. decision LLR, is obtained for each sub-channel successively. To calculate each decision LLR, we need to access the intermediate information on the factor graph. We can store all the $N \cdot \log_2 N$ intermediate values, including LLRs and partial sums on the factor graph. This approach is acceptable for short codes under SC decoding or Fano decoding. However, as the code gets longer, in particular under list decoding or stack decoding where we have a number of SC processes running in parallel, this approach will be expensive in terms of memory requirement. Also, knowing the extent of the bode updates for decoding every bit can save time by avoiding unnecessary computations. In this section, we discuss interesting properties of the SC schedule and the nodes to be updated while decoding bit i. These properties are useful in an efficient implementation of the SC-based decoding algorithms in terms of memory requirements and scheduling, in particular as hardware.

5.11.1 Updating Intermediate LLRs

The factor graph shown in Figure 5.5 has $N\log_2 N$ nodes, however, it is sufficient to update/access at most $N-1$ intermediate LLRs, indicated by λ_j (including decision LLR, indicated by λ_0) out of $N\log_2 N$ LLRs to decode any bit u_i, $0 \le i \le N-1$. Figure 5.5 illustrates the LLRs associated with the decoding of bit u_3 in tree form on the factor graph. As can be seen, there are 2^s LLRs in stage s for $s = 0, \ldots, n$.

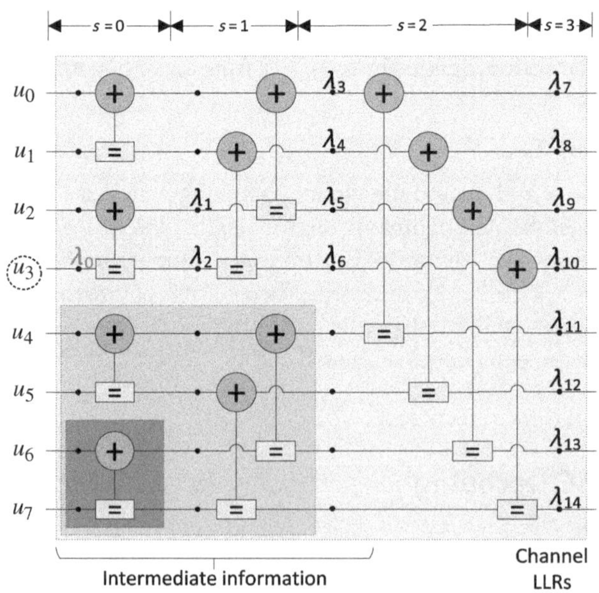

Figure 5.5 An illustrative example for updating LLRs for decoding bit u_3.

Hence, according to the geometric series, we need a total memory space of

$$\sum_{s=0}^{n} 2^s = 2^{n+1} - 1 = 2N - 1. \tag{5.31}$$

Also, assuming that each node requires one clock cycle to be processed, observe that at stage n, since half of the nodes are f and the other half are g nodes and all g and f nodes are processed in one clock cycle (CC), we require two CCs in the entire decoding. This becomes twice (2^2 CCs) in stage $n - 2$ and 2^t times in stage $n - t$. Thus, the decoding latency T_d in terms of clock cycles will be

$$T_d = \sum_{s=1}^{n} 2^s = 2(N - 1). \tag{5.32}$$

Suppose u_i is the bit that was just decoded and $\text{bin}(i) = i_{n-1}...i_0$ is the binary representation of the index i, where the least significant bit is indexed 0 and the most significant bit is indexed $n - 1$. The stages are updated from right to left (where $s = 0$). The first stage to be updated is obtained by *finding the first one* (ffo), or the position of the least significant bit set to one as

$$\eta(i) = \text{ffo}(i_{n-1}...i_0) = \begin{cases} \min\{t \in [0, n-1] : i_t = 1\}, & i > 0, \\ n - 1, & i = 0. \end{cases} \tag{5.33}$$

Example 5.17 Given a polar code of length $N = 8$. To make a decision on the value of u_3, the last stage (the deepest from left to right) in which their LLRs must be updated is

$$\eta(3) = \text{ffo}(011) = \min\{0, 1\} = 0.$$

Hence, λ_0 is computed based on λ_1, λ_2, and $\beta_0 = \hat{u}_2$ at the g node, shown by boxed "=."

5.11.2 Updating Partial Sums

The partial sums are the other set of intermediate information needed for the SC process. It turns out that we also need the same memory space for the partial sums, that is, at most $N - 1$ memory elements. We need to store 2^s bits corresponding to 2^s nodes of type **g** at stage s, which are waiting to be summed with the next decoded bit. Here, we define an operator that indicates the last stage to be updated. The last stage that its partial sums to be updated is obtained by *finding the first zero* (ffz), or the position of the least significant bit set to zero as

$$\psi(i) = \text{ffz}(i_{n-1}...i_0) = \min\{t \in [0, n-1] : i_t = 0\}. \tag{5.34}$$

It turns out that this is the only stage that consists of **g** nodes, shown by boxed "=" in Figure 5.6, in the process of updating the LLRs from stage $s = \text{ffo}(\text{bin}(i))$

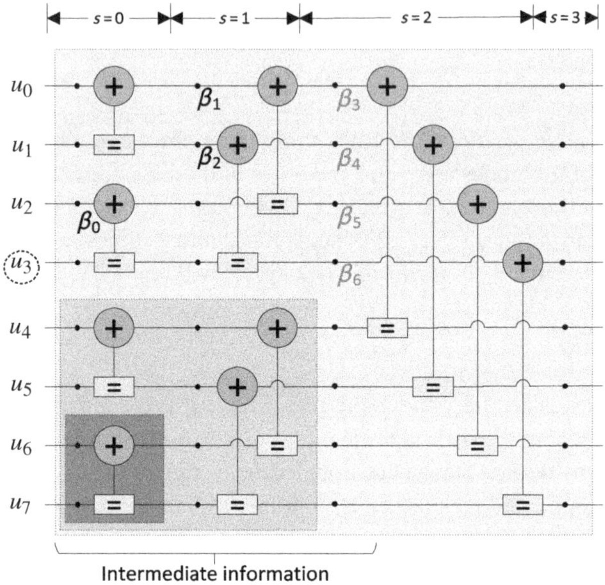

Figure 5.6 An illustrative example for updating partial sums of stage $s = 2$ after decoding u_3.

up to $s = 0$. Clearly, after decoding the last bit where there is no zero in the binary representation of the index bin$(N - 1) = 11...1$, there is no need to update the partial sums.

Example 5.18 Given a polar code of length $N = 8$. After making a decision on the value of u_3, we need to pass it backward to the factor graph. The last stage in which its partial sums should be updated is

$$\psi(3) = \text{ffz}(011) = \min\{2\} = 2.$$

Hence, starting from stage 0, where $\hat{u}_3$ and β_0 are, first the partial sums of stage 1, that is β_1 and β_2 are updated and then the partial sums of stage 2. Figure 5.6 shows $N - 1 = 8 - 1 = 7$ partial sums (β_0 to β_6) associated to u_3. The β values in gray are located in stage $\psi(3) = 2$ and will be updated by backpropagation of bit u_3 as $\beta_6 = u_3$, $\beta_5 = u_3 + \beta_0$, $\beta_4 = u_3 + \beta_2$, and $\beta_3 = u_3 + \beta_1 + \beta_0$.

You may notice that for $i \in [1, N - 1]$, we have

$$\psi(i - 1) = \eta(i). \tag{5.35}$$

That is the reason why at any bit $i \in [1, N - 1]$, the stage $\eta(i)$ where its LLRs need to be updated consists of only **g** nodes. Recall from Chapter 4 that the update relation at the g nodes requires updated partial sums to compute the output LLR. Therefore, after decoding the bit $i-1$, the partial sums of this stage are updated to be used for the **g** nodes in stage $\eta(i)$.

5.11.3 Partial Rewind: Updating LLRs and Partial Sums

In this section, we unveil properties of the SC algorithm that can help us to rewind the process efficiently. Let us begin by introducing an operator that helps us in the upcoming analysis.

Definition 5.7 The operator $\phi(j)$ finds the last zero (flz), or the position from the left side of the most significant bit set to zero in the binary representation of $j = (j_{n-1} .. j_0)_2$, where j_0 is the least significant bit

$$\phi(j) = \text{flz}(j_{n-1} .. j_0) = \begin{cases} n-1-\max\limits_{j_t=0}(t), & j < 2^n - 1, \\ n-1, & j = 2^n - 1 \end{cases} \tag{5.36}$$

for every $t \in [0, n-1]$. We denote the output of the operator $\phi(j)$ by parameter p.

Note that since the indexing is in the opposite direction when the most significant bit is set to zero, that is, $j_{n-1} = 0$, then we get $p = 0$, and when the only zero bit is $j_0 = 0$ or there is no 0-value bit, then $p = n - 1$.

Definition 5.8 ***(Set $\mathcal{Z}_p$)*** We group the bit indices $j \in [0, 1, ..., 2^n - 1]$ based on the identical $p = \phi(j)$ into n sets denoted by set $\mathcal{Z}_p$ with order $p = 0, 1, ..., n-1$, or

$$\mathcal{Z}_p = \{j \in [0, 2^n - 1] : \phi(j) = p\}. \tag{5.37}$$

Example 5.19 For $n = 3$, we can group the indices 0 to 7 into the following sets:

$$\mathcal{Z}_0 = \{0, 1, 2, 3\}, \mathcal{Z}_1 = \{4, 5\}, \text{ and } \mathcal{Z}_2 = \{6, 7\}.$$

The distribution of nonfrozen indices in set $\mathcal{A}$ among sets $\mathcal{Z}_p, p \in [0, n-1]$ depends on the code rate. Observe that as the code rate reduces, fewer nonfrozen indices will exist in low-order $\mathcal{Z}_p$, that is, $\mathcal{Z}_p$ with small p.

Properties of $\mathcal{Z}_p$: For any $n > 0$ and $p \in [0, n-1]$, set $\mathcal{Z}_p$ has the following properties:

a) The boundaries of set $\mathcal{Z}_p$ are

$$\mathcal{Z}_p = \begin{cases} \left[2^n - 2^{n-p}, 2^n - 2^{n-(p+1)} - 1\right], & 0 \le p < n-1, \\ \left[2^n - 2^{n-p}, 2^n - 1\right], & p = n-1. \end{cases} \tag{5.38}$$

b) The size of set $\mathcal{Z}_p$ is

$$|\mathcal{Z}_p| = \begin{cases} 2^{n-p-1}, & 0 \leq p < n-1, \\ 2, & p = n-1. \end{cases} \tag{5.39}$$

c) The smallest element in set $\mathcal{Z}_p$ is

$$z_p = \min(\mathcal{Z}_p) = \sum_{x=n-p}^{n-1} 2^x = 2^n - 2^{n-p}. \tag{5.40}$$

Example 5.20 For $n = 4$, we have z_p for $p = 0, 1, \ldots, n-1$ as

$$z_0 = (0000)_2 = 0, \quad z_1 = (1000)_2 = 8, \quad z_2 = (1100)_2 = 12, \quad \text{and}$$
$$z_3 = (1110)_2 = 14.$$

or based on the lower bound of $\mathcal{Z}_p$ in the properties of $\mathcal{Z}_p$ as

$$z_0 = 2^n - 2^{n-0} = 16 - 2^4 = 0, z_1 = 16 - 2^3 = 8,$$
$$z_2 = 16 - 2^2 = 12, \text{and}\, z_3 = 16 - 2 = 14.$$

Let us find the deepest updated stage while decoding any bit i within set $\mathcal{Z}_p$. For any $i \in \mathcal{Z}_p, p \in [0, n-1]$, and $z_p = \min(\mathcal{Z}_p)$, we have

$$\max_{i \in \mathcal{Z}_p}(\eta(i)) = \eta(z_p). \tag{5.41}$$

To show this, let us recall the notation $\{1\}^p + \{0\} + \{0,1\}^{n-(p+1)}$ for $i \in \mathcal{Z}_p, p < n-1$ where the operator "+" is used for concatenation, and $\{1\}^p$ is/are the most significant bits. According to (5.33), the maximum value for $\eta(i)$, that is, the largest index for the least significant bit set to one for $i \in \mathcal{Z}_p$, is obtained when we have

$$\text{bin}(i) = \{1\}^p + \{0\} + \{0\}^{n-(p+1)} = \{1\}^p + \{0\}^{n-p},$$

which is the smallest element in set $i \in \mathcal{Z}_p, p < n-1$, that is, $z_p = \min(\mathcal{Z}_p)$. For $p = n-1$, although the notation is in the form of $\{1\}^{n-1} + \{0,1\}^1$, the largest index for the least significant bit set to one is similarly obtained from $\{1\}^{n-1} + \{0\}$ which is the smallest in set $\mathcal{Z}_{n-1}$. Clearly, when $p = 0$, we have $\max(\eta(i)) = \eta(0) = n-1$ for any $i \in \mathcal{Z}_0$.

From the above discussion, we conclude that the deepest stage in which intermediate LLRs are overwritten/updated is when decoding the smallest bit index in the set $\mathcal{Z}_p$. Recall that the partial sums used in stage $\eta(z_p)$ are provided after decoding the bit with index $z_p - 1$ according to (5.35).

Now we consider updating intermediate information for i in different sets of $\mathcal{Z}_p$. For any $p, p' \in [0, n-1], p < p'$, we have

$$\eta(z_p) > \eta(z_{p'}). \tag{5.42}$$

This follows directly from (5.33) and property c of $\mathcal{Z}_p$. Note that z_p is in the form of $\{1\}^p + \{0\}^{n-p}$. It can be observed that for the smaller p, the position of the least significant bit set to one has a larger index. Therefore, $\eta(z_p)$ is larger. For any $p, p' \in [0, n-1]$, $p < p'$, we have

$$\psi(z_p - 1) > \psi(z_{p'} - 1). \tag{5.43}$$

Observe that according to (5.35), we have $\eta(z_p) = \psi(z_p - 1)$. Then based on (5.42), it follows that $\psi(z_p - 1) > \psi(z_{p'} - 1)$.

From the above discussion, we conclude that intermediate LLRs and partial sums of stage $\eta(z_p)$ are not overwritten/updated when we are decoding any bit with index $i \in \mathcal{Z}_{p'}, p < p'$. Hence, furthermore, based on the fact that the intermediate information is updated from stage $\eta(z_p)$ to stage 0, rewinding the SC algorithm from bit $i \in \mathcal{Z}_{p'}$ to bit $z_p, p < p'$ does not require any additional update of the intermediate LLRs or partial sums.

We will use these properties in Section 8.2.5 in interactive SC-based decoding algorithms for partial rewind of the SC process to avoid redecoding from scratch.

5.12 Partial Sums and Polar Constituent Codes

Partial sums as intermediate output of the encoding process have interesting properties that we discuss in this section. Denoting the encoding stages by $s = 0$ to n as shown in Figure 5.6 and partial sum at stage s by $\hat{\mathbf{u}}^{(s)}$ with length N, we get constituent codes of length 2^{n-s} at stage s by

$$\hat{\mathbf{u}}^{(s)} = \hat{\mathbf{u}}_N \left(\mathbf{I}_{2^{n-s}} \otimes \mathbf{G}_2^{\otimes s}\right), \tag{5.44}$$

which is the partial sum obtained at stage s.

Example 5.21 For $N = 8$, the partial sums at stage $s = 2$ is obtained by

$$\hat{\mathbf{u}}^{(2)} = \hat{\mathbf{u}}_8 \cdot \left(\mathbf{I}_2 \otimes \mathbf{G}_2^{\otimes 2}\right) = \hat{\mathbf{u}}_8 \cdot \begin{bmatrix} \mathbf{G}_2^{\otimes 2} & \mathbf{0} \\ \mathbf{0} & \mathbf{G}_2^{\otimes 2} \end{bmatrix} = \hat{\mathbf{u}}_8 \cdot \begin{bmatrix} 1&0&0&0&0&0&0&0 \\ 1&1&0&0&0&0&0&0 \\ 1&0&1&0&0&0&0&0 \\ 1&1&1&1&0&0&0&0 \\ 0&0&0&0&1&0&0&0 \\ 0&0&0&0&1&1&0&0 \\ 0&0&0&0&1&0&1&0 \\ 0&0&0&0&1&1&1&1 \end{bmatrix}.$$

Note that we assumed we have already decoded the entire $\hat{\mathbf{u}}_8$.

The relationship between the partial-sum vectors at stages s and t for $s < t$ denoted by $\hat{\mathbf{u}}^{(s)}$ and $\hat{\mathbf{u}}^{(t)}$, both of length N is given by

$$\hat{\mathbf{u}}^{(t)} = \hat{\mathbf{u}}^{(s)} \left(\mathbf{I}_{2^{n-t}} \otimes \mathbf{G}_2^{\otimes(t-s)} \otimes \mathbf{I}_{2^s} \right). \tag{5.45}$$

Clearly, when $t = n$ we have $\hat{\mathbf{u}}^{(n)} = \hat{\mathbf{x}}_N$ as

$$\hat{\mathbf{x}}_N = \hat{\mathbf{u}}^{(n)} = \hat{\mathbf{u}}^{(s)} \left(\mathbf{G}_2^{\otimes(n-s)} \otimes \mathbf{I}_{2^s} \right). \tag{5.46}$$

Observe from (5.46) that a polar code of length $N = 2^n$ is constitutes of 2^s polar codes of length 2^{n-s}.

Example 5.22 Given $N = 2^3 = 8$ and N-dimensional $\hat{\mathbf{u}}^{(s)}$ for $s = 1$, we have

$$\hat{\mathbf{x}}_8 = \hat{\mathbf{u}}^{(1)} \cdot \left(\mathbf{G}_2^{\otimes(3-1)} \otimes \mathbf{I}_{2^1} \right) = \hat{\mathbf{u}}^{(1)} \cdot \begin{bmatrix} 1&0&0&0&0&0&0&0 \\ 0&1&0&0&0&0&0&0 \\ 1&0&1&0&0&0&0&0 \\ 0&1&0&1&0&0&0&0 \\ 1&0&0&0&1&0&0&0 \\ 0&1&0&0&0&1&0&0 \\ 1&0&1&0&1&0&1&0 \\ 0&1&0&1&0&1&0&1 \end{bmatrix}. \tag{5.47}$$

Exercises

1 What is the error exponent of $G_2^{\otimes n}$?

2 Let us take a polar code of length $N = 128$. After making a decision on the value of $u_6 3$, we need to pass it back to the factor graph. What is the last stage where its partial sums should be updated?
Hint: Use the "ffz" operator.

3 Find the set $\mathcal{K}_i$ for $i = 15$ and $n = 7$. Assuming the minimum distance of the code is $d = 16$, how many codewords of weight 16 exist in the coset C_i.

4 Does the sum of the rows $\mathbf{g}_i, i = 15, 23, 27$ form a codeword of weight 16? If not, which additional rows should be included in the summation to adjust the weight to 16?
Hint: Find the set $\mathcal{M}(\mathcal{J})$ for J $= \{23, 27\}$ where $i = 15$.

5 Given the coset C_i for $i = 16$, does a codeword of weight 32 exist in that coset? What about codewords of weight 8? Justify your answers.
Hint: Use (5.16).

Bibliographical Notes

The main properties of polar codes were discussed by Arıkan [2009] in his seminal work, in particular, the polar transform, generator and parity check matrix, and permutation matrix. Later, Arıkan [2011] also explored the systematic polar codes, which was followed by others works such as Vangala et al. [2015] and Sarkis et al. [2016].

The minimum distance of polar codes was studied as a subcode of the Reed–Muller codes by Hussami et al. [2009] and from the coset perspective on the polar transform by Rowshan et al. [2023]. Moving forward, based on the weight structure of Reed–Muller codes by Kasami and Tokura [1970], the first closed-form formula was studied for the enumeration of minimum weight codewords of polar codes by Bardet et al. [2016] and for larger weight codewords by Drăgoi et al. [2024]. A study by Rowshan et al. [2023] decomposed the minimum weight codewords into rows of the polar transform and showed how these codewords are formed. This decomposition method also provided an alternative approach to enumerating the minimum weight codewords.

The exponent of the kernel matrices was first studied by Arıkan and Telatar [2009] followed by Korada et al. [2010] which was expanded to larger kernels and a search for a kernel matrix with larger error exponents.

In Mori and Tanaka [2010], non-binary kernel matrices were constructed based on Reed–Solomon codes and Hermitian codes that were later expanded to other algebraic geometry codes, including Suzuki codes, by Anderson and Matthews [2014], concatenated algebraic geometry codes, due to their large minimum distance and often nested structure, by Eid and Duursma [2013]. These kernels have a larger exponent than any linear binary kernel of the same dimension. Furthermore, non-linear binary kernels were studied by Presman et al. [2015] and Lin et al. [2015] which provide exponents superior to any linear binary kernel of the same number of dimensions. Yuan and Steiner [2018] proposed the construction and decoding algorithms for polar codes utilizing 2×2 non-binary kernels, demonstrating how these kernels enhance the flexibility and performance of polar codes compared to traditional binary designs, particularly in iterative decoding scenarios.

As discussed, due to the high complexity of the SC decoding algorithm for large kernels, more efficient algorithms have been proposed, such as window processing, by Trifonov [2014], Trofimiuk and Trifonov [2021], Abbasi and Viterbo [2020], and recursive trellis processing by Trifonov [2014]. The complexity of these algorithms is still exponential with respect to kernel size ℓ. Moreover, some of these approaches, such as window processing which is based on the idea of expressing LLRs for a large kernel via LLRs for the dimension-2 kernel $\boldsymbol{F}_2$, are not admitted by all kernels in an efficient way.

Shortly after, Arıkan and Telatar [2009] demonstrated that it is possible to choose $\gamma < 2^{-\Omega(2^{\beta n})}$ for any $\beta < 1/2$. This result contributed to an improved probability of decoding error of $2^{-N^{-\beta}}$. The decrease in the parameter γ with respect to n was denoted as the "rate" of (limiting) polarization.

The successive cancellation process in decoding algorithms exhibits interesting properties that are utilized for the efficient implementation of all SC-based decoding algorithms, both in hardware and software. The properties of partial sums were investigated by Berhault et al. [2015] and Fan and Tsui [2014], aiding in the development of efficient hardware designs for SC-based decoders. Additionally, Leroux et al. [2012] identified properties that are instrumental in reducing memory requirements for storing LLRs and partial sums, as well as optimizing the scheduling of the SC process. Furthermore, Rowshan and Viterbo [2022] analyzed properties of the SC process that enable partial rewinding, eliminating the need to restart the decoding algorithm from the first bit.

Bibliography

Fariba Abbasi and Emanuele Viterbo. Large kernel polar codes with efficient window decoding. *IEEE Transactions on Vehicular Technology*, 69(11):14031–14036, 2020.

Sarah E. Anderson and Gretchen L. Matthews. Exponents of polar codes using algebraic geometric code kernels. *Designs, Codes and Cryptography*, 73:699–717, 2014.

Erdal Arıkan. Channel polarization: A method for constructing capacity-achieving codes for symmetric binary-input memoryless channels. *IEEE Transactions on Information Theory*, 55(7):3051–3073, 2009.

Erdal Arıkan. A survey of Reed-Muller codes from polar coding perspective. In *2010 IEEE Information Theory Workshop on Information Theory (ITW 2010, Cairo)*, pages 1–5. IEEE, 2010.

Erdal Arıkan. Systematic polar coding. *IEEE Communications Letters*, 15(8):860–862, 2011.

Erdal Arıkan and Emre Telatar. On the rate of channel polarization. In *2009 IEEE International Symposium on Information Theory*, pages 1493–1495. IEEE, 2009.

Magali Bardet, Vlad Dragoi, Ayoub Otmani, and Jean-Pierre Tillich. Algebraic properties of polar codes from a new polynomial formalism. In *2016 IEEE International Symposium on Information Theory (ISIT)*, pages 230–234. IEEE, 2016.

Guillaume Berhault, Camille Leroux, Christophe Jego, and Dominique Dallet. Partial sums computation in polar codes decoding. In *2015 IEEE International Symposium on Circuits and Systems (ISCAS)*, pages 826–829. IEEE, 2015.

Vlad-Florin Drăgoi, Mohammad Rowshan, and Jinhong Yuan. On the closed-form weight enumeration of polar codes: 1.5d-weight codewords. *IEEE Transactions on Communications*, 72(10):5972–5987, 2024.

Abdulla Eid and Iwan Duursma. Using concatenated algebraic geometry codes in channel polarization. *arXiv preprint arXiv:1310.7159*, 2013.

You Zhe Fan and Chi-ying Tsui. An efficient partial-sum network architecture for semi-parallel polar codes decoder implementation. *IEEE Transactions on Signal Processing*, 62(12):3165–3179, 2014.

Nadine Hussami, Satish Babu Korada, and Rudiger Urbanke. Performance of polar codes for channel and source coding. In *2009 IEEE International Symposium on Information Theory*, pages 1488–1492. IEEE, 2009.

Tadao Kasami and Nobuki Tokura. On the weight structure of Reed–Muller codes. *IEEE Transactions on Information Theory*, 16(6):752–759, 1970.

Satish Babu Korada, Eren Şaşoğlu, and Rüdiger Urbanke. Polar codes: Characterization of exponent, bounds, and constructions. *IEEE Transactions on Information Theory*, 56(12):6253–6264, 2010.

Camille Leroux, Alexandre J. Raymond, Gabi Sarkis, Ido Tal, Alexander Vardy, and Warren J Gross. Hardware implementation of successive-cancellation decoders for polar codes. *Journal of Signal Processing Systems*, 69:305–315, 2012.

Hsien-Ping Lin, Shu Lin, and Khaled A. S. Abdel-Ghaffar. Linear and nonlinear binary kernels of polar codes of small dimensions with maximum exponents. *IEEE Transactions on Information Theory*, 61(10):5253–5270, 2015.

Ryuhei Mori and Toshiyuki Tanaka. Non-binary polar codes using Reed-Solomon codes and algebraic geometry codes. In *2010 IEEE Information Theory Workshop*, pages 1–5. IEEE, 2010.

Noam Presman, Ofer Shapira, Simon Litsyn, Tuvi Etzion, and Alexander Vardy. Binary polarization kernels from code decompositions. *IEEE Transactions on Information Theory*, 61(5):2227–2239, 2015.

Mohammad Rowshan and Emanuele Viterbo. Efficient partial rewind of successive cancellation-based decoders for polar codes. *IEEE Transactions on Communications*, 70(11):7160–7168, 2022.

Mohammad Rowshan, Son Hoang Dau, and Emanuele Viterbo. On the formation of min-weight codewords of polar/PAC codes and its applications. *IEEE Transactions on Information Theory*, 69(12):7627–7649, 2023.

Gabi Sarkis, Ido Tal, Pascal Giard, Alexander Vardy, Claude Thibeault, and Warren J. Gross. Flexible and low-complexity encoding and decoding of systematic polar codes. *IEEE Transactions on Communications*, 64(7):2732–2745, 2016.

Peter Trifonov. Binary successive cancellation decoding of polar codes with Reed-Solomon kernel. In *2014 IEEE International Symposium on Information Theory*, pages 2972–2976. IEEE, 2014.

Grigorii Trofimiuk and Peter Trifonov. Window processing of binary polarization kernels. *IEEE Transactions on Communications*, 69(7):4294–4305, 2021.

Harish Vangala, Yi Hong, and Emanuele Viterbo. Efficient algorithms for systematic polar encoding. *IEEE Communications Letters*, 20(1):17–20, 2015.

Peihong Yuan and Fabian Steiner. Construction and decoding algorithms for polar codes based on 2×2 non-binary kernels. In *2018 IEEE 10th International Symposium on Turbo Codes & Iterative Information Processing (ISTC)*, pages 1–5. IEEE, 2018.

6

Construction of Polar Codes

"Each problem that I solved became a rule which served afterwards to solve other problems."

— René Descartes

The construction of polar codes is critical in determining their error correction performance, computational complexity, and suitability for diverse applications. Since their introduction, polar codes have gained significant attention for their ability to achieve channel capacity under successive cancellation (SC) decoding at large block lengths. However, their practical implementation relies on efficient and accurate construction methods that identify the optimal set of information and frozen bits for a given channel and block length.

This chapter explores various methods for constructing polar codes, focusing on their trade-offs in terms of computational complexity, accuracy, and error correction performance. We explore both channel-dependent approaches, such as density evolution and Gaussian approximation, and channel-independent methods, including partial orders and polarization weights. Additionally, we discuss the construction of concatenated codes, such as CRC-polar codes and PAC codes, which have demonstrated enhanced performance in practical scenarios.

Our primary focus is on practical and low-complexity construction techniques suitable for real-world applications. While numerous advanced methods—such as search-based algorithms and machine learning approaches—exist in the literature, these are not covered in detail but are referenced in the Bibliographical notes for further exploration. By understanding the differences between the presented techniques, readers will be equipped to select and implement the most suitable construction method for their specific decoders and application needs.

Polar Codes: From Theory to Practice, First Edition. Mohammad Rowshan and Emanuele Viterbo.

6.1 Code Construction

Let W be a binary memoryless symmetric (BMS) channel. Given the code rate $R < I(W)$, polar coding is performed based on a set of $2^n R$ rows of the matrix $\mathbf{G}_{2^n} \triangleq \begin{bmatrix} 1 & 0 \\ 1 & 1 \end{bmatrix}^{\otimes n}$. This set denoted by $\mathcal{A}$ (or $\mathcal{I}$) defines the *construction* of polar codes $(2^n, 2^n R)$. This set of rows with indices in $\mathcal{A}$ forms a generator matrix $2^n R \times 2^n$ matrix used as for the polar encoding process. The approach to finding the elements of the set $\mathcal{A}$ depends on the channel W and is driven by the *channel polarization* phenomenon. In Chapter 4, we learned about channel polarization and the original method for constructing polar codes based on the Bhattacharyya parameter for the binary erasure channel (BEC).

A simple construction of polar codes is available only for the BEC. However, in practice, polar codes are expected to work over additive white Gaussian noise (AWGN) channels, where several construction methods are commonly used, such as density evolution (DE) and Gaussian approximation (GA) of density evolution. These construction methods have higher computational complexity, which scales linearly with the code length and assume that the channel state information is given (e.g., signal-to-noise ratio (SNR), noise power, erasure probability, etc.) to assess the quality of the sub-channels and select the most reliable ones. That makes these methods channel-dependent. Another approach to constructing polar codes is based on a certain partial order (PO) relation between the reliabilities of the sub-channels. These partial orders are deterministic and universal, holding for any transmission channel and any channel condition. This approach is faster and has only sublinear complexity.

Given the exact set $\mathcal{A}$ obtained using the DE method for a particular polar code for an AWGN channel, this construction is optimal only under successive cancellation decoding. In other words, to get the best performance under successive cancellation decoding, we need to accurately select the most reliable subchannels. However, the code constructed using this approach, based on the sub-channel reliability, will not necessarily have a competitive weight distribution. Hence, such a code may not perform optimally under maximum-likelihood or near-maximum-likelihood decoding algorithms.

In this chapter, we discuss various methods available for reliability-based code construction. We also discuss the importance of considering weight distribution (in particular, the minimum weight and the number of codewords with minimum weight) of the resulting code from the reliability-based approach and the need to improve it.

6.2 Channel-Dependent Code Construction

In this section, we focus mainly on the reliability-based construction, which is optimal for SC decoding.

The construction (also known as rate profile) defines the set $\mathcal{A} \subset \{1, 2, \dots, N\}$ of rows of $\mathbf{G}_N$ corresponding to the K best sub-channels $W_N^{(i)}$ whose mutual information $\{I(W_N^{(i)}), i \in \mathcal{A}\}$ is the largest among $\{I(W_N^{(i)}), i \in [N]\}$. Although the construction of polar codes is explicitly defined, the exact evaluation of $I(W_N^{(i)})$ for an additive white Gaussian noise channel is intractable, since it depends on the calculation of the parameters of channels whose output alphabets grow exponentially in code length. Note that in this section, we index the sub-channel from 1 to N to be aligned with the literature.

6.2.1 Density Evolution

As discussed in Chapter 4, the seminal work on polar codes by Arıkan proposed the use of Bhattacharyya parameter $Z(W_N^{(i)})$ as a measure of reliability over the BEC and then choosing K sub-channels with the smallest $Z(W_N^{(i)}), i \in [N]$. The block error event, $\mathcal{E}$, of the code resulting from the set $\mathcal{A}$ under SC decoding is a union over $\mathcal{A}$ of the events $\mathcal{B}_i$ in which the first bit error occurs at the i-th bit, expressed as $\mathcal{E} = \cup_{i \in \mathcal{A}} \mathcal{B}_i$. The set is defined as

$$\begin{aligned} \mathcal{B}_i &= \left\{\left(u_1^N, y_1^N\right) : \hat{u}_1^{i-1} = u_1^{i-1}, \hat{u}_i\left(y_1^N, u_1^{i-1}\right) \neq u_i\right\} \\ &\subseteq \left\{\left(u_1^N, y_1^N\right) : \hat{U}_i\left(y_1^N, u_1^{i-1}\right) \neq u_i\right\} \triangleq \mathcal{B}_i'. \end{aligned} \tag{6.1}$$

Then, the block error probability can be upper bounded by

$$P(\mathcal{E}) = \sum_{i \in \mathcal{A}} P\left(\mathcal{B}_i\right) \leq \sum_{i \in \mathcal{A}} P\left(\mathcal{B}_i'\right) \leq \sum_{i \in \mathcal{A}} Z\left(W_N^{(i)}\right). \tag{6.2}$$

Let us consider $P(\mathcal{B}_i')$ as decoding error probability in belief propagation (BP) decoding on the tree graph corresponding to the ith bit while eliminating other edges of the Tanner graph. The root and leaves of the tree correspond to the variable nodes of u_i and y_1^N, respectively. This helps to evaluate the probability of error at the root node using *density evolution* as a known approach for the LDPC codes.

The log-likelihood ratio (LLR) of ith bit, defined as

$$L_N^{(i)}\left(y_1^N, \hat{u}_i^{i-1}\right) \triangleq \log \frac{W_N^{(i)}\left(y_1^N, \hat{u}_1^{i-1} \mid 0\right)}{W_N^{(i)}\left(y_1^N, \hat{u}_1^{i-1} \mid 1\right)}, \tag{6.3}$$

is calculated recursively the intermediate LLRs from leaves to the root of the tree by the following updating rules:

$$L_N^{(2i-1)}\left(y_1^N, \hat{u}_1^{2i-2}\right) = 2\tanh^{-1}\left(\tanh\left(L_{N/2}^{(i)}\left(y_1^{N/2}, \hat{u}_{1,e}^{2i-2} \oplus \hat{u}_{1,o}^{2i-2}\right)/2\right)\right.$$
$$\left.\times \tanh\left(L_{N/2}^{(i)}\left(y_{N/2+1}^{N}, \hat{u}_{1,e}^{2i-2}\right)/2\right)\right) \tag{6.4}$$

$$L_N^{(2i)}(y_1^N, \hat{u}_1^{2i-1}) = L_{N/2}^{(i)}\left(y_{N/2+1}^{N}, \hat{u}_{1,e}^{2i-2}\right)$$
$$+(-1)^{\hat{u}_{2i-1}} L_{N/2}^{(i)}\left(y_1^{N/2}, \hat{u}_{1,e}^{2i-2} \oplus \hat{u}_{1,o}^{2i-2}\right), \tag{6.5}$$

where $\hat{u}_{1,e}^{i}$ and $\hat{u}_{1,o}^{i}$ are subvectors that consist of elements of $\hat{u}_1^i$ with even and odd indices, respectively, and channel LLR:

$$L_1^{(i)}(y_i) = \log \frac{W(y_i \mid 0)}{W(y_i \mid 1)}. \tag{6.6}$$

For a symmetric B-DMC with probability density functions (PDFs) $a_w(x)$ of the channel LLRs in (6.6) for the all-zero codeword, we have

$$P\left(\mathcal{B}_i'\right) = \mathfrak{E}\left(\mathrm{a}_N^i\right),$$

where

$$\mathfrak{E}(\mathrm{a}) := \lim_{\epsilon \to +0}\left(\int_{-\infty}^{-\epsilon} \mathrm{a}(x)\mathrm{d}x + \frac{1}{2}\int_{-\epsilon}^{+\epsilon} \mathrm{a}(x)\mathrm{d}x\right), \tag{6.7}$$

and $\mathrm{a}_N^i, i \in [1, N]$ are recursively obtained using

$$\mathrm{a}_{2N}^{2i} = \mathrm{a}_N^i \circledast \mathrm{a}_N^i, \quad \mathrm{a}_{2N}^{2i-1} = \mathrm{a}_N^i \boxast \mathrm{a}_N^i, \quad \mathrm{a}_1^1 = \mathrm{a}_w, \tag{6.8}$$

and $\circledast$ and $\boxast$ denote the convolutions of LLR density functions corresponding to (6.4) and (6.5), respectively. Recall that the addition of independent random variables, here LLRs, implies the convolution of their densities. Then, a polar code can be constructed by choosing a set $\mathcal{A}$, where $|\mathcal{A}| = N \cdot R$, that minimizes $\sum_{i\in\mathcal{A}} P(\mathcal{B}_i')$ as

$$\mathcal{A} = \mathrm{argmin}_{\mathcal{A}} \sum_{i\in\mathcal{A}} \mathfrak{E}(\mathrm{a}_N^i). \tag{6.9}$$

6.2.2 Density Evolution with Gaussian Approximation

To reduce the complexity of DE, several approximate methods have been proposed, which are based on the upper bound and the lower bound on the error probability of the sub-channels.

The success of the Gaussian approximation for sparse graph codes suggests the use of a similar approach for polar codes to approximate the density evolution of LLRs throughout the tree graph. The intuition behind this approximation is that the channels $W_N^{(i)}$ behave like binary AWGN channels with varying noise levels

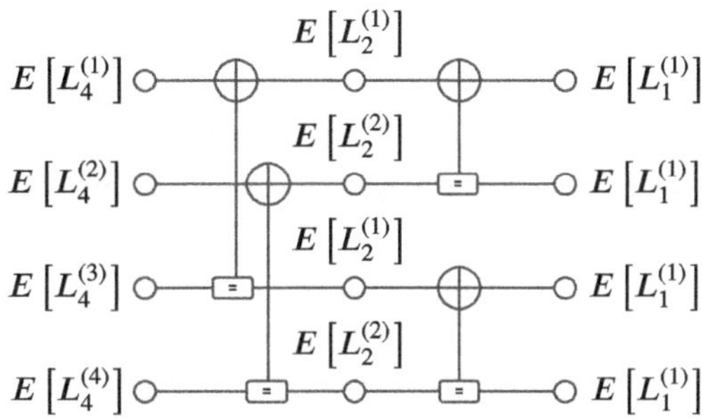

Figure 6.1 Mean LLRs through check nodes and variable nodes.

when N is sufficiently large. Therefore, we only need to monitor the noise variances of these channels. For AWGN channels, the distribution of channel LLR in (6.6) is Gaussian, that is, $L_1^{(i)}(y_i) \sim \mathcal{N}(m, 2m)$ with mean $m = \frac{2}{\sigma_n^2}$ and variance $\frac{4}{\sigma_n^2}$, where $\frac{1}{\sigma_n^2}$ is the channel SNR. We use a design parameter called *design-SNR*, denoted by SNR_d. Given the design-SNR SNR_d in dB, we can obtain the variance of the Gaussian distribution as

$$\sigma_n^2 = \frac{1}{2\frac{K}{N}10^{\text{SNR}_d/10}}, \tag{6.10}$$

where the term $10^{\text{SNR}_d/10}$ is used for converting design-SNR from dB to linear form. The listing in Algorithm 6.1 gives a procedure to construct a polar code using the density evolution method with Gaussian approximation.

To simplify computations, intermediate LLRs in (6.5) and (6.4) are considered Gaussian random variables where the relationship between the mean (expected value) E and the variance V is $V[L_N^{(i)}] = 2E[L_N^{(i)}]$. As a result, considering the check and variable nodes of degree $d_c = d_v = 3$ in the factor graph of polar codes in Figure 6.1, the convolution operations in (6.8) are reduced to

$$E\left[L_N^{(2i-1)}\right] = \phi^{-1}\left(1 - \left(1 - \phi\left(E\left[L_{N/2}^{(i)}\right]\right)\right)^2\right) \tag{6.11}$$

$$E\left[L_N^{(2i)}\right] = 2E\left[L_{N/2}^{(i)}\right], \tag{6.12}$$

where function $\phi(x)$ for $x \in [0, \infty)$ is defined as

$$\phi(x) = \begin{cases} 1 - \frac{1}{\sqrt{4\pi x}}\int_{\mathbb{R}} \tanh\frac{u}{2}e^{-\frac{(u-x)^2}{4x}}du, & \text{if } x > 0 \\ 1, & \text{if } x = 0 \end{cases}. \tag{6.13}$$

To reduce the complexity of calculating $\phi(x)$, an approximation with acceptable accuracy is used. For $x < 10$, one can use the following curve fitting:

$$\phi(x) \sim e^{\alpha x^{\gamma}+\beta},$$

Algorithm 6.1: Density Evolution w/Gaussian Approximation (DEGA) Method

---

input : N, K, mean channel LLR $m = \frac{2}{\sigma_n^2} = 4\frac{K}{N}10^{\mathrm{SNR}_d/10}$

output: The indices of the most reliable sub-channels $\mathcal{A}$

$\mathbf{Z} \leftarrow [m \cdots m]$ `// Initialization with an` N`-length vector`

for $\ell \leftarrow 1$ **to** $\log_2 N$ **do**

    $S \leftarrow 2^{\ell}$ `// The number of tree branches at layer` ℓ

    for $j \leftarrow 0$ **to** $\frac{S}{2} - 1$ **do**

        $z \leftarrow \mathbf{Z}[j]$

        $\mathbf{Z}[j] \leftarrow \phi^{-1}\left(1 - (1 - \phi(z))^2\right)$ `// Upper branch/sub-channel`

        $\mathbf{Z}\left[\frac{S}{2} + j\right] \leftarrow 2z$ `// Lower branch/sub-channel`

$\mathcal{A} \leftarrow$ Find the indices of the largest elements in vector $\mathbf{Z}$

return $\mathcal{A}$

---

where $\alpha = -0.4527$, $\beta = 0.0218$, and $\gamma = 0.86$. However, for $x \geq 10$, we can use the average of the upper and lower bounds for $\phi(x)$ given below:

$$\sqrt{\frac{\pi}{x}} e^{-\frac{x}{4}} \left(1 - \frac{3}{x}\right) < \phi(x) < \sqrt{\frac{\pi}{x}} e^{-\frac{x}{4}} \left(1 + \frac{1}{7x}\right), x > 0. \tag{6.14}$$

Example 6.1 Construct the polar code $(8, 4)$ using the density evolution with Gaussian approximation (DEGA) method over the AWGN channel with $\mathrm{SNR}_d = 2$. The variance σ_n^2 is calculated using (6.10) which gives $\sigma_n^2 = 0.6310$. Starting from $E\left\{L_1^{(1)}\right\} = \frac{2}{\sigma_n^2} = 3.1698$, the recursion of the mean channel LLR is executed using (6.11) and (6.2.2) and results in the tree shown in Figure 6.2. Ranging the

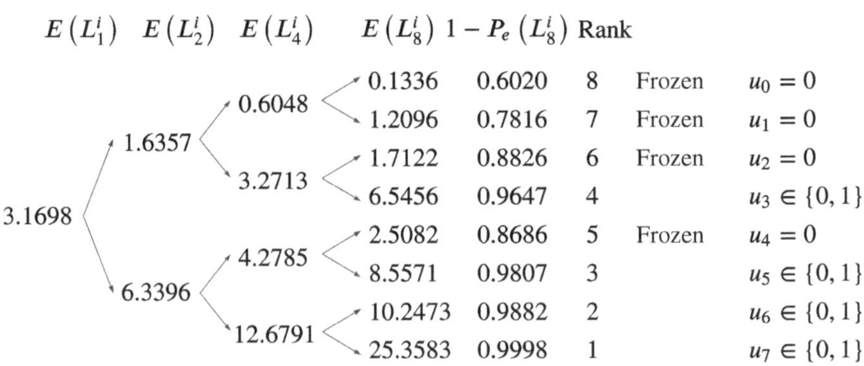

Figure 6.2 The polar code $(8, 4)$ constructed using the DE/GA method over AWGN channel with $\mathrm{SNR}_d = 2$.

sub-channels, four subchannels with the highest reliabilities (corresponding to the largest evolved LLRs) are selected to transmit information $u_i, i \in \mathcal{A} = \{3, 5, 6, 7\}$ and the rest transmit a known value, by default $u_i = 0, i \in \mathcal{A}^c = \{0, 1, 2, 4\}$.

6.3 Channel-Independent Code Construction

This section discusses methods used to partially and fully order the reliabilities of sub-channels, $W_N^{(i)}, i \in [0, N-1]$, to construct polar codes, independent of the underlying channel W and its parameter(s).

6.3.1 Universal Partial Order (UPO)

In this section, the notion of partial order (PO) is introduced and its inherent recursive structure is analyzed. There exists a certain *partial order* relation between the reliabilities of the sub-channels in polar coding. They are called "partial" because the available relations cannot form a fully ordered integer sequence corresponding to the indices of all N sub-channels. The partial orders are deterministic and universal, that is, the relation holds for any transmission channel. Hence, they are called the universal partial order (UPO).

The partial order relies on the binary expansion of the sub-channel index i. Let us denote the binary expansion of i by $\text{bin}(i) = (b_{n-1} \cdots b_1 b_0), b_k \in \{0, 1\}$, where the b_0 is the least significant bit (LSB).

As discussed in Chapter 4, the vector channel $W_N : \{0, 1\}^N \to \mathcal{Y}^N$, which is the result of combining N copies of a given binary discrete memoryless channel (B-DMC) W in a recursive manner, can be split back into N sub-channels $W_N^{(i)} : \{0, 1\} \to \mathcal{Y}^N \times \{0, 1\}^{i-1}, 0 \leq i \leq N-1$. The sub-channel $W_N^{(i)}$ can be defined as

$$W_N^{(i)} = ((W^{b_0})^{b_2} \cdots)^{b_{n-1}}, \quad 0 \leq i < N, \tag{6.15}$$

where $b_{n-1} \cdots b_1 b_0 \in \{0, 1\}$ is the binary expansion of i denoted by $\text{bin}(i)$ and $W^{b_k}, b_k \in \{0, 1\}, 0 \leq k < n$ taking either W^0 or W^1 are defined as

$$W^0(y_0, y_1 | x_1) = \frac{1}{2} \sum_{x_1 \in \mathcal{X}} W(y_0 | x_0 \oplus x_1) W(y_1 | x_1), \tag{6.16}$$

$$W^1(y_0, y_1, x_0 | x_2) = \frac{1}{2} W(y_0 | x_0 \oplus x_1) W(y_1 | x_1). \tag{6.17}$$

Note that the sub-channel W^0 and W^1 are the results of splitting the basic vector channel W_2 into a pair of channels $W^0 : \mathcal{X} \to \mathcal{Y}^2$ and $W^1 : \mathcal{X} \to \mathcal{X} \times \mathcal{Y}^2$ where $\mathcal{X} \in \{0, 1\}$.

Now, we start with the relation between two sub-channels obtained by one-step channel transformation for two copies of binary memoryless symmetric (BMS) channel W. We say that the sub-channel W^1 is stochastically upgraded with respect to W^0 and we denote it by

$$W^0 \preccurlyeq W^1. \tag{6.18}$$

This relation can be shown by the Bhattacharyya parameters of the channels W^0, W^1, and W as

$$Z(W)\sqrt{2 - Z(W)^2} \leqslant Z(W^0) \leqslant 2Z(W) - Z(W)^2, \tag{6.19}$$

$$Z(W^1) = Z(W)^2. \tag{6.20}$$

Now, given a sub-channel $W_N^{(i)}$. This sub-channel is stochastically upgraded to $W_N^{(j)}$ if j is obtained by

- Swapping a more significant 0 with a less significant 1 (also known as left-swap operation) in the binary expansion of i. That is,

  $$W^{(c^n 0 b^l 1 a^m)} \preccurlyeq W^{(c^n 1 b^l 0 a^m)}, \tag{6.21}$$

  where $a^n \in \{0,1\}^n, b^l \in \{0,1\}^l, c^m \in \{0,1\}^m$, and $n, l, m \geq 0$. Such a swapping pattern can happen on one pair of bits or multiple pairs. For example,

  $$W^{(\underline{1},1,\underline{0},0)} \quad \preccurlyeq \quad W^{(\underline{0},1,\underline{1},0)}, \tag{6.22}$$

  $$W^{(\underline{1,1},\underline{0,0})} \quad \preccurlyeq \quad W^{(\underline{0,0},\underline{1,1})}. \tag{6.23}$$

  Observe that channels $W_N^{(i)}$ and $W_N^{(j)}$ have binary indexes i and j with identical Hamming weights, that is $w_H(\text{bin}(i)) = w_H(\text{bin}(j))$.
- Flipping a 0 to 1 (also known as addition operation) in the binary expansion of i. That is,

  $$W^{b^m 0^n a^l} \preccurlyeq W^{b^m 1^n a^l} \tag{6.24}$$

  This kind of flipping pattern can happen on one bit or multiple bits. For example,

  $$W^{(0,\underline{0},1,1)} \quad \preccurlyeq \quad W^{(0,\underline{1},1,1)}, \tag{6.25}$$

  $$W^{(\underline{0,0},1,1)} \quad \preccurlyeq \quad W^{(\underline{1,1},1,1)}. \tag{6.26}$$

  Observe that if $W^{b^m} \preccurlyeq W^{c^m}$, then flipping low significant bits will not upgrade the sub-channel. That is,

  $$W^{b^m a^k 1^n} \preccurlyeq W^{c^m a^k 0^n}. \tag{6.27}$$

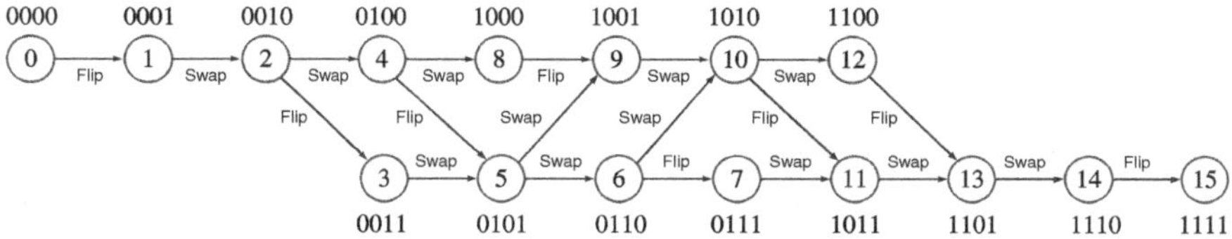

Figure 6.3 Hasse diagram of the partial order for $N = 16$ sub-channels.

Observe that channels $W_N^{(i)}$ and $W_N^{(j)}$ have binary indexes i and j with different Hamming weights, that is $w_H(\text{bin}(i)) < w_H(\text{bin}(j))$.

Example 6.2 Figure 6.3 illustrates the chain of partial orders for $N = 16$ sub-channels in a Hasse diagram. The sub-channel indices are provided in decimal and binary. The right arrows indicate the $x \preccurlyeq y$ relations. They are also labeled by the two operations of bit flipping from "0" to "1" and left-swapping of a "1" on the binary expansion of the index x to arrive at y.

Note that with this method, we cannot compare all the sub-channels and construct a code. Nevertheless, knowing the quality of a sub-channel $W_N^{(i)}$ gives us information about the quality of all sub-channels $W_N^{(j)}$ that are comparable to $W_N^{(i)}$ based on the right-swap operation. Hence, PO can be used in simplifying the code construction of polar codes.

6.3.1.1 Nested and Symmetric Properties

The partial order of sub-channels has two properties as the following that can be useful:

- Nested property: The partial orders obtained for a polar code of length N remain unchanged for the code length of $2N$. Observe that the binary expansion of any sub-channel $i \in [0, 2^n)$ in the form of $\text{bin}(i) = b_{n-1} \cdots {}_1 b_0$ can be expressed as $0b_{n-1} \cdots {}_1 b_0$ in length $2N = 2^{n+1}$, where $b_n = 0$.
- Symmetric property: For a polar code of length N, the partial order of $x \preccurlyeq y$ has a twin order of $(N - 1 - x) \succcurlyeq (N - 1 - y)$. Knowing that the binary expansion of $N - 1 - i$ is the two's complement of the binary expansion i, the flipping of 0s to 1s and vice versa causes upgrading or degradation and hence inverses the relation between x and y regardless of their membership in intervals $[0, N/2 - 1]$ and/or $[N/2, N - 1]$.

Given the nested and symmetric properties of the partial order of sub-channels, we give an example to show the recursive structure of the partial order. Let us denote the partial order relation $x \preccurlyeq y$ of two sub-channels $x, y \in [0, N - 1]$ by $\{x, y\}$. We show the set of partial orders that exist between the pairs with the adjacent sub-channels of a code with length $N = 2^n$ by $\mathbf{UPO_n}$. Obviously, due to the nested property, the partial orders in level n are inherited from level $n - 1$.

Example 6.3 The recursive structure of partial orders for polar codes of length $N = 2, 4, 8, 16$ using the notation $\{x, y\}$ is listed below. Every set $\mathbf{UPO}_n, n > 1$ is divided into three subsets as follows:

- The first set lists the relations inherited from $\mathbf{UPO}_{n-1}$ to construct $\mathbf{UPO}_n$.
- The last set represents the symmetric order relations corresponding to the first set. The order pairs in this set are the copy of $\mathbf{UPO}_{n-1}$ while the indices change to $\{x + N/2, y + N/2\}$.
- The middle set represents the new orders formed to construct $\mathbf{UPO}_n$ from $\mathbf{UPO}_{n-1}$.

$$\begin{aligned}
\mathbf{UPO}_1 &= \{\{0,1\}\}\,, \\
\mathbf{UPO}_2 &= \{\{0,1\}\} \cup \{\{1,2\}\} \cup \{\{2,3\}\}\,, \\
\mathbf{UPO}_3 &= \{\{0,1\},\{1,2\},\{2,3\}\} \cup \{\{2,4\},\{3,5\}\} \cup \{\{4,5\},\{5,6\},\{6,7\}\}\,, \\
\mathbf{UPO}_4 &= \{\{0,1\},\{1,2\},\{2,3\},\{2,4\},\{3,5\},\{4,5\},\{5,6\},\{6,7\}\} \\
&\quad \cup \{\{4,8\},\{5,9\},\{6,10\},\{7,11\}\} \cup \{\{8,9\},\{9,10\},\{10,11\}, \\
&\quad \{10,12\},\{11,13\},\{12,13\},\{13,14\},\{14,15\}\}\,.
\end{aligned}$$

Figure 6.4 shows the recursive structure corresponding to the $\mathbf{UPO}_n, n = 2, 3, 4$ sets above, arriving at the Hasse diagram in Example 6.2. The dotted arrows represent the new order relations between the orders inherited from $\mathbf{UPO}_{n-1}$ (indicated by solid arrows) and the corresponding symmetric orders (indicated by dashed arrows) in $\mathbf{UPO}_n$.

6.3.2 Polarization Weight (PW)

In this section, the concept of β-expansion and the polarization weight algorithm are discussed. The recursive structure of PO allows constructing polar codes by applying β-expansion at each recursive step given β is carefully chosen.

Let us first introduce a function called *polarization weight* that maps a sub-channel index i with binary expansion $\text{bin}(i) = (i_{n-1} \ldots i_1 i_0)$ to a reliability measure. This measure is calculated based on a nonintegral representation of a number i with base β as follows:

$$w_P(i) = \sum_{k=0}^{n-1} b_k \beta^k, \tag{6.28}$$

where β is a parameter that can be chosen carefully for every code length to get a reliability order close to or the same as the channel-dependent method. It is

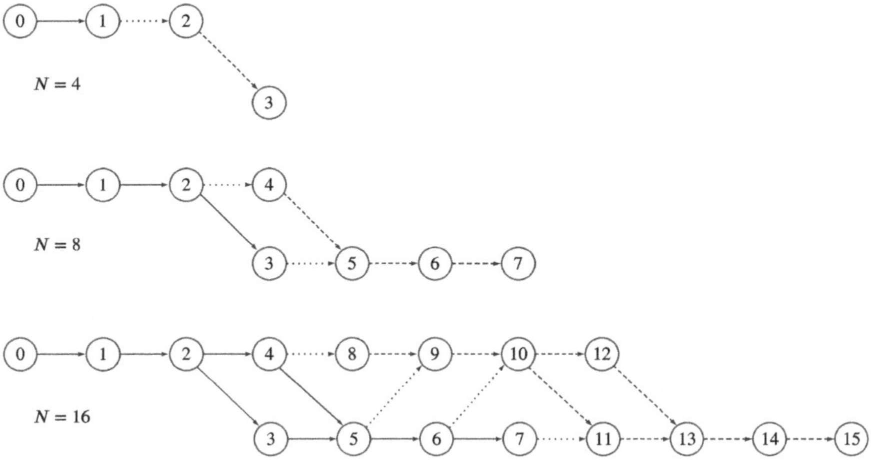

Figure 6.4 Recursive structure of partial orders for $N = 4, 8, 16$ sub-channels.

observed that $\beta = 2^{1/4}$ can be used for most of the block length with an acceptable reliability order for sub-channels.

Example 6.4 Suppose we want to find the reliability order of sub-channels of a polar code with length $N = 2^4 = 16$. First, we find the β-expansion of all the sub-channel indices. For $\beta = 2^{1/4}$, we have $\beta^0 = 1, \beta^1 = 1.1892, \beta^2 = 1.4142, and\ \beta^3 = 1.6818$. Then, by plugging in the binary expansion of sub-channels indices from $i = (b_3b_2b_1b_0)_2 = 0$ to 15, we can get their polarization weight $w_P(i) = b_3\beta^3 + b_2\beta^2 + b_1\beta^1 + b_0\beta^0$ as follows:

$$\mathbf{w} = \{0.000\ 1.000\ 1.189\ 2.189\ 1.414\ 2.414\ 2.603\ 3.603$$
$$1.682\ 2.682\ 2.871\ 3.871\ 3.096\ 4.096\ 4.285\ 5.285\},$$

where a larger value of $w_P(i)$ means the sub-channel with index i has higher reliability. Now, we can obtain the order of sub-channels' reliability as

$$\textbf{Reliability Order} = \{0\ 1\ 2\ 4\ 8\ 3\ 5\ 6\ 9\ 10\ 12\ 7\ 11\ 13\ 14\ 15\}.$$

Obviously, the indices of nonfrozen bits (i.e., good sub-channels) of the code $(16, K)$ are selected from the last K elements in the set above as they have the highest reliability.

Observe that the nested property of polar codes discussed for partial order holds for polarization weight as well. The reason is similar. Any arbitrary binary expansion $(b_{n-1} \dots b_1b_0)$ in length N can be expressed as $(0b_{n-1} \dots b_1b_0)$ in length $2N$. Clearly, since $b_n = 0$, the polarization weight $w_P(i), i \in [0, N-1]$ does not change by increasing the code length.

Table 6.1 The 5G polar sequence table.

i	0	1	2	3	4	5	6	7	$\cdots$	1018	1019	1020	1021	1022	1023
Q_i^{1024}	0	1	2	4	8	16	32	3	$\cdots$	1007	1015	1019	1021	1022	1023
$W(Q_i^{1024})$	0	1	2	3	4	5	6	7	$\cdots$	1018	1019	1020	1021	1022	1023

6.4 5G Code Construction

Given a block length $N_{\max}$, the reliability order of the sub-channel index $Q_i^{N_{\max}}, i \in [0, N_{\max}-1]$, where $0 \leq Q_i^{N_{\max}} \leq N_{\max}-1$ is denoted by $W(Q_i^{N_{\max}})$. By sorting the reliability of sub-channels in ascending order, that is $W(Q_0^{N_{\max}}) < W(Q_1^{N_{\max}}) \cdots < W(Q_{N_{\max}-1}^{N_{\max}})$, we get a sequence of indices called *polar sequence* and denoted by $\mathbf{Q}_0^{N_{\max}-1}$.

For any block length $N < N_{\max}$, where N is a power of two, the *polar sequence* $\mathbf{Q}_0^{N-1} = \{Q_0^N, Q_1^N, \ldots, Q_{N-2}^{N-1}\}$ is a subset of the polar sequence $\mathbf{Q}_0^{N_{\max}-1}$ with every element $Q_i^N < N$ sorted in ascending order of reliability. The sequence $[0, N-1]$ can be split into two sub-sequences of information indices $\overline{\mathbf{Q}}_I^N$, where $\left|\overline{\mathbf{Q}}_I^N\right| = K + n_{PC}$, and frozen indices $\overline{\mathbf{Q}}_F^N$, where $\left|\overline{\mathbf{Q}}_F^N\right| = N - \left|\overline{\mathbf{Q}}_I^N\right|$, and n_{PC} denotes the number of parity check bits.

In the technical specification TS 38.212 of 3GPP's Release 17, a polar sequence table has been provided that gives the mapping $W : Q_i^{N_{\max}} \rightarrow [0, N_{\max}-1]$ determining the reliability order of every $Q_i^{N_{\max}}, i \in [0, N_{\max}-1]$. Table 6.1 partially shows the elements of the table. A MATLAB script including the entire sequence for $N_{\max} = 1024$ is provided in Appendix C. This script also illustrates how to use this sequence for rate-profiling, that is, placing the information bits at most reliable positions. Note that we need to exclude the rate matching pattern, the set of coordinates used for puncturing, shortening, or extending/repeating (see Chapter 10, for the set $\overline{\mathbf{Q}}_I^N$).

6.5 Code Construction for ML Decoding

The code constructions discussed so far are based on the most reliable sub-channels. This approach is suitable for greedy search decoding procedures such as successive cancellation (SC) decoding, where the final decision on the value of every bit is made locally without considering the partial or whole sequence. Recall that in SC decoding, the assumption is that all the previous bits are decoded correctly when decoding every bit. Hence, due to the sequential nature of SC decoding, a single error can fail the decoding as the error is propagated to the rest of the bits and corrupt them. Observe that selection of the most reliable sub-channels can play an important role in the error correction performance of such a decoder.

Example: Let us take two sets $\mathcal{A}_1$ and $\mathcal{A}_2$ such that they differ only by one element, say $12 \in \mathcal{A}_1$ while $9 \in \mathcal{A}_2$ are those two different elements and $W_N^{(12)} \preccurlyeq W_N^{(9)}$. As the block error rate is

$$P_{SC}(\mathcal{A}) = 1 - \prod_{i \in \mathcal{A}} (1 - p_{e,i}), \tag{6.29}$$

where $p_{e,i}$ is the probability of error at sub-channel i at a particular noise power or SNR assuming that bits 0 to $i-1$ are decoded successfully. Note that $p_{e,i} = 0$ for any $i \in \mathcal{A}^c$. Then, we expect that the code with construction based on $\mathcal{A}_2$ outperforms the code based on $\mathcal{A}_1$ due to $p_{e,9} \leq p_{e,12}$.

As the example above showed, it is paramount in such a decoding process to select the most reliable sub-channels. However, good performance of a code under (near) maximum-likelihood (ML) decoding algorithms, such as SC list deciding (see Chapter 6), sequential decoding, or sphere decoding (see Chapter 7), depends on the weight distribution of the underlying code.

To understand the relationship between the weight distribution and the error correction performance, let us recall the *union-Bhattacharyya bound* in Section 2.2 of Chapter 2, which is the simplest upper bound on the block error rate (BLER) of a linear code with rate R and weight distribution $\{A_{d_{\min}}, A_{d_{\min}+1}, \dots, A_N\}$. The error probability of decoding a received sequence from an additive white Gaussian noise (AWGN) channel under a soft-decision ML decoder is

$$P_{BL} \leq \sum_{d=d_{\min}}^{N} A_d Q\left(\sqrt{2dRE_b/N_0}\right), \tag{6.30}$$

where E_b/N_0 is the SNR per information bit (SNR_b) and Q-function is defined as

$$Q(\alpha) = \frac{1}{\sqrt{2\pi}} \int_{\alpha}^{\infty} e^{-x^2/2} dx.$$

It can be seen that $Q(\alpha)$ decreases exponentially as α increases. As $\alpha \propto \sqrt{d}$, then the term $Q\left(\sqrt{2d_{\min} RE_b/N_0}\right)$ is the dominant term in the summation. Hence, the first term

$$A_{d_{\min}} Q\left(\sqrt{2d_{\min} RE_b/N_0}\right), \tag{6.31}$$

can be used as a good approximation for the upper bound at high SNR_b or low noise regimes as this bound is loose and even diverges in low-SNR regimes. Moreover, since $A_{d_{\min}}$ is the coefficient of the first term, it is also called the *error coefficient*. Note that the improved variants of the conventional union bound (based

Table 6.2 Comparison of $d_{\min}$ and $A_{d_{\min}}$ of different codes.

Coding scheme	$d_{\min}$	$A_{d_{\min}}$
Polar code (128,64)	8	4
– Appending 8 CRC bits: CRC-Polar (128,64)	8	20
– Employing convolutional pretransform: PAC code (128,64) with polar rate-profile	8	4
Reed–Muller (RM) code (128,64)	16	94488
– Employing convolutional pretransform: PAC code (128,64) with RM rate-profile	16	3120

on truncation and tightening) also rely on the weight distribution of the code, in particular, the error coefficient and the number of codewords with small weights. Hence, any reduction in A_d with small d, in particular, $A_{d_{\min}}$, would lower the upper bound. Therefore, an improvement in the performance of the linear code should be expected.

Example 6.5 Suppose that we have the code (128,64). Table 6.2 illustrates the characteristics of this code in various scenarios.

Among the codes listed in Table 6.2, PAC code (128,64) with RM rate-profile outperforms the rest under (near) ML list decoding such as sequential decoders or list decoder with a very large list size for higher minimum distance and smaller number of minimum weight codewords. This is not the case when code length increases because the impact of including the sub-channels with low reliability becomes significant and deteriorating while channel reliability is not considered when constructing codes with weight-based RM rule.

This section discusses methods to improve the error coefficient of polar code design based on the reliability of the sub-channels.

6.5.1 RM-Polar Codes

To have control over the minimum distance of the code resulting from the reliability-based construction process, we can employ the rule used in the construction of Reed-Muller codes. More precisely, we can employ both the weight-based RM rule and the reliability-based Arıkan rule to construct a polar-like code. That is, the set $\mathcal{A}$ is divided into two as

$$\mathcal{A} = \mathcal{A}^{RM} \cup \mathcal{A}^{Polar}. \tag{6.32}$$

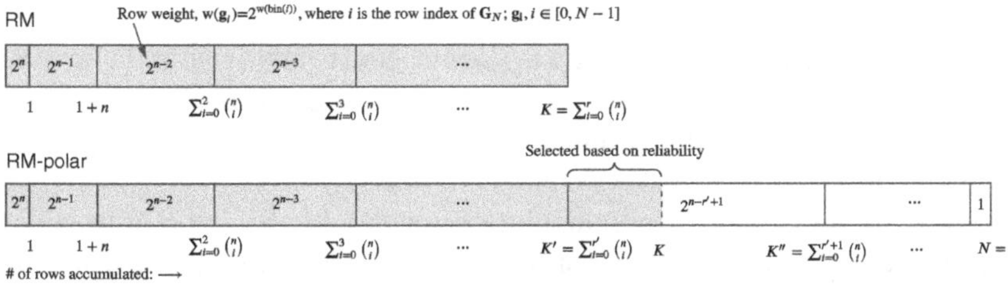

Figure 6.5 Comparison of RM construction and RMxPolar Construction, a construction based on RM rule and Arıkan rule, which is flexible for the choice of K.

Given the blocklength $N = 2^n$ and code dimension K, we find $r' = \arg\max_{r} \sum_{j=0}^{r} \binom{n}{j} \leq K$ and denote $K' = \sum_{j=0}^{r'} \binom{n}{j}$. Then, we find the indices of K' rows of $\mathbf{G}_N$ with weights $\mathrm{w}(\mathbf{g}_i) \geq 2^{n-r'}$. That is,

$$\mathcal{A}^{RM} = \left\{ i \in [0, N-1] : 2^{\mathrm{w}(\mathrm{bin}(i))} \geq 2^{n-r'} \right\},$$

where $K' = |\mathcal{A}^{RM}|$. To find $\mathcal{A}^{Polar}$, we exclude the indices in $\mathcal{A}^{RM}$ from the reliability sequence of length N and select the remaining $K - K'$ indices from the most reliable indices remaining in the sequence with row weight $2^{n-(r'+1)}$. That is,

$$\mathcal{A}^{Polar} = \underset{\substack{\mathcal{S} \subseteq \left([0,N-1] \backslash \mathcal{A}^{RM}\right) \\ |\mathcal{S}|=K-|\mathcal{A}^{RM}|}}{\arg\max} \sum_{\substack{i \in \mathcal{S} \\ \mathrm{w}(\mathrm{bin}(i))=n-r'-1}} L_N^{(i)},$$

where $L_N^{(i)}$ is the average LLR of sub-channel i which can be found from the methods discussed in Section 6.2.2. Observe that this approach gives rth-order RM code (r, n) when $K = \sum_{j=0}^{r} \binom{n}{j}$ and therefore $\mathcal{A}^{Polar} = \emptyset$. If we use the 5G reliability sequence in this process, we call it *5G-RM* construction, otherwise (crossed) *RM-polar* construction can be denoted by RMxPolar codes, due to intersection with RM codes. Figure 6.5 illustrates the selection of K indices with this approach.

A different approach for constructing RM-Polar codes is to constrain the selection of most reliable sub-channels with row weights of $\mathrm{w}(\mathbf{g}_i) \geq 2^{n-r'-1}$, where set $\mathcal{A}$ is defined as below:

$$\mathcal{A} = \underset{\substack{\mathcal{S} \subset [0,N-1] \\ |\mathcal{S}|=K}}{\arg\max} \sum_{\substack{i \in \mathcal{S} \\ \mathrm{w}(\mathrm{bin}(i)) \geq n-r'-1}} L_N^{(i)},$$

To differentiate from the RMxPolar codes, we refer to this code as *constrained RM-Polar* codes. Note that the set $\mathcal{A}$ in this approach is most likely different from (6.32), because there might be sub-channels with row weight(s) larger than the minimum weight $\mathrm{w}_{\min}$ excluded from $\mathcal{A}$ and instead include more sub-channels with row weight w_{min}. This results in a code with a higher number of cosets $C_i(\mathcal{A})$, where $w(bin(i)) = \log_2(w_{min})$ and, consequently, containing more minimum weight codewords. The RMxPolar codes outperform the constrained approach at short code lengths, while the constrained RM-Polar codes outperform polar codes due to a slightly better weight distribution.

According to the Union bound, the block error rate is a function of the minimum distance d_{min} and the number of small-weight codewords, in particular, the number of minimum weight $A_{w_{min}}$, also known as the error coefficient. Hence, a code design approach that considers both the sub-channel reliability and the *error coefficient* is of interest; however, optimizing both would be highly complex and would require performing an exhaustive search.

6.6 Pre-transformed Polar Codes

In Chapter 2, we learned about the classic concatenation in which two coding schemes are used as inner and outer codes. That is, the information bits are first encoded by an outer coding scheme and an inner code. The decoding is naturally performed in reverse order.

6.6.1 CRC-Polar Codes

CRC polar codes are an example of this kind of concatenation. In this concatenation, the information vector $\mathbf{u}$ is first systematically encoded using a cyclic code with generator polynomial $g(x)$, as described in Chapter 2. This is followed by polar coding as the inner code.

As we learned in Chapter 2, this particular concatenation has the capability of error detection. Then, when we have a list of candidates for the decoding solution, the parity bits in the systematic cyclic coding, which are also called cyclic redundancy check (CRC) bits, can be used to find the solution among the candidates that aligns with the CRC bits. Later, we learn about other decoding algorithms in addition to SC decoding, where CRC is their essential component. For example, when CRC is used with SC list decoding, therefore, in the context of polar coding, the decoding algorithm is widely referred to as CRC-aided SC list decoding. However, it is still an SC list decoding that is used for CRC-polar concatenated codes.

Based on the detection capability of the outer code, we can use CRC for early detection of decoding failure. In this arrangement, the information bits are divided into several segments, each of which is encoded separately by the outer code. During the decoding process, if the estimated partial information sequence does not pass the corresponding CRC, the decoding is terminated. This can reduce computational and time complexity. However, we need to understand that separate outer encoding of the segments will result in a loss of rate.

Example 6.6 Given an information sequence of length 256 bits and CRC generator polynomials of degrees 12 and 8. Compare the effective rate of inner polar coding of length 512 in two scenarios: (1) outer encoding of the entire sequence once, (2) dividing the sequence into two segments, and performing the outer encoding for each segment separately.

Ans: Let us use the generator polynomial of degree 12, which appends 12 CRC bits to the 256 information bits by systematic cyclic encoding. The effective rate will be

$$R_{\text{eff}} = \frac{k + n_{\text{crc}}}{N} = \frac{256 + 12}{512} = 0.5234375.$$

Then, we do the outer encoding with two segments using polynomial of degree 8. This results in the effective rate of

$$R_{\text{eff}} = \frac{k + n_{\text{crc}}}{N} = \frac{256 + 2 \times 8}{512} = 0.53125.$$

If we use polynomial of degree 12. This results in the effective rate of

$$R_{\text{eff}} = \frac{k + n_{\text{crc}}}{N} = \frac{256 + 2 \times 12}{512} = 0.546875.$$

Clearly, the rate increases even though we use a polynomial with a smaller degree. As we learned in Chapter 2, smaller degree polynomials may result in false detection.

However, rate loss has its consequences. If we increase the code rate from R_1 to R_2, according to equation (1.39) in Chapter 1, to keep the channel capacity and BLER constant, the SNR should also increase by

$$10 \cdot \log_{10}\left(\frac{R_2}{R_1}\right) = 10 \cdot \log_{10}\left(\frac{0.546875}{0.5234375}\right) = 0.19 \text{ dB}.$$

Hence, to justify the use of CRC, the power gain due to the utilization of CRC should overcome the power loss due to rate loss.

6.6.1.1 CRC Bits in 5G

Given the information bits $(a_0, a_1, \ldots, a_{A-1})$, the CRC bits are calculated using one of the cyclic generator polynomials provided in Table 6.3. As Figure 6.6 shows, 24-bit CRC is used in the transmission of downlink control information (DCI) and the others are used in uplink transmission. For more information about the conceptual channels such as, please see Appendix A.

Table 6.3 Cyclic generator polynomials for CRC calculation in 5G polar coding.

	Generator polynomials $g(D)$
CRC-24C	$D^{24}+D^{23}+D^{21}+D^{20}+D^{17}+D^{15}+D^{13}+D^{12}+D^{8}+D^{4}+D^{2}+D+1$
CRC-11	$D^{11}+D^{10}+D^{9}+D^{5}+1$
CRC-6	$D^{6}+D^{5}+1$

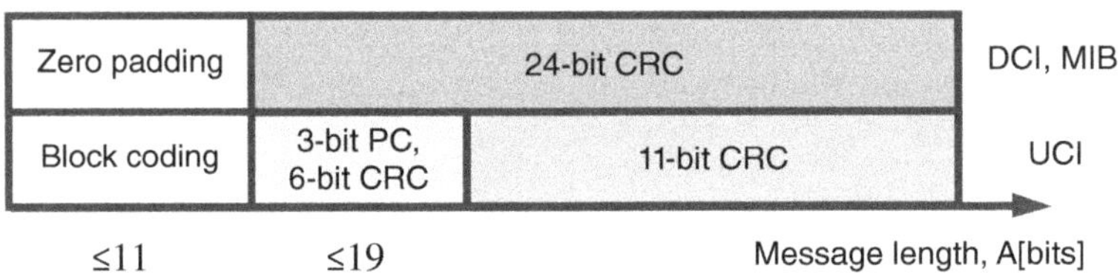

Figure 6.6 Polar coding in 5G NR.

Encoding is performed in a systematic form (see Section 2.6 in Chapter 2), which results in the following CRC bits:

$$p(D) = \text{Rem}\left(\frac{D^{n-k}a(D)}{g(D)}\right), \tag{6.33}$$

where $a(D)$ is the information polynomial and Rem(.) gives the remainder of the polynomial division. Note that (6.33) is identical to (2.37) in Chapter 2. Here, r is the CRC length which we also denote by *crcLen*. The resulting CRC bits $(p_0, p_1, \dots, p_{r-1})$ are appended to the information bits. In polynomial form, the information bits and the CRC bits together are represented by the polynomial

$$b(D) = a_0D^{A+r-1} + a_1D^{A+r-2} + \cdots + a_{A-1}D^{r} + p_0D^{r-1} + p_1D^{r-2} + \cdots$$
$$+ p_{r-2}D^{1} + p_{r-1}.$$

Note that $\text{Rem}\left(\frac{b(D)}{g(D)}\right) = 0$.

6.6.2 Parity-Check (PC)-Polar Codes

While cyclic redundancy check (CRC) bits are indeed a form of parity check bits, it is worth noting that, in this context, parity check bits are generated as a linear combination of a few information bits and transmitted as frozen bits. Consequently, they are often referred to as *dynamic frozen* bits, since the values of these frozen bits replaced by parity bits are not static. Unfortunately, there is no (near-) optimal low-complexity approach to define *PC functions* that combines the information bits and returns the parity bits (preferably located on frozen bits to preserve the code rate):

$$f_{PC} : u_i, i \in \mathcal{A} \rightarrow u_j, j \in [0, N-1].$$

6.6.2.1 PC Bits in 5G

Here, we discuss the heuristic approach utilized in the 5G standard. Given the dependency among sub-channels, decoding errors tend to propagate across channels. Therefore, an efficient parity check (PC) function should combine relatively independent bit errors. Empirical observations reveal that the most common error patterns occur between bit positions spaced apart by powers of two, such as 1, 2, 4, and 8, due to the recursive nature of the polar kernel. In contrast, errors propagate less frequently between positions spaced five bits apart. Thus, an effective strategy for defining PC functions involves combining bit positions with a spacing of five positions between them. This spacing can be conveniently implemented using a circular shift register (CSR).

The coordinates of the PC bits collected in the set Q_{PC}^N are determined as follows. Given the sequence $\overline{\mathbf{Q}}_I^N$, the number of parity bits n_{PC}, the minimum Hamming weight of the code, $w_{\min}$, and the reliability sequence corresponding to the row weights $w_{\min}$ and $2w_{\min}$ in $\overline{\mathbf{Q}}_I^N$, that is,

$$\overline{\mathbf{Q}}_{w_{\min}}^N = \left\{ i \in \overline{\mathbf{Q}}_I^N \mid \mathrm{w}(\mathbf{g}_i) = w_{\min} \right\},$$

and

$$\overline{\mathbf{Q}}_{2w_{\min}}^N = \left\{ i \in \overline{\mathbf{Q}}_I^N \mid \mathrm{w}(\mathbf{g}_i) = 2w_{\min} \right\}.$$

Let us denote $n_{w_{\min}} = |\overline{\mathbf{Q}}_{w_{\min}}^N|$. Then, if we have $n_{PC} \leq n_{w_{\min}}$, we select all n_{PC} coordinates of PC bits from $\overline{\mathbf{Q}}_{w_{\min}}^N$ in the descending reliability order. Otherwise, in addition to selecting all $\overline{\mathbf{Q}}_{w_{\min}}^N$, we select the rest of the required coordinates, that is, $n_{PC} - n_{w_{\min}}$, from $\overline{\mathbf{Q}}_{2w_{\min}}^N$ in the descending reliability order.

The calculation of the PC bits is performed through a cyclic shift register of length 5, initialized to 0. Each PC bit is calculated as the XOR of the message bits assigned to preceding subchannels, modulo 5, excluding the previously calculated parity check bits. To summarize, a PC bit u_i, with $i \in Q_{PC}^N$, is calculated as

$$u_i = \bigoplus_{j=\lfloor i_{PC}/5 \rfloor}^{q-1} u_{5j+p},$$

where $q = \lfloor i/5 \rfloor, p = i \bmod 5$ and $i_{PC} \in Q_{PC}^N$ is the highest index smaller than i for which $i_{PC} \bmod 5 = p$. If no such index exists, $i_{PC} = 0$.

6.6.3 PAC Codes

A different approach to pre-transformation in polar coding is to perform forward convolution similar to convolutional encoding in Chapter 2 but with rate-1. The resulting concatenated codes are called *polarization-adjusted convolutional (PAC)* codes. As mentioned in Chapter 4, although the motivation behind polar codes was to build a synthetic channel to boost the cutoff rate inspired by classical concatenated schemes, concatenation was not employed at that stage due to the performance of the polar codes and the availability of the low-complexity decoding procedure. Nevertheless, it was later demonstrated that a rate-1 convolutional pre-transformation can improve the block error rate (BLER) for short codes under sequential decoding.

In PAC coding, the information bits $\boldsymbol{d} = (d_0, d_1, ..., d_{K-1})$ are first mapped to a vector $\boldsymbol{v} = (v_0, v_1, ..., v_{N-1})$ using a rate-profile defined by set $\mathcal{A}$. This set includes the indices of the positions where the information bits are placed in the input vector to the convolutional transform (CT). The rule for forming this set does not have to be the same as that for polar codes. The bit values in the remaining positions in $\boldsymbol{v}$ are set to 0. The constraint $v_{\mathcal{A}^c} = 0$ simply leads to an irregular decoding tree. Note that (1) the outputs of CT for the indices in $\mathcal{A}^c$, which enter the polar transform, are no longer fixed, that is known a priori—unlike in polar coding. (2) In contrast to convolutional coding, in which usually $R_c < 1$, here we use a one-to-one transform $\boldsymbol{P}$, hence the vectors $\boldsymbol{v}$ and $\boldsymbol{u}$ in $\boldsymbol{u} = \boldsymbol{vP}$ have the same dimension.

The relation between the input bits v_{i-m}^{i} and the output bit u_i, at time-step i, is obtained as a binary convolution by

$$u_i = \sum_{j=0}^{m} p_j v_{i-j}, \tag{6.34}$$

where $p_i \in \{0, 1\}$ and m is the number of previous input bits stored in a shift register (m is also known as *memory size*). By representing bit sequences as polynomials in the delay variable D representing a time step in the encoder, an output sequence $x(D)$ is obtained as $p(D)v(D)$, where $p(D) = \sum_{j=0}^{m} p_j D^j$ is the *generator polynomial*. The coefficients of the generator polynomial in the context of convolutional codes are represented in octal notation. For example, the commonly used $\mathbf{p} = [1\ 0\ 1\ 1\ 0\ 1\ 1]$ is represented by 133.

The convolution operation can be represented in the form of an upper triangular matrix where the rows of the *pre-transformation matrix*, denoted as $\mathbf{P}$ with dimensions $N \times N$, are formed by shifting right the vector $\mathbf{p} = (p_0, p_1, \dots p_s)$ by one position in each row. Note that $p_0 = p_s = 1$ by convention. Then, we can obtain $\mathbf{u}$ by matrix multiplication as $\mathbf{u} = \mathbf{vP}$, where

$$\mathbf{P} = \begin{bmatrix} p_0 & p_1 \cdots\cdots p_s & 0 & \cdots & \cdots & 0 \\ 0 & p_0 & p_1 \cdots\cdots p_s & \ddots & & \vdots \\ \vdots & \ddots & \ddots & \ddots & \ddots & 0 \\ \vdots & & \ddots & p_0 & p_1 \cdots\cdots & p_s \\ \vdots & & & \ddots & \ddots & \vdots \\ \vdots & & & \ddots & p_0 & p_1 \\ 0 & \cdots & \cdots & \cdots & 0 & p_0 \end{bmatrix} \tag{6.35}$$

Due to this precoding, we would have $u_i \in \{0, 1\}$ for $i \in \mathcal{I}^c$, indicating that u_i corresponding to a frozen $v_i = 0, i \in \mathcal{A}^c$ may no longer be fixed. Consequently, the vector $\mathbf{u}$ is mapped to codeword $\mathbf{x} = \mathbf{u}\mathbf{G}_N$. Overall, we obtain $\mathbf{x} = \mathbf{v}\mathbf{P}\mathbf{G}_N$.

After the convolutional transform, the vector $\mathbf{u}$ is mapped to $\mathbf{x}$ by employing the polar transform $\mathbf{G}_N$; therefore, $\boldsymbol{x} = \mathbf{u}\mathbf{G}_N$. Algorithm 6.2 demonstrates the encoding process in PAC coding.

6.6.3.1 Minimum Distance of PAC Codes

In convolutional pre-transformation of $\mathbf{v}$ to $\mathbf{u}$, we have $i = \min\{\text{supp}(\mathbf{u})\}$, the coordinate of the first nonzero element, after precoding by $\mathbf{p}$ remains the same. That is, given any input vector $\boldsymbol{v}$, after precoding by $\mathbf{u} = \boldsymbol{v}\mathbf{P}$, we will have

$$\min\{\text{supp}(\mathbf{u})\} = \min\{\text{supp}(\boldsymbol{v})\}.$$

According to (6.34), it is easy to see that $u_k = 0$ for every $k < i$ and $u_i = v_i$. On the other hand, according to (5.16) in Chapter 5 which shows the lower bound for the weight of codewords in every coset C_i where i is the coordinate of the first nonzero element in $\mathbf{u}$, we can conclude that $d_{PAC} \geq \min_{i\in\mathcal{A}} \text{w}(\mathbf{g}_i)$. Hence, as $d_P = \min_{i\in\mathcal{A}}$, we have $d_{PAC} \geq d_P$.

Note that as a result of precoding, we may not have a row $\mathbf{g}_i$ alone as a codeword, and depending on $\mathbf{p}$ and i, there is a possibility of combining/summing it with one or more other rows with indices larger than i resulting codewords with weight equal to or larger than $w(\mathbf{g}_i)$, according to (5.16).

In Section 6.5, we observed how the number of codewords of minimum weight as a result of precoding is reduced relative to the polar codes. Interestingly, this is not the case for all cosets $C_i, i \in \mathcal{B}$ of a code. There are sets for which precoding, regardless of the choice of polynomial $\mathbf{p}$, has no impact on them. That is, $A_{i,w_{\min}}(\mathbf{G}, \mathcal{A}) = A_{i,w_{\min}}(\mathbf{PG}, \mathcal{A})$. This phenomenon occurs due to the distribution of

Algorithm 6.2: PAC Encoding

```
input  : profiled information bits v, g
output: the codeword x
u ← convTrans(v, g)
x ← uG_N                                        // Like polar encoder
return x;
subroutine convTrans (v, g):
    cState[1,...,|g| − 1] ← [0,...,0]           // currState
    for i ← 0 to |v| − 1 do
        (u_i, cState) ← conv1bTrans(v_i, cState, g)
    return u;
subroutine conv1bTrans (v, currState, g):
    u ← v · g_0
    for j ← 1 to |g| do
        if g_j = 1 then
            u ← u ⊕ currState[j − 1];

    nextState ← [v_i] + currState[1,...,|g| − 2]
    return (u, nextState);
```

frozen rows I^c, in particular, in the cosets C_i with large i. By characterizing these cosets in the following lemma, the enumeration of PAC codes becomes easier.

6.6.3.2 Invariant Coset

Here we define cosets where the pre-transformation leaves the number of minimum weight codewords invariant, that is unchanged relative to before pre-transformation. For any coset C_i where

1) $\mathcal{F} = \mathcal{A}^c \cap (i, N-1] = \emptyset$, or
2) $\mathcal{F} = \mathcal{A}^c \cap (i, N-1] = \mathcal{F}^{\mathcal{J}}$,

or in other words, any cosets C_i where there is no frozen row $\mathbf{g}_f$ for $f \in \mathcal{A}^c \cap (i, N-1]$ such that $|\text{supp}(f)\backslash\text{supp}(i)| > 1$, we get $A_{i,w_{\min}} = 2^{|\mathcal{K}_i|}$ in PAC coding, independently of polynomial $\mathbf{p}$. That is, the number of minimum weight codewords is invariant or remains unchanged: $A_{i,w_{\min}}(\mathbf{G}, \mathcal{A}) = A_{i,w_{\min}}(\mathbf{PG}, \mathcal{A})$.

Hence, we divide cosets $C_i, i \in \mathcal{B}$ in terms of the capability of providing a smaller error coefficient into two groups as follows:

$$C_i \rightarrow \begin{cases} \text{variant} & \text{if } \exists f \in \mathcal{A}^c \cap (i, N-1] : |\text{supp}(f)\backslash\text{supp}(i)| > 1, \\ \text{invariant} & \text{otherwise.} \end{cases} \tag{6.36}$$

Table 6.4 The error coefficient of cosets $C_i, i \in \mathcal{B}$ of polar/PAC code (64,16,16).

i	$A_{i,w_{\min}}(G, \mathcal{A})$	$A_{i,w_{\min}}(PG, \mathcal{A})$	Coset type
30	64	0	Variant
45	64	0	Variant
46	32	0	Variant
51	64	64	Invariant
53	32	32	Invariant
54	16	16	Invariant
57	16	16	Invariant
58	8	8	Invariant
60	4	4	Invariant

Hence, we collect the coset leaders of invariant cosets in set $\mathcal{B}^*$

$$\mathcal{B}^* = \{i \in \mathcal{B} : C_i \text{ is invariant}\}. \tag{6.37}$$

Example 6.7 Let us take the code (64,16,16). Table 6.4 lists the error coefficient of every coset C_i for $i \in \mathcal{B} = \{30, 45, 46, 51, 53, 54, 57, 58, 60\}$ by precoding based on polynomial $\mathbf{p} = [1\ 0\ 1\ 1]$. As can be seen, cosets C_i for $i \in \{30, 45, 46\}$ are variant cosets and could remove all the min-weight codewords by precoding, though in general, this is not always the case, but the precoding can reduce it.

The variant cosets include some frozen row(s) $\mathbf{g}_j$, where $|\text{supp}(j)\backslash\text{supp}(i)| > 1$. For instance, coset C_{46} includes row $\mathbf{g}_{49}$, coset C_{45} includes row $\mathbf{g}_{50}$, and coset C_{30} includes rows $\mathbf{g}_{33}$ and $\mathbf{g}_{49}$, all satisfying the condition $|\text{supp}(j)\backslash\text{supp}(i)| > 1$.

Among the invariant cosets, cosets C_{57}, C_{58}, and C_{60} follow the first condition, that is $\mathcal{F} = \mathcal{A}^c \cap (i, N-1] = \emptyset$ meaning they do not cover any frozen rows. The reset of the invariant cosets, namely C_{51}, C_{53}, and C_{54}, includes one or both frozen rows of $\mathbf{g}_{52}$ and $\mathbf{g}_{56}$ which follow the second condition, that is $\mathcal{F} = \mathcal{A}^c \cap (i, N-1] = \mathcal{F}^{\mathcal{J}}$ meaning they only cover frozen rows $\mathbf{g}_f$ where $|\text{supp}(f)\backslash\text{supp}(i)| = 1$.

Due to limitations that exist for the error coefficient of PAC codes as a result of invariant cosets, we define a lower bound for their error coefficient $A_{w_{\min}}(\mathcal{A})$.

6.6.3.3 Lower Bound for $A_{w_{\min}}(PG, \mathcal{A})$

Given a polar code (N, K, d_P) with the information set $\mathcal{A}$, the convolutional precoding by generator polynomial $\mathbf{p}$ results in a PAC code with the number of minimum weight codewords

$$A_{w_{\min}}(\mathbf{PG}, \mathcal{A}) \geq \sum_{i \in \mathcal{B}^*} 2^{|\mathcal{K}_i|}. \tag{6.38}$$

Exercises

1 Given $\beta = 2^{1/4}$ and block length $N = 32$. Find the polarization weight of the sub-channel $i = 8$.

2 Form the reliability chain of sub-channels with decimal indices 2,3,5,9 for $n = 4$. Among them, which sub-channels are comparable with sub-channel 11? Show their reliability relations.

3 Given two codes with $A_{d_{\min}} = 94488$ and 3120. Assuming the number of codewords for other weights is not much different for these two codes, which one would perform better under an ML decoding algorithm?

4 Suppose we append 9-bit CRC for detection purposes to the messages of length $k = 15$, which are encoded into blocks of length $N = 64$. Calculate the effective code rate and compare it with k/N. How much should the minimum power gain due to use of these CRC bits (in a list decoding for instance) at a practical range of BLER be to justify the utilization of CRC?

Bibliographic Notes

The first method for the construction of polar codes was suggested by Arıkan [2009] in his seminal work for BECs. Later, Mori and Tanaka [2009] employed density evolution by Richardson and Urbanke [2008] to construct polar codes over the AWGN channels. The reduced complexity of this approach based on the Gaussian approximation for the evolution by Chung et al. [2001], which was originally used for LDPC codes, was used by Korada et al. [2010] and Trifonov [2012]. The main equations for the convolution operations at the check and variable nodes in the factor graph are provided in Chung et al. [2001, Equations (6), (4)]. The updating rules for LLRs at the check and variable nodes have resulted from the contributions of Hartmann and Rudolph [1976], Battail et al. [1979], and Hagenauer et al. [1996].

The reliability order of the sub-channels is channel-dependent. It was shown by Bardet et al. [2016] and Schürch [2016] that the sub-channels can be partially ordered regardless of the channel condition. They were further expanded and generalized by Wu and Siegel [2019]. In another attempt, He et al. [2017] explored a simple approach for a low-complexity ordering method based on beta expansion, which was used in the 5G standard, as demonstrated in the technical report by Huawei [2016].

By the introduction of PAC codes as a class of concatenated codes by Arıkan [2019] and studies showing the impact of pre-transformation on the error coefficient, the number of minimum weight codewords by Rowshan and Viterbo [2021] and Rowshan and Yuan [2023], it was further observed that using two rules of weight and reliability borrowed from polar codes and RM codes, respectively, for short codes where the damage to the over reliability of the super-channel is limited can outperform polar codes constructed solely based on the reliability. Rowshan et al. [2023] also showed that the error coefficient of the polar codes can be significantly improved by a simple modification.

There are several concatenations of polar codes with other codes. The most effective and popular concatenated codes are CRC-polar codes by Niu and Chen [2012] and Tal and Vardy [2015] codes, parity-check (PC-) polar codes using dynamic frozen bits by Trifonov and Miloslavskaya [2013], Zhang et al. [2018], and PAC codes by Arıkan [2019]. They are all considered pre-transformed polar codes, as the input to the polar transform undergoes a pre-transformation. The design of pre-transformed polar codes has been studied in several works by Zhang et al. [2017] and Trifonov [2020].

Given the competitive performance of PAC codes in short block lengths, a significant number of works have focused on this. This direction is still under development, they reveal interesting ideas. We will review them in the following. Rowshan and Viterbo [2021] and Rowshan and Yuan [2023] showed that the precoding stage can reduce the number of minimum weight codewords as a result of the inclusion of frozen rows in $\boldsymbol{G}_N$. Hence, precoding can potentially improve the block error rate of polar codes. Furthermore, reverse PAC coding by Gu et al. [2022, 2023] and sparsely pre-transformed polar codes by Choi and Lee [2024] have recently been proposed that can overcome the limitations of forward precoding to further reduce the number of minimum weight codewords. Among other precoding schemes and variants of PAC codes, we can name tail-biting PAC codes by Feng et al. [2022], row-merged polar codes by Gelincik et al. [2022] and Zunker et al. [2024], spatially coupled PAC codes by Ying et al. [2022], modified polar codes by Cai et al. [2023], parity check PAC codes by YU et al. [2024], and other schemes by Mishra and Kim [2020], Choi and Lee [2023], Liu et al. [2023a], and Wan et al. [2023]. Numerous construction methods for PAC codes, predominantly based on search, have also been suggested by Abdullah and Mow [2023], Mishra et al. [2023], Liu et al. [2023b], and Chiu and Su [2023]. It was proposed by Rowshan et al. [2023] to use the reliability-based rate-profile of polar codes as it is easy to construct, and instead to modify it to drastically reduce the number of minimum weight codewords. Recently, inspired by the concatenation of CRC bits in

CRC-polar coding, shifted-profile PAC codes were proposed by Gu et al. [2025] and demonstrated that they can outperform CRC-polar codes in short and medium block-lengths.

An approach by Rowshan et al. [2022, 2023] was suggested that simply improves the error coefficient of an available reliability-based construction, which was discussed above. Also, as shown by Rowshan and Viterbo [2019], we can modify the reliability-based construction by bit-swapping to reduce the possibility of eliminating the correct path in the list decoding. This shows that one can tailor the construction of polar code for a specific decoding algorithm. This insight was also used by Rowshan and Viterbo [2018] to partition the code block and to use different list sizes for each partition (depending on the possibility of correct path elimination) in order to reduce computational complexity. Among the various approaches to designing polar codes, which are predominantly based on search or machine learning, Yao et al. [2024] proposed a trellis-based search method for detecting polar codes. However, a limitation of search-based approaches is that they cannot be effectively applied in adaptive coding and modulation schemes, as they are optimized for a specific code.

To enhance the efficiency of transport blocks (TB) in communication standards, Wu et al. [2018a,2018b,2021] introduced a technique involving the coupling of consecutive systematic code blocks (CBs) within a TB by sharing a portion of the information bits. This approach is referred to as *partially information coupled* (PIC) polar codes. In a slightly different approach, Wang et al. [2018] proposed transmitting the coupled portions of the information bits either as frozen bits or appending them to the information bits in the subsequent CB.

Different from the above, there are innovative coding schemes combining polar codes with repetition techniques to enhance performance in challenging communication scenarios. Abbasi et al. [2021b] introduced "polar-coded repetition," a method that concatenates outer polar codes with inner repetition codes to improve efficiency in low-capacity channels while maintaining manageable complexity. Building on this, Abbasi et al. [2021a] adapted this concept into "hybrid non-binary repeated polar codes" tailored for low SNR regimes, using a two-stage hybrid polar code with binary and nonbinary kernels to boost reliability. This idea was further refined by Abbasi et al. [2022a], which details the design and performance of hybrid non-binary repeated polar codes, emphasizing their applicability in wireless systems under varying conditions. Finally, Abbasi et al. [2022b] provided a comprehensive analysis of polar-coded repetition, optimizing it for general low-capacity settings and demonstrating its superiority over traditional repetition schemes.

Bibliography

Fariba Abbasi, Hessam Mahdavifar, and Emanuele Viterbo. Hybrid non-binary repeated polar codes for low-SNR regime. In *2021 IEEE International Symposium on Information Theory (ISIT)*, pages 1742–1747. IEEE, 2021a.

Fariba Abbasi, Hessam Mahdavifar, and Emanuele Viterbo. Polar coded repetition for low-capacity channels. In *2020 IEEE Information Theory Workshop (ITW)*, pages 1–5. IEEE, 2021b.

Fariba Abbasi, Hessam Mahdavifar, and Emanuele Viterbo. Hybrid non-binary repeated polar codes. *IEEE Transactions on Wireless Communications*, 21(9):7582–7594, 2022a.

Fariba Abbasi, Hessam Mahdavifar, and Emanuele Viterbo. Polar coded repetition. *IEEE Transactions on Communications*, 70(10):6399–6409, 2022b.

Murad Abdullah and Wai Ho Mow. New search for the polarization-adjusted convolutional codes with respect to the AFER-optimality criterion. In *2023 IEEE International Symposium on Information Theory (ISIT)*, pages 1723–1728. IEEE, 2023.

Erdal Arıkan. Channel polarization: A method for constructing capacity-achieving codes for symmetric binary-input memoryless channels. *IEEE Transactions on Information Theory*, 55(7):3051–3073, 2009.

Erdal Arıkan. From sequential decoding to channel polarization and back again. *arXiv preprint arXiv:1908.09594*, 2019.

Magali Bardet, Vlad Dragoi, Ayoub Otmani, and Jean-Pierre Tillich. Algebraic properties of polar codes from a new polynomial formalism. In *2016 IEEE International Symposium on Information Theory (ISIT)*, pages 230–234. IEEE, 2016.

Gerard Battail, Martine Decouvelaere, and Philippe Godlewski. Replication decoding. *IEEE Transactions on Information Theory*, 25(3):332–345, 1979.

Zuoxin Cai, Li Chen, Wenxin Liu, and Huazi Zhang. Modified PAC codes. In *2023 IEEE International Symposium on Information Theory (ISIT)*, pages 1908–1913. IEEE, 2023.

Mao-Ching Chiu and Yi-Sheng Su. Design of polar codes and PAC codes for SCL decoding. *IEEE Transactions on Communications*, 71(5):2587–2601, 2023.

Geon Choi and Namyoon Lee. Deep polar codes. *arXiv preprint arXiv:2308.03004*, 2023.

Geon Choi and Namyoon Lee. Sparsely pre-transformed polar codes for low-complexity SCL decoding. In *2024 IEEE International Symposium on Information Theory (ISIT)*, pages 25–30. IEEE, 2024.

Sae-Young Chung, Thomas J Richardson, and Rüdiger L Urbanke. Analysis of sum-product decoding of low-density parity-check codes using a Gaussian approximation. *IEEE Transactions on Information Theory*, 47(2):657–670, 2001.

Bowen Feng, Yi Yang, Jian Jiao, and Qinyu Zhang. On tail-biting polarization-adjusted convolutional (TB-PAC) codes and small-sizes list decoding. *IEEE Communications Letters*, 27(2):433–437, 2022.

Samet Gelincik, Philippe Mary, Anne Savard, and Jean-Yves Baudais. A pre-transformation method to increase the minimum distance of polar-like codes. *arXiv preprint arXiv:2202.04366*, 2022.

Xinyi Gu, Mohammad Rowshan, and Jinhong Yuan. Selective reverse PAC coding for sphere decoding. *arXiv preprint arXiv:2212.00254*, 2022.

Xinyi Gu, Mohammad Rowshan, and Jinhong Yuan. Improved convolutional precoder for PAC codes. In *GLOBECOM 2023-2023 IEEE Global Communications Conference*, pages 1836–1841. IEEE, 2023.

Xinyi Gu, Mohammad Rowshan, and Jinhong Yuan. PAC codes meet CRC-polar codes. *arXiv preprint arXiv:2501.18080*, 2025.

Joachim Hagenauer, Elke Offer, and Lutz Papke. Iterative decoding of binary block and convolutional codes. *IEEE Transactions on Information Theory*, 42(2):429–445, 1996.

Carlos R P Hartmann and Luther D Rudolph. An optimum symbol-by-symbol decoding rule for linear codes. *IEEE Transactions on Information Theory*, 22(5):514–517, 1976.

Gaoning He, Jean-Claude Belfiore, Ingmar Land, Ganghua Yang, Xiaocheng Liu, Ying Chen, Rong Li, Jun Wang, Yiqun Ge, Ran Zhang, and Wen Tong. Beta-expansion: A theoretical framework for fast and recursive construction of polar codes. In *GLOBECOM 2017-2017 IEEE Global Communications Conference*, pages 1–6. IEEE, 2017.

HiSilicon Huawei. Polar code design and rate matching. In *3GPP TSG RAN WG1 Meeting*, volume 86, pages R1–167209, 2016.

Satish Babu Korada, Andrea Montanari, Emre Telatar, and Rüdiger Urbanke. An empirical scaling law for polar codes. In *2010 IEEE International Symposium on Information Theory*, pages 884–888. IEEE, 2010.

Aolin Liu, Bowen Feng, Chulong Liang, Jin Xu, and Qinyu Zhang. A novel hamming check concatenated polarization-adjusted convolutional (PAC) codes. In *2023 IEEE/CIC International Conference on Communications in China (ICCC)*, pages 1–5. IEEE, 2023a.

Wenxin Liu, Li Chen, and Xingcheng Liu. A weighted sum based construction of PAC codes. *IEEE Communications Letters*, 27(1):28–31, 2023b.

Samir Kumar Mishra and KwangChul Kim. Selectively precoded polar codes. *arXiv preprint arXiv:2011.04930*, 2020.

Samir Kumar Mishra, Digvijay Katyal, and Sarvesha Anegundi Ganapathi. A heuristic algorithm for rate-profiling of polarization adjusted convolutional (PAC) codes. *Authorea Preprints*, 2023.

Ryuhei Mori and Toshiyuki Tanaka. Performance of polar codes with the construction using density evolution. *IEEE Communications Letters*, 13(7):519–521, 2009.

Kai Niu and Kai Chen. CRC-aided decoding of polar codes. *IEEE Communications Letters*, 16(10):1668–1671, 2012.

Tom Richardson and Ruediger Urbanke. *Modern coding theory*. Cambridge University Press, 2008.

Mohammad Rowshan and Emanuele Viterbo. Stepped list decoding for polar codes. In *2018 IEEE 10th International Symposium on Turbo Codes & Iterative Information Processing (ISTC)*, pages 1–5. IEEE, 2018.

Mohammad Rowshan and Emanuele Viterbo. How to modify polar codes for list decoding. In *2019 IEEE International Symposium on Information Theory (ISIT)*, pages 1772–1776. IEEE, 2019.

Mohammad Rowshan and Emanuele Viterbo. On convolutional precoding in PAC codes. In *2021 IEEE Globecom Workshops (GC Wkshps)*, pages 1–6. IEEE, 2021.

Mohammad Rowshan and Jinhong Yuan. On the minimum weight codewords of PAC codes: The impact of pre-transformation. *IEEE Journal on Selected Areas in Information Theory*, 4:487–498, 2023.

Mohammad Rowshan, Son Hoang Dau, and Emanuele Viterbo. Improving the error coefficient of polar codes. In *2022 IEEE Information Theory Workshop (ITW)*, pages 249–254. IEEE, 2022.

Mohammad Rowshan, Son Hoang Dau, and Emanuele Viterbo. On the formation of min-weight codewords of polar/PAC codes and its applications. *IEEE Transactions on Information Theory*, 7(12):7627–7649, 2023.

Christian Schürch. A partial order for the synthesized channels of a polar code. In *2016 IEEE International Symposium on Information Theory (ISIT)*, pages 220–224. IEEE, 2016.

Ido Tal and Alexander Vardy. List decoding of polar codes. *IEEE Transactions on Information Theory*, 61(5):2213–2226, 2015.

Peter Trifonov. Efficient design and decoding of polar codes. *IEEE Transactions on Communications*, 60(11):3221–3227, 2012.

Peter Trifonov. Randomized polar subcodes with optimized error coefficient. *IEEE Transactions on Communications*, 68(11):6714–6722, 2020.

Peter Trifonov and Vera Miloslavskaya. Polar codes with dynamic frozen symbols and their decoding by directed search. In *2013 IEEE Information Theory Workshop (ITW)*, pages 1–5. IEEE, 2013.

Heping Wan, Joonyoung Cho, and Charlie Jianzhong Zhang. Polar codes with enhanced weight distribution. In *2023 IEEE Globecom Workshops (GC Wkshps)*, pages 1237–1242, 2023.

Kai-Hsin Wang, Wei Hou, Shan Lu, Ping-Yuen Wu, Yeong-Luh Ueng, and Jun Cheng. Improving polar codes by spatial coupling. In *2018 International Symposium on Information Theory and Its Applications (ISITA)*, pages 432–436. IEEE, 2018.

Wei Wu and Paul H Siegel. Generalized partial orders for polar code bit-channels. *IEEE Transactions on Information Theory*, 65(11):7114–7130, 2019.

Xiaowei Wu, Lei Yang, Yixuan Xie, and Jinhong Yuan. Partially information coupled polar codes. *IEEE Access*, 6:63689–63702, 2018a.

Xiaowei Wu, Lei Yang, and Jinhong Yuan. Information coupled polar codes. In *2018 IEEE International Symposium on Information Theory (ISIT)*, pages 861–865. IEEE, 2018b.

Xiaowei Wu, Min Qiu, and Jinhong Yuan. Partially information coupled bit-interleaved polar coded modulation. *IEEE Transactions on Communications*, 69(10):6409–6423, 2021.

Xinyuanmeng Yao, Xiangping Zheng, and Xiao Ma. Trellis-based construction of polar codes for SCL decoding. In *2024 IEEE International Symposium on Information Theory (ISIT)*, pages 2210–2215. IEEE, 2024.

Siyu Ying, Shan Jin, and Zhaoyang Zhang. A spatially coupled PAC coding scheme and its list decoding. In *2022 IEEE/CIC International Conference on Communications in China (ICCC Workshops)*, pages 82–87. IEEE, 2022.

Qingping Yu, You Zhang, Renze Luo, Longye Wang, and L I Xingwang. Parity-check polarization-adjusted convolutional coding. *IEICE Transactions on Fundamentals of Electronics, Communications and Computer Sciences*, 107(2):187–191, 2024.

Qingshuang Zhang, Aijun Liu, Xiaofei Pan, and Kegang Pan. CRC code design for list decoding of polar codes. *IEEE Communications Letters*, 21(6):1229–1232, 2017.

Huazi Zhang, Rong Li, Jian Wang, Shengchen Dai, Gongzheng Zhang, Ying Chen, Hejia Luo, and Jun Wang. Parity-check polar coding for 5G and beyond. In *2018 IEEE International Conference on Communications (ICC)*, pages 1–7. IEEE, 2018.

Andreas Zunker, Marvin Geiselhart, Lucas Johannsen, Claus Kestel, Stephan ten Brink, Timo Vogt, and Norbert Wehn. Row-merged polar codes: Analysis, design, and decoder implementation. *IEEE Transactions on Communications*, 73(1):39–53, 2024.

7

Monomial Codes and Permutations

It is better to solve one problem five different ways than to solve five problems one way.

— George Pólya

In this chapter, we introduce an algebraic formalism to represent polar codes, placing them within the broader family of decreasing monomial codes. By exploring the relationship between monomial codes and linear block codes, we uncover how polar codes and Reed–Muller codes are special instances of this larger family. This algebraic perspective provides a unified framework for analyzing these codes, enabling deeper insight into their structure and properties.

We delve into the role of permutation or automorphism groups and their linear action on monomials, which is essential for understanding permutation decoding techniques, particularly for polar codes. Additionally, we investigate the partial order of subchannels, which plays a critical role in the construction of polar codes, and provide a systematic approach for enumerating minimum weight codewords in terms of monomials. This chapter equips readers with the knowledge to implement permutation decoding of polar codes and to appreciate the rich interplay between algebraic representations and practical decoding methods.

As a prerequisite, recall that we used the concept of monomials earlier in Chapter 2 to represent Reed–Muller codes. Reviewing Section 2.5.1 is strongly recommended before proceeding with this chapter.

7.1 Monomial Codes

A monomial is a single-term algebraic expression indicating the product of any subset of variables in $\mathbf{x} = (x_0, \ldots, x_{m-1})$. Assuming $\mathbf{i} = (i_0, \ldots, i_{m-1})$, where

Polar Codes: From Theory to Practice, First Edition. Mohammad Rowshan and Emanuele Viterbo.

$i_j \in \{0,1\}$, an m-variate monomial is denoted as

$$\mathbf{x}^{\mathbf{i}} = \prod_{j=0}^{m-1} x_j^{i_j} = x_0^{i_0} \cdots x_{m-1}^{i_{m-1}}. \tag{7.1}$$

For simplicity, we denote a monomial by $f = \mathbf{x}^{\mathbf{i}}$. Moreover, let us denote the degree of a monomial $\mathbf{x}^{\mathbf{i}}$ as $\deg(\mathbf{x}^{\mathbf{i}}) = \mathrm{w}(\mathbf{i})$ and the set of all monomials by

$$\mathcal{M}_m \stackrel{\text{def}}{=} \left\{ x_0^{i_0} \cdots x_{m-1}^{i_{m-1}} \mid \left(i_0, \ldots, i_{m-1}\right) \in \mathbb{F}_2^m \right\}. \tag{7.2}$$

For any monomial f of degree s, for $1 \leq s \leq m$, denoted as $f = x_{l_1} \ldots x_{l_s}$, where $0 \leq l_1 \leq l_2 \cdots \leq l_s \leq m-1$, the *support of monomial* is denoted as $\mathrm{ind}(f) = \{l_1, \ldots, l_s\}$. Observe that the support of a monomial includes all j, where $f_j = 1$. Degree induces a ranking on any monomial set $\mathcal{I} \subseteq \mathcal{M}_m$, that is, $\mathcal{I} = \bigcup_{j=0}^{m} \mathcal{I}_j$, where $\mathcal{I}_j = \{f \in \mathcal{I} \mid \deg(f) = j\}$.

Example 7.1 For $m = 3$, we have

$$\mathcal{M}_3 = \{1, x_0, x_1, x_2, x_0x_1, x_0x_2, x_1x_2, x_0x_1x_2\}.$$

Consider $\mathcal{I} = \mathcal{M}_3 \backslash \{1\}$, then we have

$$\mathcal{I} = \mathcal{I}_1 \cup \mathcal{I}_2 \cup \mathcal{I}_3,$$

where $\mathcal{I}_1 = \{x_0, x_1, x_2\}, \mathcal{I}_2 = \{x_0x_1, x_0x_2, x_1x_2\}$, and $\mathcal{I}_3 = \{x_0x_1x_2\}$. The degree and support of $f = x_0x_1x_2$ are $\deg(f) = 3$ and $\mathrm{ind}(f) = \{0, 1, 2\}$, respectively.

Now, any subset $\mathcal{I} \subseteq \mathcal{M}_m$ forms a *generating set* for a $(n = 2^m, k = |\mathcal{I}|)$ monomial code C. This monomial code C as a linear code is the vector subspace $C(\mathcal{I}) \subseteq \mathbb{F}_2^m$ generated by the span of the row vectors as a *basis* resulting from the *evaluation* of the monomials in $\mathcal{I}$. That is,

$$C(\mathcal{I}) = \mathrm{span}(\{\mathrm{ev}(f) \mid f \in \mathcal{I}\}), \tag{7.3}$$

where $\mathrm{ev}(f)$ is the evaluation of monomial f for all binary entries of $f = (f_0, \ldots, f_{m-1}) \in \mathbb{F}_2^n$.

7.1.1 Evaluation of a Monomial

Let $f = \mathbf{x}^{\mathbf{i}}$ be a monomial where each variable x_j can take the value $a_j \in \{0,1\}$, and each variable appears with an exponent $i_j \in \{0,1\}$. The evaluation of f, denoted by $\mathrm{ev}(f)$, is defined as the following function over the binary inputs $\mathbf{a} = (a_{m-1}, \ldots, a_1, a_0) \in \{0,1\}^m$:

$$\mathrm{ev}(f)(\mathbf{a}) = \prod_{j=0}^{m-1} a_j^{i_j}, \tag{7.4}$$

where each term $a_j^{i_j}$ evaluates as follows:

- If $i_j = 1$, then $a_j^{i_j} = a_j$.
- If $i_j = 0$, then $a_j^{i_j} = 1$ (i.e. the variable x_j does not affect the evaluation).

In other words, the monomial evaluates to 1 if and only if every variable with an exponent of 1 is assigned the value 1; otherwise, it evaluates to 0. Note that we use a convention for the order of entries $(a_{m-1}, \dots, a_0)$, in which the leftmost entry is the most significant bit (as the index of entries indicates) in the binary representation of $\mathbf{i}$. In general, this evaluation process captures how monomials behave over binary inputs, highlighting which combinations of variables yield a non-zero (or 1) result based on the exponents in the monomial.

The number of non-zero elements in ev(f) is written as $|\text{ev}(f)|$ or simply $|f|_m$ and is called the weight of f.

Example 7.2 For evaluation of $f = x_0x_1x_2$ in Example 7.1, we consider all $\mathbf{a} = (a_2, a_1, a_0) \in \{0, 1\}^3$; that is, we substitute all

$$(a_2, a_1, a_0) \in \{(1,1,1), (1,1,0), (1,0,1), (1,0,0), (0,1,1), (0,1,0), (0,0,1), (0,0,0)\},$$

in f and collect all the results in the same order in a vector as follows:

$$\text{ev}(f) = (1 \ \ 0 \ \ 0 \ \ 0 \ \ 0 \ \ 0 \ \ 0 \ \ 0).$$

Clearly, $x_0x_1x_2$ gives 1 only for $a_0 = 1, a_1 = 1, a_2 = 1$, corresponding to the first element of the vector above. The weight of f is $|\text{ev}(f)| = 1$.

The minimum distance of a monomial code is then

$$d = \min_{\mathbf{c} \in \mathcal{C}(\mathcal{I})} \text{w}(\mathbf{c}) = \min_{f \in \mathcal{I}} \text{w}(\text{ev}(f)) = 2^{m-r^+(\mathcal{C})}, \tag{7.5}$$

where $r^+(\mathcal{C}) = \max_{f \in \mathcal{I}} \deg(f)$.

7.1.2 Reed–Muller Codes

We learned about Reed–Muller codes in Chapter 2. With the formalism defined above, we can redefine the Reed–Muller code $\mathcal{R}(r, m)$ as

$$\mathcal{R}(r, m) \stackrel{\text{def}}{=} \text{span}\left(\left\{\text{ev}(f) \mid f \in \mathcal{M}_m, \text{w}(f) \leqslant r\right\}\right). \tag{7.6}$$

Hence, Reed–Muller code $\mathcal{R}(r, m)$ is a monomial code that has a generating set of

$$\mathcal{I} = \left\{f \in \mathcal{M}_m \mid \text{w}(f) \leqslant r\right\}.$$

7.2 Polar Codes as Monomial Codes

As we learned in Chapter 4, polar codes of length $N = 2^m$ are constructed based on the mth Kronecker power, denoted by $(.)^{\otimes m}$, of binary Walsh–Hadamard matrix $\mathbf{G}_2 = \begin{bmatrix} 1 & 0 \\ 1 & 1 \end{bmatrix}$, that is, $\boldsymbol{G}_N = \mathbf{G}_2^{\otimes m}$, which we called it *polar transform*. Recall that we denoted the polar transform by rows $\mathbf{g}_i, i = [0, N-1]$ as $\boldsymbol{G}_N = [\mathbf{g}_0, \mathbf{g}_1, \dots, \mathbf{g}_{N-1}]^T$, where operator T in $[\cdot]^T$ denotes the transpose of the matrix. The generator matrix of a polar code is formed by selecting a set of rows of $\boldsymbol{G}_N$. We use $\mathcal{A}$ to denote the set of indices of these rows and $C(\mathcal{A})$ to denote the linear code generated by the set of rows of $\boldsymbol{G}_N$ indexed by $\mathcal{A}$. Note that $\mathcal{A} \subseteq [0, N-1]$. We learned in Chapters 4 and 6 that the characterization of the information set $\mathcal{A}$ of polar codes is based on the channel polarization theorem by Arıkan and the concept of *synthetic channel reliability*. A polar code of length $N = 2^m$ is constructed by selecting a set $\mathcal{A}$ of indices $i \in [0, N-1]$ with high reliability. The indices in $\mathcal{A}$ are dedicated to information bits, while the rest of the synthetic channels with the indices in $\mathcal{A}^c \triangleq [0, N-1]\backslash\mathcal{A}$ are used to transmit a known value, "0" by default, which are called *frozen bits*. Regardless of the method used to form the set $\mathcal{A}$ for a polar code, every synthetic channel represented by the index i, denoted by $W_{2^m}^{(i)}$ where $i \in \mathcal{A}$, must be more reliable than any synthetic channel in $\mathcal{A}^c$.

Converting all decimal indices in the information set $\mathcal{A}$ to their binary representations, the complements of these binary numbers give $\mathbf{i}$ and their corresponding monomials $\mathbf{x}^{\mathbf{i}}$. Then, we will have a generating set $\mathcal{I}$. Hence, polar codes with the information set $\mathcal{A}$ can also be considered as monomial codes $C(\mathcal{I})$, where $\mathcal{I} \subset \mathcal{M}_m$ is the generating set. The rows of matrix $\mathcal{G}_N$ represent all possible evaluations of monomials over $\mathbb{F}_2^m$. Observe that the lth row of $\boldsymbol{G}_N$ corresponding to monomial f, where

$$l = \sum_{j \in [0, n-1] \backslash \mathrm{ind}(f)} 2^j,$$

can be obtained by ev(f). Hence, there exists a monomial $f \in \mathcal{I}$ for every row $\mathbf{g}_i$, where $i \in \mathcal{A}$ such that $\mathrm{ev}(f) = \mathbf{g}_i$

Example 7.3 Consider the polar code (8,4) with the information set $\mathcal{A} = \{3, 5, 6, 7\}$. To find the equivalent generating set $\mathcal{I}$, we find the binary representation of elements in $\mathcal{A}$ and their complements. Then, the corresponding monomials are found as follows:

$j \in \mathcal{A}$	3	5	6	7
bin(j)	011	101	110	111
bin$(j)^c$: $\mathbf{i} = (i_2, i_1, i_0)$	(1,0,0)	(0,1,0)	(0,0,1)	(0,0,0)
$\mathbf{x}^{\mathbf{i}} = f \in \mathcal{I}$	x_2	x_1	x_0	1

Hence, the generating set is $\mathcal{I} = \{1, x_0, x_1, x_2\}$.

Example 7.4 Let $m = 2$. The evaluation ev(f) of all monomials $f \in \mathcal{M}_n$ and their corresponding rows in $\boldsymbol{G}_4$ is as follows:

$\mathbf{g}_j$	bin(j)	bin(j)^c	(a_1, a_0):	(1,1)	(1,0)	(0,1)	(0,0)
$\mathbf{g}_0$	00	11	ev(x_0x_1)	1	0	0	0
$\mathbf{g}_1$	01	10	ev(x_1)	1	1	0	0
$\mathbf{g}_2$	10	01	ev(x_0)	1	0	1	0
$\mathbf{g}_3$	11	00	ev(1)	1	1	1	1

where

- x_0x_1 evaluates to 1 only when both $x_0 = 1$ and $x_1 = 1$, otherwise it evaluates to 0,
- x_1 evaluates to 1 whenever $x_1 = 1$, regardless of the value of x_0,
- x_0 evaluates to 1 whenever $x_0 = 1$, regardless of the value of x_1, and
- 1 evaluates to 1 for all input values.

7.2.1 Codeword as a Polynomial

In (7.3), we defined a monomial code as a row span for the evaluation of monomials in the generating set of the code. Hence, a codeword $\mathbf{c}\left(x_0, x_1, \dots, x_{m-1}\right)$ can be represented as a polynomial that is the sum of a subset of monomials from the generating set $\mathcal{I}$. Formally, let $\mathcal{S} \subseteq \mathcal{I}$ be a subset of monomials. The corresponding codeword $\mathbf{c}\left(x_0, x_1, \dots, x_{m-1}\right)$ is defined as the sum of these monomials:

$$\mathbf{c}\left(x_0, x_1, \dots, x_{m-1}\right) = \sum_{\mathbf{x}^{\mathbf{i}} \in \mathcal{S}} \mathbf{x}^{\mathbf{i}},$$

where $\mathbf{x}^{\mathbf{b}} = x_0^{i_0} x_1^{i_1} \cdots x_{m-1}^{i_{m-1}}$, and each $\mathbf{x}^{\mathbf{i}}$ is a monomial in the subset $\mathcal{S}$.

Example 7.5 Suppose that we have three variables x_0, x_1, x_2, and a generating set of monomials:

$$\mathcal{I} = \left\{x_0x_1, x_1x_2, x_0\right\}.$$

Now, if we take a subset $\mathcal{S} = \left\{x_0x_1, x_0\right\} \subseteq \mathcal{I}$, the corresponding codeword is:

$$\mathbf{c}\left(x_0, x_1, x_2\right) = x_0x_1 + x_0.$$

This codeword corresponds to $\mathbf{c} = \mathbf{g}_4 \oplus \mathbf{g}_6$, the sum of rows 4 and 6 of $\boldsymbol{G}_8$.

Since every monomial is also a codeword, the code $C(\mathcal{I})$ has $\sum_{\ell=0}^{|\mathcal{I}|} \binom{|\mathcal{I}|}{\ell} = 2^{|\mathcal{I}|}$ codewords in total, where $\binom{|\mathcal{I}|}{0}$ corresponds to the all-zero codeword, $\binom{|\mathcal{I}|}{1}$ represents all individual monomials as a codewords, and $|\mathcal{I}| = K$.

7.3 Decreasing Monomial Codes

Polar codes and Reed–Muller codes belong to a large family of codes called Decreasing Monomial Codes. The monomial order was partially characterized, which is called (universal) partial order in the context of sub-channel relation in terms of reliability. The fundamental algebraic relation between Reed–Muller codes and polar codes goes beyond their monomial structure. It is well known that both the polar and Reed–Muller generating sets ($\mathcal{I}$) obey a certain order relation denoted by $\preceq$. We investigated this in Chapter 5 as well. Here, we look at it from the monomial perspective. Note that the direction of the order for monomials is the opposite of what we observed in Chapter 5.

7.3.1 Partial Order

We establish the relation $\preceq$ and $\preceq_w$ between two monomials f and g with the same degree and different degrees in the following. Note that "w" in $\preceq_w$ stands for "weak" (as in weak order) compared to $\preceq$ in the sense that any pair of monomials f, g that satisfy the relation $f \preceq_w g$, also satisfy the relation $f \preceq g$, by definition.

Definition 7.1 Let m be a positive integer and $f, g \in \mathcal{M}_m$. Then, we have

- Divisibility order: $f \preceq_w g$ if and only if $f|g$, which means that every variable in f appears in g.
- Shifted order: $f \preceq_{sh} g$ for $\deg(f) = \deg(g) = s$ if $\forall\ 1 \leq \ell \leq s \quad i_\ell \leq j_\ell$, where $f = x_{i_1} \dots x_{i_s}$, $g = x_{j_1} \dots x_{j_s}$ and $i_1 \leq i_2 \leq \cdots \leq i_s$ and $j_1 \leq j_2 \leq \cdots \leq j_s$ (assuming we order the indices for comparison), and
- General order: $f \preceq g$ if there exists $g^* \in \mathcal{M}_m$ such that $f \preceq_{sh} g^* \preceq_w g$, which allows a monomial g to be transformed into some g^* such that $f \preceq_{sh} g^*$ and $g^* \mid g$.

One might have $g^* = g$ in the definition of $\preceq$. In such cases if $f \preceq g$ and $\deg(f) = \deg(g)$, we have $\preceq = \preceq_{sh}$. The notation $f \preceq_{sh} g$ comes from the fact that one could obtain g from f by positively shifting some of the variables in f. For example $x_2x_3 \preceq_{sh} x_2x_6$ since x_6 is a shift by 3 positions of x_3. Note that there is a chain relation on the variables, i.e. $x_0 \preceq x_1 \preceq \cdots \preceq x_{m-1}$. Also, the $\preceq$ is an order relation that is partial, e.g. x_3x_4 and x_1x_5 are not comparable with respect to $\preceq$.

Example 7.6 Let us consider two monomials $f = x_0x_2$ and $g = x_0x_1x_2$. Here, $f \mid g$, since all variables in f (i.e. x_0 and x_2) also appear in g. Therefore, we write $f \preceq_w g$. If we take $f = x_1x_2$ and $g = x_0x_1x_2$, then $f \preceq_w g$ still holds because x_1x_2 divides $x_0x_1x_2$.

Now, let's take $f = x_0x_2$ (where $i_1 = 0$ and $i_2 = 2$) and $g = x_1x_2$ (where $j_1 = 1$ and $j_2 = 2$). Here, for all ℓ, we check the condition $i_\ell \leq j_\ell$: $i_1 = 0 \leq j_1 = 1$ and $i_2 = 2 = j_2 = 2$. Since the condition is satisfied, we conclude that $f \preceq_{sh} g$. However, if we take

$f = x_1x_2$ and $g = x_0x_2$, then the condition does not hold because $i_1 = 1 \npreceq j_1 = 0$. Hence, $f \npreceq_{sh} g$, but the opposite order $g \preceq_{sh} f$ holds.

Lastly, let's take $f = x_0x_1$ and $g = x_0x_2x_3$. We can set $g^* = x_0x_2$, which is a monomial that divides g (i.e. $g^* \mid g$, or g^*(weak g). Now, we check if $f \preceq_{sh} g^*$: comparing $i_2 = 1$ for f and $j_2 = 2$ for g^*, and since the condition $i_\ell \leq j_\ell$ is satisfied for all ℓ, we have $f \preceq_{sh} g^*$. Therefore, the general ordering $f \preceq g$ holds.

Definition 7.2 A set $\mathcal{I} \subseteq \mathcal{M}_m$ is *decreasing* if and only if ($f \in \mathcal{I}$ and $g \preceq f$) implies $g \in \mathcal{I}$.

If we map every monomial $f \in \mathcal{M}_m$ to the corresponding synthetic channel denoted by W_m^f in the context of polar codes, the relationship between each pair of synthetic channels in terms of channel reliability can be established as follows. Let f and g be two monomials such that $f \preceq_w g$, then we have $W_m^g \preceq_d W_m^f$. Here, "d" in $\preceq_d$ indicates that the channel W_m^g is a degradation of W_m^f. Now, as a polar code is generated by the set of monomials $\mathcal{I} \subset \mathcal{M}_m$, if $g \in \mathcal{I}$, then it implies that f also belongs to $\mathcal{I}$. We can use this fact to represent a generating set of a decreasing monomial code using only a few monomials rather than a complete set of monomials. We show this in the following example.

Example 7.7 For $m = 4$, we can find a chain relation between the monomials $f \in \mathcal{M}_m$. Figure 7.1 demonstrates this relation. However, as can be seen, the relation or ordering cannot be defined between all pairs. These pairs are shown by rectangles. Hence, this order of monomials is called the partial order.

Example 7.8 For $m = 7$, find all 19 monomials in the generating set $\mathcal{I}$ of a decreasing monomial code represented by $\mathcal{I} = \{f : f \preceq x_1x_6\}$.

We can apply the shifted and divisibility orders in Definition 7.1 to find all $f \preceq x_1x_6$, as shown in Figure 7.2.

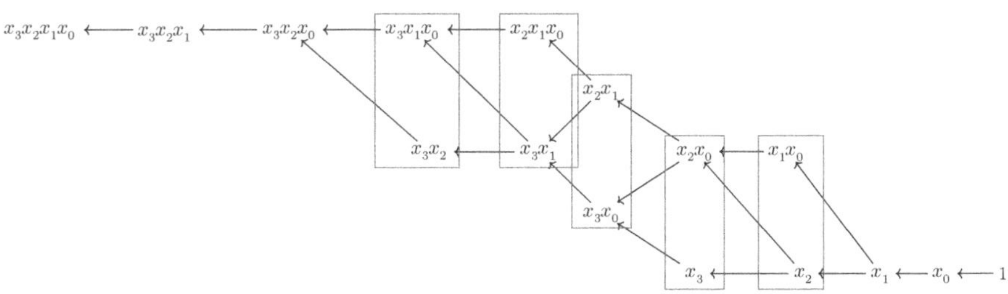

Figure 7.1 Partial order for all monomials $f \in \mathcal{M}_4$. The relation between the monomials within the rectangles is undefined.

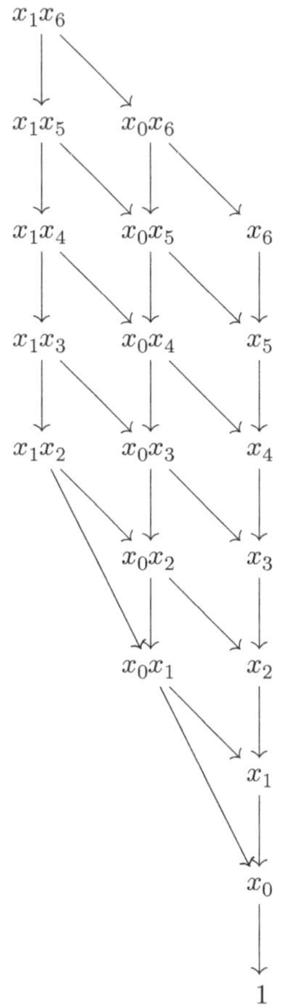

Figure 7.2 All monomials $f \preceq x_1x_6$ obtained using the shifted and divisibility orders in Example 7.8.

7.4 Permutation Group

A permutation π is a bijective function from a set of elements to itself. Assuming that the set of elements consists of the positions (indices) of a codeword, a permutation π of length N is a mapping

$$\pi : \{0, 1, \dots, N-1\} \to \{0, 1, \dots, N-1\},$$

where $\pi(i)$ denotes the position where the element i is mapped under π. Here, we adapt the cycle notation for permutations. In cycle notation, a permutation π is written as a product of disjoint cycles, where each cycle, denoted by the list of their elements in the order in which they are permuted enclosed with parentheses, indicates the positions being permuted in a cyclic manner. For example,

$$\pi = (i_1\ i_2\ \dots\ i_m)$$

means that $\pi(i_1) = i_2, \pi(i_2) = i_3, \dots,$ and $\pi(i_m) = i_1$. Any element not mentioned in the cycle is mapped to itself (i.e. $\pi(i) = i$ for those elements). Alternatively, a permutation can be denoted by two-line notation as

$$\pi = \begin{pmatrix} i_1 & i_2 & i_3 & \cdots & i_m \\ \pi(i_1) & \pi(i_2) & \pi(i_3) & \cdots & \pi(i_m) \end{pmatrix}.$$

Permutations can also be represented using square matrices. Each row and column of a permutation matrix contains exactly one entry of 1, while all other entries are 0. The 1's in the matrix indicate the positions where the elements are permuted such that π(column index) = row index.

Example 7.9 For a permutation $\pi = (2\ 3)$ acting on a 4-element set $\{0, 1, 2, 3\}$, we have

- $\pi(2) = 3$,
- $\pi(3) = 2$,
- $\pi(0) = 0$,
- $\pi(1) = 1$

Here, (2 3) is a transposition, meaning it only swaps two elements (positions 2 and 3) and leaves the others unchanged. Now, let $\mathbf{c} = (c_0, c_1, c_2, c_3)$ be a codeword of length 4. The permutation $\pi = (2\ 3)$ applied to $\mathbf{c}$ would result in a new codeword $\pi(\mathbf{c}) = (c_0, c_2, c_1, c_3)$, where the second and third positions are swapped. The corresponding permutation matrix R_π is

$$R_\pi = \begin{bmatrix} 1\ 0\ 0\ 0 \\ 0\ 0\ 1\ 0 \\ 0\ 1\ 0\ 0 \\ 0\ 0\ 0\ 1 \end{bmatrix}.$$

Now, consider the permutation $\pi = (1\ 3)(0\ 2)$ on the set $\{0, 1, 2, 3\}$. This means:

- $\pi(1) = 3$,
- $\pi(3) = 1$,
- $\pi(0) = 2$,
- $\pi(2) = 0$.

In this case, π consists of two disjoint cycles: (1) The first cycle (13) swaps positions 1 and 3. (2) The second cycle (0 2) swaps positions 0 and 2. Applying π to a codeword $\mathbf{c} = (c_0, c_1, c_2, c_3)$, the result is:

$$\pi(\mathrm{c}) = (c_2, c_3, c_0, c_1).$$

7.4.1 Symmetric Permutation Group

The set of all possible permutations of m elements forms a group called the *symmetric group*, denoted by $S(m)$ or S_m. Observe that the permutations in the symmetric group satisfy the closure and associativity conditions under the operation of permutation composition or permutation matrix multiplication and contain an identity element and an inverse element. Note that the term symmetric group may seem misleading at first, because it might imply that the group is related to symmetry in the geometric sense. However, the name is historically rooted in the concept of symmetry in a different context.

Given a permutation $\pi \in S(N)$ and an N-length vector $\mathbf{c} = \left(c_1, \dots, c_N\right)$, we denote the vector obtained from the permutation of $\mathbf{c}$ according to π as

$$\pi(\mathbf{c}) \triangleq \left(c_{\pi(1)}, \dots, c_{\pi(N)}\right).$$

The set of permutations that map codewords of a code C to other codewords (not necessarily different), i.e. the set of permutations that leaves the code invariant, forms the *automorphism group* of the code C, which is denoted by $\mathrm{Aut}(C)$. Therefore, a permutation π is an automorphism of code C if and only if for every N-length codeword $\mathbf{c} \in C$, we have $\pi(\mathbf{c}) \in C$. In other words,

$$\mathrm{Aut}(C) = \{\pi \in S(N) \mid \pi(\mathbf{c}) \in C,\ \forall \mathbf{c} \in C\}.$$

Example 7.10 Consider the code C as follows:

$$C = \{(0,0,0,0), (1,1,0,0), (0,0,1,1), (1,1,1,1)\}.$$

This is a code where swapping the first two positions with the last two positions results in another valid codeword. Now, define the permutation $\pi = (12)(34)$, which swaps the first two positions with the last two.

Applying π to each codeword:

- $\pi(0,0,0,0) = (0,0,0,0) \in C$,
- $\pi(1,1,0,0) = (0,0,1,1) \in C$,
- $\pi(0,0,1,1) = (1,1,0,0) \in C$,
- $\pi(1,1,1,1) = (1,1,1,1) \in C$.

Thus, $\pi = (12)(34)$ is an automorphism of the code C, and the automorphism group $\mathrm{Aut}(C)$ for this code includes the identity permutation and π.

7.4.2 Lower Triangular Affine (LTA) Permutation Group

In contrast to symmetric permutations, affine transformations combine linear transformations with translations. The term “affine” comes from the Latin word “affinis,” which means “related” or “connected.” Here, translation refers to the process of shifting the coordinates by adding a constant vector. Lower triangular affine (LTA) and general affine (GA) transformations are examples of affine transformations. The difference between LTA and GA transformations is that the transformation matrix in LTA, unlike GA, is in the lower triangular form.

The automorphisms of decreasing monomial codes are induced by the lower triangular affine (LTA) group. Let us formally define the affine transform and the

LTA transform. A bijective affine transformation over $\mathbb{F}_2^m$ is represented by a pair $(\mathbf{A}, \mathbf{b})$, where $\mathbf{A} = (a_{i,j})$ is an $m \times m$ invertible matrix lying in the general linear group GL$(m, 2)$ and $\mathbf{b}$ in $\mathbb{F}_2^m$. Clearly, m and 2 refer to the dimension and the binary field $\mathbb{F}_2$ of the transform, respectively. The action of $(\mathbf{A}, \mathbf{b})$ on a monomial $g = \prod_{i \in \text{ind}(g)} x_i$ denoted by $(\mathbf{A}, \mathbf{b}) \cdot g$ replaces each variable x_i of g by a variable y_i as

$$y_i = x_i + \sum_{j=0}^{i-1} a_{i,j} x_j + b_i, \tag{7.7}$$

where $a_{i,j}$ and b_i are in $\mathbb{F}_2$. For decreasing monomial codes, a lower triangular affine transformation denoted by LTA$(m, 2)$ is used, where $\mathbf{A} \in$ GL$(m, 2)$ is a lower triangular binary matrix with $a_{i,i} = 1$ and $a_{i,j} = 0$ whenever $j > i$. Therefore, the lower triangular affine group LTA$(m, 2)$ can be expressed as the following mapping from $\mathbb{F}_2^m$ to itself.

$$\mathbf{x} \to \mathbf{A}\mathbf{x} + \boldsymbol{b}, \tag{7.8}$$

where the matrix multiplication represents linear maps, and vector addition represents translations. Note that the variables in (7.8) represent code bit indices, and, thus, affine transformations represent code bit permutations.

Example 7.11 Let us consider $m = 3$ variables for the affine mapping $\mathbf{x} \to \mathbf{A}\mathbf{x} + \mathbf{b}$ of an input vector $\mathbf{x}$ and define a lower triangular matrix $\mathbf{A}$ and a translation vector $\mathbf{b}$ for introducing the affine shift, as follows:

$$\mathbf{A} = \begin{bmatrix} 1 & 0 & 0 \\ 1 & 1 & 0 \\ 0 & 1 & 1 \end{bmatrix}, \quad \mathbf{b} = \begin{bmatrix} 1 \\ 0 \\ 1 \end{bmatrix}.$$

To apply the Affine Transformation, we first compute

$$\mathbf{A}\mathbf{x} = \begin{bmatrix} 1 & 0 & 0 \\ 1 & 1 & 0 \\ 0 & 1 & 1 \end{bmatrix} \begin{bmatrix} x_0 \\ x_1 \\ x_2 \end{bmatrix} = \begin{bmatrix} x_0 \\ x_0 + x_1 \\ x_1 + x_2 \end{bmatrix},$$

and then we add the translation vector

$$\mathbf{x}' = \mathbf{A}\mathbf{x} + \mathbf{b} = \begin{bmatrix} x_0 \\ x_0 + x_1 \\ x_1 + x_2 \end{bmatrix} + \begin{bmatrix} 1 \\ 0 \\ 1 \end{bmatrix} = \begin{bmatrix} x_0 + 1 \\ x_0 + x_1 \\ x_1 + x_2 + 1 \end{bmatrix}.$$

Thus, the transformation maps $\mathbf{x} = (x_0, x_1, x_2)$ to $\mathbf{x}' = (x_0 + 1, x_0 + x_1, x_1 + x_2 + 1)$.

Example 7.12 Given $f = x_0 x_2$ and the specific LTA transform in Example 7.11, to show how the monomial f is transformed, we treat the monomial f as the binary vector $\mathbf{x} = [x_0, x_1, x_2] = [1, 0, 1]$, where each bit corresponds to the exponent of the variables x_0, x_1, x_2. Then, we substitute $\mathbf{x}$ in $\mathbf{x}' = [x_0+1, x_0+x_1, x_1+x_2+1]$ and perform modulo-2 addition. The transformed vector is $\mathbf{x}' = [0,1,0]$, which corresponds to the monomial x_1. In other words, the LTA transformation has mapped the original monomial $f = x_0x_2$ to x_1, and the corresponding LTA permutation is

$$\pi(x_0x_2) = x_1.$$

7.4.3 Orbit of a Monomial, Orb(*f*)

The set of polynomials resulting from the action of LTA$(m,2)$ on a monomial is collected in a set named an orbit.

Definition 7.3 The orbit of a monomial $f = \mathbf{x}$ under the action of LTA$(m,2)$ is all its possible destinations under the group action and is defined as the set of polynomials

$$\text{LTA}(m,2)\cdot f = \{(\mathbf{A},\mathbf{b})\cdot f | (\mathbf{A},\mathbf{b}) \in \text{LTA}(m,2)\} = \{\mathbf{A}\mathbf{x}+\mathbf{b} | (\mathbf{A},\mathbf{b}) \in \text{LTA}(m,2)\}.$$

Example 7.13 Given $m = 3$ and the monomial $f = x_1$, find the orbit of this monomial, Orb(f).

To find the orbit, we need to consider all possible LTA permutations within the group permutation LTA$(m,2)$. We know that any transformation $\mathbf{A}\cdot\mathbf{x}+\mathbf{b}$ could keep x_1 unchanged or introduce linear combinations involving x_0 and/or x_2.

For example,

- With $\mathbf{A} = \begin{bmatrix} 1 & 0 & 0 \\ 1 & 1 & 0 \\ 0 & 0 & 1 \end{bmatrix}$ and $\mathbf{b} = (0,0,0)$, we get $f' = x_1$.
- With $\mathbf{A} = \begin{bmatrix} 1 & 0 & 0 \\ 1 & 1 & 0 \\ 1 & 1 & 1 \end{bmatrix}$ and $\mathbf{b} = (1,1,1)$, we get $f' = x_0 + x_1 + 1$.
- With $\mathbf{A} = \begin{bmatrix} 1 & 0 & 0 \\ 0 & 1 & 0 \\ 0 & 0 & 1 \end{bmatrix}$ and $\mathbf{b} = (0,1,0)$, we get $f' = x_1 + 1$.

In general, the orbit of $f = x_1$ is

$$\text{Orb}(x_1) = \left\{x_1 + a_{1,0}x_0 + a_{1,2}x_2 + b_1 \mid a_{1,0}, a_{1,2}, b_1 \in \{0,1\}\right\}.$$

Observe that $|\text{Orb}(x_1)| = 2^3$ because we have three variables $a_{1,0}, a_{1,2}, b_1$, where each can take two possible values of 0 and 1. Thus, the orbit of x_1, as demonstrated in Figure 7.3, includes

$$\text{Orb}\left(x_1\right) = \{x_1, x_1+1, x_0+x_1, x_0+x_1+1, x_1+x_2, x_1+x_2 \\ +1, x_0+x_1+x_2, x_0+x_1+x_2+1\}.$$

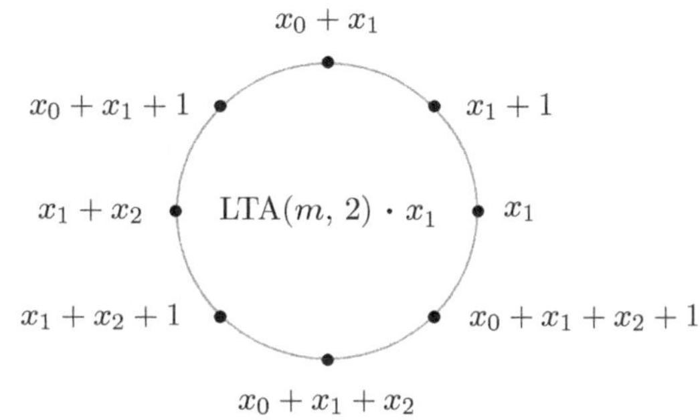

Figure 7.3 Visualization of the orbit of monomial $f = x_1$, Orb(x_1) in Example 7.13.

Example 7.14 Given $m = 2$ and the monomial $f = x_0x_1$, find the orbit of this monomial, Orb(f). In general, the orbit of $f = x_0x_1$ is

$$\text{Orb}(x_0x_1) = \{(x_0 + a_{0,1}x_1 + b_0)(x_1 + a_{1,0}x_0 + b_1) \mid a_{0,1}, a_{1,0}, b_0, b_1 \in \{0, 1\}\}.$$

Observe that $|\text{Orb}(x_0x_1)| = 2^4$ because we have four variables $a_{0,1}, a_{1,0}, b_0, b_1$, where each can take two possible values of 0 and 1. To find all the polynomial elements of the orbit, we need to multiply all possible two terms $(x_0 + a_{0,1}x_1 + b_0)$ and $(x_1 + a_{1,0}x_0 + b_1)$ and simplify them. Recall that $x_j^2 = x_j$ in the binary field. For instance, when $a_{0,1} = a_{1,0} = b_0 = b_1 = 0$, we get the monomial x_0x_1 unchanged. However, when $a_{0,1} = a_{1,0} = b_1 = 0, b_0 = 1$, it results in the polynomial $x_0x_1 + x_1$.

Since LTA$(m, 2)$ acts as a permutation on ev(f), all elements in LTA$(m, 2) \cdot f$ have the same Hamming weight. This fact will be useful when estimating the number of minimum weight codewords of a decreasing monomial code.

7.4.4 Restricted LTA Transform/Permutation Group

As we learned, to find the orbit of a monomial, we need to use the action of LTA$(m, 2)$ on monomials g in the set $\mathcal{I}_r$, where r is the maximum degree of monomials in a decreasing monomial set $\mathcal{I}$. As the following example shows, if we consider the complete LTA permutation group as defined by LTA$(m, 2) \cdot g$ for all $(\mathbf{A}, \mathbf{b})$, we will overcount the number of polynomials $P \in$ LTA$(m, 2) \cdot g$ equivalent to the minimum weight codewords. To avoid this, we define a particular subgroup of LTA$(m, 2)$.

Definition 7.4 For any $g, f \in \mathcal{M}_m$, define LTA$(m, 2)_g \cdot f$ as the subgroup of $(\mathbf{A}, \mathbf{b}) \in$ LTA$(m, 2)$ by the following restrictions:

- $b_i = 0$ if $i \notin$ ind(f): This restricts the vector $\mathbf{b}$, forcing certain entries to be zero, based on the indices of **f**.

- $a_{i,j} = 0$ for all $i \notin \text{ind}(f)$. That is, $a_{i,j} = 0$ for all $i \notin \text{ind}(f)$ and $j \in [0, i-1]$. Hence, rows of $\mathbf{A}$ with indices not in ind(f) must be zero below the diagonal.
- $a_{i,j} = 0$ for all $j \in \text{ind(g)}$. That is, $a_{i,j} \in \{0, 1\}$ for all $i \in \text{ind(g)}$ and $j \in [0, i-1] \backslash \text{ind(g)}$. Hence, the columns of $\mathbf{A}$ with indices in ind(g) must be zero below the diagonal.

Observe that for every $x_i, i \in \text{ind}(f)$, the last restriction in the definition above changes (7.7) to

$$y_i = x_i + \sum_{j=0, j \notin \text{ind(g)}}^{i-1} a_{i,j} x_j + b_i, \tag{7.9}$$

hence, the LTA transform of $f = x_{i_1} x_{i_2} \cdots x_{i_s}$ works as

$$\text{LTA}(m, 2)_g \cdot f = \prod_{k=1}^{s} \left(x_{i_k} + \sum_{j < i_k, j \notin \text{ind(g)}} a_{i_k, j} x_j + b_{i_k} \right). \tag{7.10}$$

Example 7.15 Consider the complete permutation group $(\mathbf{A}, \mathbf{b}) \in \text{LTA}(m, 2)$ with

$$\mathbf{A} = \begin{bmatrix} 1 & 0 & 0 \\ a_{1,0} & 1 & 0 \\ a_{2,0} & a_{2,1} & 1 \end{bmatrix}, \quad \mathbf{b} = \begin{bmatrix} b_0 \\ b_1 \\ b_2 \end{bmatrix}$$

that results in

$$\mathbf{y} = (x_0 + b_0, x_1 + a_{1,0} x_0 + b_1, x_2 + a_{2,1} x_1 + a_{2,0} x_0 + b_2),$$

where $a_{i,j}, b_i \in \{0, 1\}$.

Now, given $g = x_0 x_1$ with the support ind(g) $= \{0, 1\}$, we apply the following restrictions on LTA(m, 2) for $\text{LTA}(m, 2)_g \cdot g$:

- $b_i = 0$ for all $i \notin \text{ind(g)}$, i.e. $b_2 = 0$.
- $a_{i,j} = 0$ for all $i \notin \text{ind(g)}$ and $a_{i,j} = 0$ for all $j \in \text{ind(g)}$, i.e. $a_{2,0} = a_{2,1} = 0$, $a_{1,0} = 0$. Note that all $a_{i,j}$ are the elements under the diagonal; hence, $a_{i,j} \in \{0, 1\}$ for every $i \in \text{ind}\, g$ and $j \in [0, i-1] \backslash \text{ind(g)}$.

Then, the corresponding $\text{LTA}(m, 2)_g \cdot g$ which results in $\mathbf{y} = (x_0 + b_0, x_1 + b_1, 0)$ is

$$\mathbf{A} = \begin{bmatrix} 1 & 0 & 0 \\ 0 & 1 & 0 \\ 0 & 0 & 1 \end{bmatrix}, \quad \mathbf{b} = \begin{bmatrix} b_0 \\ b_1 \\ 0 \end{bmatrix}.$$

This transformation is constrained by the support of the monomial $g = x_0 x_1$, which modifies the matrix and vector accordingly.

Example 7.16 Let $g = x_0 x_1$, then by definition, we have $\text{LTA}(m, 2) \cdot g = \{(x_0 + b_0)(x_1 + a_{1,0}x_0 + b_1) | b_0, a_{1,0}, b_1 \in \mathbb{F}_2\}$. The complete group has four distinct elements; however, since $b_0, a_{1,0}, b_1$ can take two values in $\{0, 1\}$, we expected $2^3 = 8$ elements. Four repeated elements leave g invariant, namely

$$(x_0 + 1)(x_1 + x_0 + 1) = (x_0 + 1)(x_1 + 1) = x_0 x_1 + x_1 + x_0 + 1,$$

$$(x_0 + 1)(x_1 + x_0) = (x_0 + 1)x_1 = x_0 x_1 + x_1,$$

$$x_0(x_1 + x_0) = x_0(x_1 + 1) = x_0 x_1 + x_0,$$

$$x_0(x_1 + x_0 + 1) = x_0 x_1.$$

As can be seen, since $0 \in \text{ind}(g)$, when $a_{1,0} = 1$, the obtained polynomials are identical to another one in the group. Recall that $x_j^2 = x_j$ in $\mathbb{F}_2$. On the other hand, the subgroup $\text{LTA}(m, 2)_g$ gives only 2^2 distinct elements since, by definition, $a_{1,0} = 0$. The elements of the subgroup are $(x_0 + 1)(x_1 + 1), (x_0 + 1)x_1, x_0(x_1 + 1)$, and $x_0 x_1$.

Example 7.17 If we let $g = x_0 x_4$ and $f = x_1 x_2 x_5$, then any polynomial from $\text{LTA}(m, 2)_g \cdot f$ can be written as

$$(x_1 + b_1)(x_2 + a_{2,1}x_1 + b_2)(x_5 + a_{5,3}x_3 + a_{5,2}x_2 + a_{5,1}x_1 + b_5),$$

where the corresponding $(\mathbf{A}, \mathbf{b})$ of the LTA transform are given below:

$$\mathbf{A} = \begin{bmatrix} 1 & 0 & 0 & 0 & 0 & 0 \\ 0 & 1 & 0 & 0 & 0 & 0 \\ 0 & a_{2,1} & 1 & 0 & 0 & 0 \\ 0 & 0 & 0 & 1 & 0 & 0 \\ 0 & 0 & 0 & 0 & 1 & 0 \\ 0 & a_{5,1} & a_{5,2} & a_{5,3} & 0 & 1 \end{bmatrix}, \quad \mathbf{b} = \begin{bmatrix} 0 \\ b_1 \\ b_2 \\ 0 \\ 0 \\ b_5 \end{bmatrix}.$$

7.4.5 Block Lower Triangular Affine (BLTA) Permutation Group

The lower-triangular affine (LTA) group is only a part of the automorphism group, which is solely based on the partial order of the synthetic channels of the code. However, depending on the design, polar codes can have a richer set of automorphisms. Hence, the LTA is extended to a larger subgroup of the general affine group (GA), namely the block lower triangular affine (BLTA) group. In fact, for decreasing monomial codes, all automorphisms that can be expressed as affine transformations are equal to BLTA.

The BLTA transformation is in the form of (7.8), where the matrix $\boldsymbol{A}$, shown in Figure 7.4, is a block lower triangular (BLT) matrix, with the diagonal block sizes of $\mathbf{s} = [s_1, \ldots, s_t]$ such that $s_1 + \cdots + s_t = m$. Diagonal blocks $\boldsymbol{D}_{k,k}$ are invertible

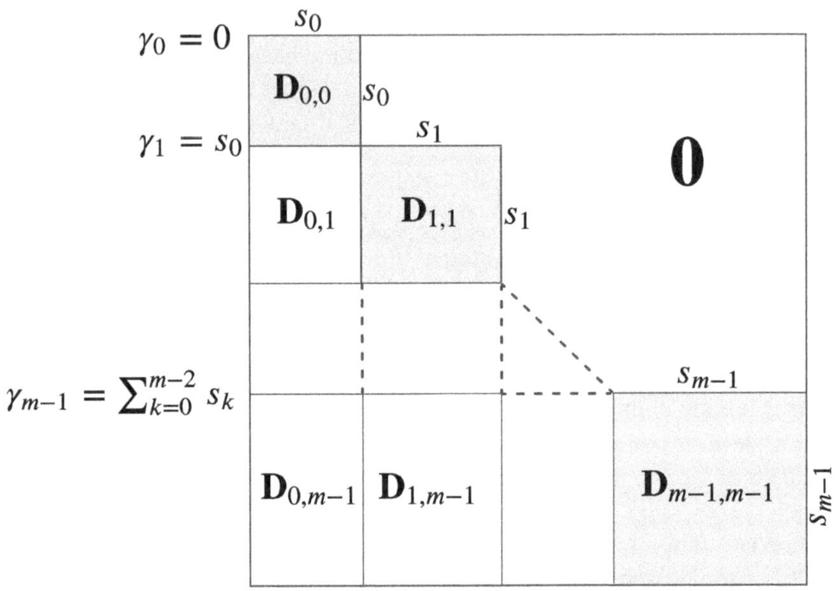

Figure 7.4 The general structure of a BLT matrix $\boldsymbol{A}$ where block $\boldsymbol{D}_{k,l}, 0 \leq k \leq m-1$ has size $s_l \times s_k$ and starting row index $\gamma_k = \sum_{j=0}^{k-1} s_j$.

binary random square matrices of size $s_k \times s_k$, where the diagonal elements are nonzero, while the off-diagonal blocks $\boldsymbol{D}_{k,l}$, where $k < l$ are binary random rectangular matrices.

Let us review the properties of block lower-triangular matrices from linear algebra:

- The product of two block lower-triangular matrices is another block lower-triangular matrix.
- A block lower-triangular matrix is non-singular if and only if all of its diagonal blocks are non-singular. This means that each square matrix $\boldsymbol{D}_{k,k}$ must have a non-zero determinant (to have an inverse). If any of the diagonal blocks $\boldsymbol{D}_{k,k}$ is singular, then the block lower-triangular matrix as a whole is singular and does not have an inverse.
- The inverse of a block lower-triangular matrix $(\mathbf{A}, \mathbf{b})$ is another block lower-triangular matrix, where each block is the inverse of the corresponding block in the original matrix.

Note that the block sizes are preserved in the inverse of a BLT matrix or in the product of two BLT matrices.

For a set of elements (here, non-singular BLT matrices) to form a group, they must satisfy several conditions, including closure, associativity, identity element, and inverses. Considering the aforementioned properties of BLT matrices, observe that:

- Any two BLT matrices produce another BLT matrix. Hence, the closure condition is satisfied.
- The inverse of any (non-singular) BLT matrix is another (non-singular) BLT matrix. Hence, the inverse condition is satisfied.
- The associativity and identity conditions hold for matrix multiplication in general, regardless of the specific form of the matrices. Note that the identity matrix can also be considered a BLT matrix.

Hence, non-singular BLT matrices form a group.

Example 7.18 The following is a non-singular BLT matrix with diagonal block sizes (structure) of $\mathbf{s} = [4, 1, 1, 3]$.

$$A = \left[\begin{array}{cccc|c|c|ccc} 1 & 0 & 0 & 0 & 0 & 0 & 0 & 0 & 0 \\ 1 & 1 & 0 & 0 & 0 & 0 & 0 & 0 & 0 \\ 1 & 0 & 1 & 0 & 0 & 0 & 0 & 0 & 0 \\ 0 & 1 & 1 & 1 & 0 & 0 & 0 & 0 & 0 \\ \hline 0 & 0 & 0 & 0 & 1 & 0 & 0 & 0 & 0 \\ \hline 0 & 0 & 0 & 0 & 0 & 1 & 0 & 0 & 0 \\ \hline 0 & 0 & 0 & 0 & 0 & 0 & 1 & 0 & 0 \\ 0 & 0 & 0 & 0 & 0 & 0 & 1 & 1 & 0 \\ 0 & 0 & 0 & 0 & 0 & 0 & 1 & 1 & 1 \end{array}\right].$$

The matrix consists of four following blocks:

- $\boldsymbol{D}_{0,0}$: A 4×4 block in the top-left.
- $\boldsymbol{D}_{1,1}$: A 1×1 block on the diagonal in the middle-left.
- $\boldsymbol{D}_{2,2}$: A 1×1 block on the diagonal in the middle.
- $\boldsymbol{D}_{3,3}$: A 3×3 block in the bottom-right.

Let's calculate the determinants of the diagonal blocks: Since the determinant of a lower triangular matrix is the product of the diagonal elements, we have $\det\left(\boldsymbol{D}_{0,0}\right) = \det\left(\boldsymbol{D}_{1,1}\right) = \det\left(\boldsymbol{D}_{2,2}\right) = \det\left(\boldsymbol{D}_{3,3}\right) = 1$. Hence, since all diagonal blocks are non-singular (i.e. their determinants are non-zero), the entire matrix A is non-singular.

The block structure corresponds to groups of variables that are "coupled" together, meaning that variables within the same block can influence each other, but variables in one block do not directly influence variables in other blocks (since the off-diagonal blocks are zero). For instance, in this matrix, the first four variables (associated with the 4×4 block) can interact, but they don't interact directly with the variables associated with the 3×3 block.

The structure of this matrix allows us to solve equations involving A by solving smaller systems of equations block by block, starting from the bottom-right corner and moving upward (back-substitution). The inverse of a BLT matrix can also be computed block-wise. Since the matrix is non-singular, the inverse exists and can be constructed by inverting each diagonal block individually.

7.5 Stabilizer Group

A permutation $\pi \in S(m)$ stabilizes a set of monomials $\mathcal{I} \subseteq \mathcal{M}_m$ if and only if

$$\mathbf{x}^{\mathbf{i}} \in \mathcal{I} \Longrightarrow \mathbf{x}^{\pi(\mathbf{i})} \in \mathcal{I}. \tag{7.11}$$

This means that applying the permutation π to the exponents of the monomials in $\mathcal{I}$ must result in another monomial that is still in $\mathcal{I}$. The key observation here is that a permutation induces an automorphism of a monomial code if and only if it stabilizes the generating set of the code. The set of all such stabilizers forms a subgroup in $\mathfrak{S}_m$. We will call this subgroup the stabilizer of $\mathcal{I}$ and denote it $\mathrm{Stab}(\mathcal{I})$. In other words, $\mathrm{Stab}(\mathcal{I})$ is the set of all permutations on x_i that induce permutation automorphisms of $C(\mathcal{I})$:

$$\forall \pi \in \mathfrak{S}_m \quad \pi \in \mathrm{Stab}(\mathcal{I}) \Leftrightarrow \alpha_\pi \in \mathrm{Aut}(C(\mathcal{I})),$$

where α_π is the codeword position permutation induced by π.

Example 7.19 Consider $m = 3$ variables x_0, x_1 and x_2, and a set of monomials $\mathcal{I} = \{x_0x_1, x_1x_2\}$. We check two example permutations π from the symmetric group $S(3)$.

Let us consider the permutation $\pi = (0\ 1)$, which swaps x_0 and x_1, leaving x_2 unchanged. Observe that $\pi(x_0x_1) = x_1x_0 = x_0x_1$, which is still in $\mathcal{I}$. However, we have $\pi(x_1x_2) = x_0x_2$, which is not in $\mathcal{I}$. Hence, this permutation does not stabilize $\mathcal{I}$.

Now, let us check the identity permutation $\pi = \mathrm{id}$, which leaves all variables unchanged. Clearly, this permutation stabilizes $\mathcal{I}$. Therefore, the identity permutation is part of the stabilizer group. We say that the stabilizer group is trivial, i.e. $\mathrm{Stab}(\mathcal{I}) = \{\mathrm{id}\}$.

7.5.1 Stabilizer Group Structure

In general, the stabilizer group $\mathrm{Stab}(\mathcal{I})$ of a set of monomials $\mathcal{I} \subset \mathcal{M}_m$ can be non-trivial, and it will have the structure described:

- The variables $x_0, x_1, \ldots, x_{m-1}$ are partitioned into disjoint intervals of sizes m_i, $\sum m_i = m$. Some intervals may include only one variable, which are called trivial intervals.

- Permutations are allowed within these intervals but not between them.
- For any non-trivial interval (intervals with more than one variable), there exists a permutation transposing elements within that interval, ensuring the stabilization of the monomial set.

The stabilizer group Stab($\mathcal{I}$) is then a direct product of symmetric groups acting on each interval as

$$\text{Stab}(\mathcal{I}) = S_1 \times S_2 \times \cdots, \quad S_i \leq \mathfrak{S}_{m_i},$$

where each S_i is a subgroup of $\mathfrak{S}_{m_i}$.

Example 7.20 Suppose $\mathcal{I} = \{x_0x_1, x_1x_2\}$, and we partition the variables into two intervals: $\{x_0\}$ and $\{x_1, x_2\}$. Then, the stabilizer group would be the product of two groups:

$$\text{Stab}(\mathcal{G}) = S_1 \times S_2.$$

Here, S_1 is the trivial group (because there's only one variable in the first interval, so there's no room for any permutation), and S_2 would contain the permutations of x_1 and x_2, $\pi = (1\ 2)$.

7.6 Permutation Ensemble Decoding

We learned in Section 7.4 that an automorphism π of a code C is a permutation that maps any codeword $\mathbf{c} \in C$ to another codeword $\pi(\mathbf{c}) \in C$. The automorphism group Aut(C) of a code C is the set containing all automorphisms of code C. This is the basis of the decoder we want to introduce in this section. A permutation ensemble decoder consists of M parallel successive cancellation (SC)-based decoders (with fixed throughput, such as the SC decoder and SC list decoder) or a belief propagation (BP)-based decoder, where each decoder operates on a permuted version of the received codeword. The permutations are chosen from the automorphism group of the code. The codewords resulting from the parallel decoders are de-interleaved π^{-1}, and the most likely candidate is selected. The structure of a permutation ensemble decoding is shown in Figure 7.5, where $\mu(\hat{\mathbf{c}}_i)$ gives the modulated sequence since $\mathbf{y}$ is modulated.

The M different permutations π_j for $j \in \{1, 2, \ldots, M\}$ are randomly sampled (or carefully selected) from the automorphism group and used in the interleavers and de-interleavers. The received vector $\mathbf{y}$ is accordingly permuted with the M different permutations π_j leading to M permuted noisy codewords $\mathbf{y}'_j$, where $j \in \{1, 2, \ldots, M\}$. Every $\mathbf{y}'_j$-vector is decoded using an independent decoder, resulting in the estimated codeword estimated codewords $\hat{\mathbf{x}}_j$ after the de-interleaving

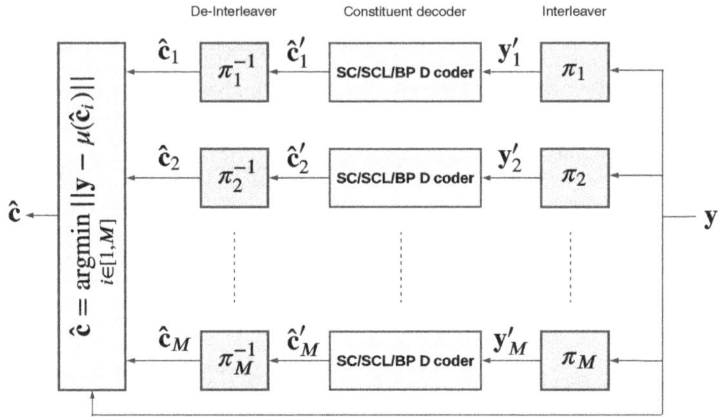

Figure 7.5 Permutation ensemble decoding with M constituent decoders.

stage. Denoting the decoding function by Dec(·) that maps $\mathbf{y}_j'$ to $\hat{\mathbf{x}}_j'$, we can write the jth decoding process including interleaving and de-interleaving as

$$\hat{\mathbf{x}}_j = \pi_j^{-1}\left(\text{Dec}\left(\pi_j(\mathbf{y})\right)\right). \tag{7.12}$$

This decoding process has advantages and disadvantages. Although the structure of the decoder seems similar to the list decoder, it has a lower latency (essentially equivalent to the latency of a single decoder), as the parallel component decoders do not exchange information during the decoding process. The disadvantage comes from the large automorphism group of polar codes, knowing that the choice of permutation determines the error correction performance of the ensemble decoder.

Now, let us look at the motivation behind the permutation ensemble decoding. Suppose we flip the signs of the elements of the received log-likelihood ratio (LLR) vector $\mathbf{L}$ according to some codeword $\mathbf{x} \in C$, that is, $\mathbf{L} \cdot (-1)^{\mathbf{x}}$, the codeword estimated by a linear decoder will shift exactly by $\mathbf{x}$, that is, a linear decoder does not favor any codeword over another. In other words, a linear decoder gives

$$\text{Dec}\left(\mathbf{L} \cdot (-1)^{\mathbf{x}}\right) = \text{Dec}(\mathbf{L}) \oplus \mathbf{x}, \tag{7.13}$$

which means the decoding behavior is only dependent on the noise induced by the channel (here the sign of LLR elements), rather than the choice of the transmitted codeword $\mathbf{x}$. Thus, the permutation in (7.12) corresponds to the permutation of the noise. As a result, one can observe that the permutation ensemble decoding is based on the assumption that suboptimal decoders behave differently to noise realizations in different permutations, hoping that one of the permutations in the ensemble gives the ML estimate of the codeword.

Example 7.21 We revisit Example 4.3. We assume that the message data $\mathbf{d} = [1,\ 1,\ 0,\ 1]$ is sent. First, the message is rate-profiled into $\mathbf{u} = [0,\ 0,\ 0,\ 1,\ 0,\ 1,\ 0,\ 1]$ and encoded as $\mathbf{c} = [1, 1, 0, 0, 0, 0, 1, 1]$. With BPSK modulation, the codeword becomes $\mathbf{x} = [-1, -1,\ 1,\ 1,\ 1,\ 1, -1, -1]$.

Let $\mathbf{x}$ be transmitted over the AWGN channel with $E_b/N_0 = 2$ dB, resulting in the received sequence $\mathbf{y} = [-0.4167, -0.6357,\ 1.6604, -0.5944,\ 0.6982,\ 1.6727, -1.7881, -1.4540]$.

Let us walk through ensemble decoding using four decoders with distinct LTA permutations for code length $N = 8$, corresponding to $m = 3$ variables. The affine group acts on the binary representation of the positions (indices) of codeword symbols. Here, we use two-line notation for permutations.

LTA permutations are defined as $\pi_i(x) = A_i x + \mathbf{b}_i \bmod 2$, where x is the binary index (0 to 7), A_i is a 3×3 invertible matrix, and $\mathbf{b}_i$ is a 3-bit vector. We'll define four distinct permutations below:

Binary Indices: $0 = [0, 0, 0]^T, 1 = [1, 0, 0]^T, 2 = [0, 1, 0]^T, 3 = [1, 1, 0]^T, 4 = [0, 0, 1]^T, 5 = [1, 0, 1]^T, 6 = [0, 1, 1]^T, 7 = [1, 1, 1]^T$.

1) π_1: Identity

$$A_1 = \begin{bmatrix} 1 & 0 & 0 \\ 0 & 1 & 0 \\ 0 & 0 & 1 \end{bmatrix}, \mathbf{b}_1 = \begin{bmatrix} 0 \\ 0 \\ 0 \end{bmatrix}.$$

$$\pi_1(x) = x, \text{ so } \pi_1 = (0, 1, 2, 3, 4, 5, 6, 7), \text{ Two-line: } \begin{pmatrix} 0 & 1 & 2 & 3 & 4 & 5 & 6 & 7 \\ 0 & 1 & 2 & 3 & 4 & 5 & 6 & 7 \end{pmatrix}.$$

$$\mathbf{y}_1 = \mathbf{y} = [-0.4167, -0.6357, 1.6604, -0.5944, 0.6982, 1.6727, -1.7881, -1.4540].$$

2) π_2: Shift and Swap

$$A_2 = \begin{bmatrix} 1 & 0 & 0 \\ 1 & 1 & 0 \\ 0 & 1 & 1 \end{bmatrix}, \mathbf{b}_2 = \begin{bmatrix} 0 \\ 1 \\ 0 \end{bmatrix}.$$

Compute: $0 \to 2, 1 \to 1, 2 \to 4, 3 \to 7, 4 \to 6, 5 \to 5, 6 \to 0, 7 \to 3$.

$$\pi_2 = (2, 1, 4, 7, 6, 5, 0, 3), \text{ Two-line: } \begin{pmatrix} 0 & 1 & 2 & 3 & 4 & 5 & 6 & 7 \\ 2 & 1 & 4 & 7 & 6 & 5 & 0 & 3 \end{pmatrix}.$$

Inverse: $\pi_2^{-1} = (6, 1, 0, 7, 2, 5, 4, 3)$.

$$\mathbf{y}_2 = [-1.7881, -0.6357, -0.4167, -1.4540, 1.6604, 1.6727, 0.6982, -0.5944].$$

3) π_3: Shift and Inversion

$$A_3 = \begin{bmatrix} 1 & 0 & 0 \\ 1 & 1 & 0 \\ 1 & 0 & 1 \end{bmatrix}, \mathbf{b}_3 = \begin{bmatrix} 1 \\ 0 \\ 1 \end{bmatrix}.$$

Compute: $0 \to 5, 1 \to 2, 2 \to 7, 3 \to 0, 4 \to 1, 5 \to 6, 6 \to 3, 7 \to 4$.

$\pi_3 = (5, 2, 7, 0, 1, 6, 3, 4)$, Two-line: $\begin{pmatrix} 0 & 1 & 2 & 3 & 4 & 5 & 6 & 7 \\ 5 & 2 & 7 & 0 & 1 & 6 & 3 & 4 \end{pmatrix}$.

Inverse: $\pi_3^{-1} = (3, 4, 1, 6, 7, 0, 5, 2)$.

$\mathbf{y}_3 = [-0.5944, 0.6982, -0.6357, -1.7881, -1.4540, -0.4167, 1.6727, 1.6604]$.

4) π_4: Full Swap

$$A_4 = \begin{bmatrix} 1 & 0 & 0 \\ 0 & 1 & 0 \\ 1 & 1 & 1 \end{bmatrix}, \mathbf{b}_4 = \begin{bmatrix} 1 \\ 1 \\ 0 \end{bmatrix}.$$

Compute: $0 \to 3, 1 \to 6, 2 \to 5, 3 \to 0, 4 \to 7, 5 \to 2, 6 \to 1, 7 \to 4$.

$\pi_4 = (3, 6, 5, 0, 7, 2, 1, 4)$, Two-line: $\begin{pmatrix} 0 & 1 & 2 & 3 & 4 & 5 & 6 & 7 \\ 3 & 6 & 5 & 0 & 7 & 2 & 1 & 4 \end{pmatrix}$.

Inverse: $\pi_4^{-1} = (3, 6, 5, 0, 7, 2, 1, 4)$ (same, as it is self-inverse).

$\mathrm{y}_4 = [-0.5944, -1.7881, 1.6727, -0.4167, -1.4540, 1.6604, -0.6357, 0.6982]$.

Now, we use four decoders as follows:

Decoder 1:
- $\hat{\mathbf{u}}_1 = [0, 0, 0, 1, 0, 1, 0, 1]$.
- Encode: $\hat{\mathbf{c}}_1 = \hat{\mathbf{u}}_1 F^{\otimes 3} = [1, 1, 0, 0, 0, 0, 1, 1]$ (matches transmitted).
- De-permute: $\pi_1^{-1} = (0, 1, 2, 3, 4, 5, 6, 7)$, $\hat{\mathbf{c}}_{1,\text{dep}} = [1, 1, 0, 0, 0, 0, 1, 1]$.
- BPSK: $\hat{\mathbf{x}}_{1,\text{dep}} = [-1, -1, 1, 1, 1, 1, -1, -1]$.
- Distance: $d_E^2 = (0.5833)^2 + (0.3643)^2 + (0.6604)^2 + (-1.5944)^2 + (-0.3018)^2 + (0.6727)^2\ (-0.7881)^2 + (-0.454)^2 = 7.4518$.

Decoder 2:
- $\hat{\mathbf{u}}_2 = [0, 0, 0, 1, 0, 0, 0, 1]$.
- Encode: $\hat{\mathbf{C}}_2 = [1, 1, 0, 0, 0, 0, 0, 1]$.
- De-permute: $\pi_2^{-1} = (6, 1, 0, 7, 2, 5, 4, 3)$, $\hat{\mathbf{c}}_{2,\text{dep}} = [0, 1, 1, 1, 0, 0, 0, 0]$.
- BPSK: $\hat{\mathbf{x}}_{2,\text{dep}} = [1, -1, -1, -1, 1, 1, 1, 1]$.
- Distance: $d_E^2 = (-1.4167)^2 + (0.3643)^2 + (2.6604)^2 + (0.4056)^2 + (-0.3018)^2 + (0.6727)^2 + (-2.7881)^2 + (-2.454)^2 = 22.6295$.

Decoder 3:
- $\hat{\mathbf{u}}_3 = [0, 0, 0, 1, 0, 1, 0, 0]$.
- Encode: $\hat{\mathbf{c}}_3 = [1, 1, 0, 0, 0, 0, 1, 0]$ (note: earlier error corrected).
- De-permute: $\pi_3^{-1} = (3, 4, 1, 6, 7, 0, 5, 2)$, $\hat{\mathbf{c}}_{3,\text{dep}} = [0, 0, 1, 1, 0, 1, 0, 0]$.
- BPSK: $\hat{\mathbf{x}}_{3,\text{dep}} = [1, 1, -1, -1, 1, -1, 1, 1]$.
- Distance: $d_E^2 = (-1.4167)^2 + (-1.6357)^2 + (2.6604)^2 + (0.4056)^2 + (-0.3018)^2 + (2.6727)^2 + (-0.7881)^2 + (-2.454)^2 = 22.1295$.

Decoder 4:

- $\hat{\mathbf{u}}_4 = [0,0,0,1,0,0,1,0]$.
- Encode: $\hat{\mathbf{c}}_4 = [1,1,0,0,0,0,1,0]$.
- De-permute: $\pi_4^{-1} = (3,6,5,0,7,2,1,4), \hat{\mathbf{c}}_{4,\text{dep}} = [0,1,0,1,0,0,1,0]$.
- BPSK: $\hat{\mathbf{x}}_{4,\text{dep}} = [1,-1,1,-1,1,1,-1,1]$.
- Distance: $d_E^2 = (-1.4167)^2 + (0.3643)^2 + (0.6604)^2 + (0.4056)^2 + (-0.3018)^2 + (0.6727)^2 + (-0.7881)^2 + (-2.454)^2 = 10.9058$.

Summary of results

Decoder	$\hat{\mathbf{c}}_{i,\text{dep}}$	$\hat{\mathbf{x}}_{i,\text{dep}}$	d_E^2
1	$[1,1,0,0,0,0,1,1]$	$[-1,-1,1,1,1,1,-1,-1]$	7.4518
2	$[0,1,1,1,0,0,0,0]$	$[1,-1,-1,-1,1,1,1,1]$	22.6295
3	$[0,0,1,1,0,1,0,0]$	$[1,1,-1,-1,1,-1,1,1]$	22.1295
4	$[0,1,0,1,0,0,1,0]$	$[1,-1,1,-1,1,1,-1,1]$	10.9058

As can be seen, the closest solution is from Decoder 1.

7.7 Enumeration of Min-Weight Codewords

In Chapter 5, we discussed a method for enumeration of minimum weight codewords of polar codes, characterizing the formation of these codewords based on $\boldsymbol{G}_N$-rows. In this section, we see how the minimum-weight codewords are counted using the orbit of maximum-degree monomials.

Now, we look at the size of the orbit $\text{LTA}(m,2)_f \cdot f$ denoted by $|\text{LTA}(m,2)_f \cdot f|$. We can break down the action of $\text{LTA}(m,2)_f$ on f into two operations:

- Translation: Every variable x_i in the monomial can be translated by a scalar $b_i \in \mathbb{F}_2$ as $x_i + b_i$. Hence, a monomial f of degree $\deg(f)$ admits as many translations as possible for all its variables, which is equal to $2^{\deg(f)}$.
- Linear mapping: Every variable x_i can be mapped to a "new variable" (y_i), which is a linear combination of variable(s) x_j, where $j \in [0, i-1]\backslash\text{ind}(f)$ as $y_i = x_i + \sum_{j=0, j\notin \text{ind}(f)}^{i-1} a_{i,j} x_j$. The extra variables considered in y_i express the degree of freedom we have on x_i. This will be denoted by $\lambda_f(i) = |\{j \in [0, i-1] \mid j \notin \text{ind}(f)\}|$ and represents the maximum number of variables in the group action on the variable x_i. The total number of free variables on all x_i in the support of f will be $|\lambda_f(f)| = \sum_{i\in\text{ind}(f)} \lambda_f(i)$. Then, the total possible actions for all the variables in support of f is $2^{|\lambda_f|}$ (since we are defined over $\mathbb{F}_2$).

Example 7.22 Let $f = x_0x_2x_4$. Then, $\lambda_f = (4-2, 2-1, 0-0) = (2, 1, 0)$, which corresponds to degree of freedom 2 for x_4 (x_3, x_1 are free variables for x_4), and degree of freedom 1 for x_2 (x_1 is a free variable for x_2).

Let $g = x_2x_4$. Then, $\lambda_g = (3, 2)$ because x_3, x_1, x_0 are free variables for x_4 and x_1, x_0 are free variables for x_2.

We also have $\lambda_f(g) = (2, 1)$ since $g|f$ and $\text{ind}(g) = \{2, 4\}$, which correspond to the first and second indices in ind(f).

In general, we can have $\text{LTA}(m, 2)_f \cdot g$, where f and g might be different. In this case, $|\lambda_f(g)| = \sum_{i\in\text{ind}(g)} \lambda_f(i)$. When $g = f$, we shall simplify the notation and use λ_f. Therefore, the cardinality of the orbit of monomial f under the action of LTA(m, 2) is

$$|\text{LTA}(m, 2)_f \cdot f| = 2^{\deg(f)+|\lambda_f|}. \tag{7.14}$$

In Example 7.23, we illustrate how this counting procedure works. We know that any codeword of minimum weight $C(\mathcal{I})$ is of weight 2^{m-r} and, more significant, any $\mathbf{c} \in C(\mathcal{I})$ with $\text{w}(\mathbf{c}) = 2^{m-r}$ has the following form $\mathbf{c} = \text{ev}(y_1 \dots y_r)$, where y_i is the linearly independent forms.

1) Any monomial $f \in \mathcal{I}_r$ will define an orbit under $\text{LTA}(m, 2)_f$ in which all polynomials evaluate to a minimum weight codeword.
2) We know how to count the cardinal of the orbit $\text{LTA}(m, 2)_f \cdot f = |\text{LTA}(m, 2)_f|$.
3) For any pair of monomials $f, g \in \mathcal{I}_r$ the orbits $\text{LTA}(m, 2)_f \cdot f$ and $\text{LTA}(m, 2)_g \cdot g$ are disjoint.
4) Finally, sum over all monomials $f \in \mathcal{I}_r$.

$$|W_{w_{\min}}| = \sum_{f\in\mathcal{I}_r} 2^{r+|\lambda_f|}. \tag{7.15}$$

Observe that every $f \in \mathcal{I}_r$ represents a coset leader for coset $C_{\bar{f}}$, and the orbit represents the set of core rows where their row combinations along with the balancing rows result in minimum weight codewords.

Example 7.23 Let $m = 8$ and $\mathcal{I}$ be a decreasing monomial set with $r = 4$. Suppose $f \in \mathcal{I}_r = \{x_0x_1x_2x_3, x_0x_1x_2x_4, x_0x_1x_2x_5, x_0x_1x_3x_4\}$, then the table below illustrates the procedure of finding the total number of codewords of $C(\mathcal{I})$ with weight $w_{\min}$, i.e. $|W_{w_{\min}}|$.

ind(f)	$\{0, 1, 2, 3\}$	$\{0, 1, 2, 4\}$	$\{0, 1, 2, 5\}$	$\{0, 1, 3, 4\}$
λ_f	$(0, 0, 0, 0)$	$(1, 0, 0, 0)$	$(2, 0, 0, 0)$	$(1, 1, 0, 0)$
$\|\text{LTA}(m, 2)_f \cdot f\|$	2^4	2^5	2^6	2^6
$\|W_{w_{\min}}\|$		176		

Linking the combinations of rows discussed in Chapter 5 to the corresponding polynomials discussed here, we can say that a coset formed by a leading row $\mathbf{g}_{\text{ind}(\bar{f})}$ for $f \in \mathcal{I}_r$ can be mapped to the orbit resulting from the action of the permutation group LTA$(m, 2)_f$ on f. We demonstrate this in the following example.

Example 7.24 Let us take the monomial $f = x_1x_0$ and $m = 3$. Similarly to Example 7.23, we find that the size of the orbit LTA$(m, 2)_f \cdot f$ is $2^{\deg(f)+|\lambda_f|} = 2^{2+0} = 4$ where $\lambda_f = (0\,0)$. Polynomials corresponding to the $w_{\min}$-weight codewords on the orbit can be found by changing the variables of the monomial f, where every x_i is replaced by the variable $y_i = x_i + \sum_{j=0, j\notin \text{ind}(f)}^{i-1} b_{ij}x_j + \varepsilon_i$. The resulting polynomials $\prod_{i\in\text{ind}(f)} y_i$ are listed below.

Polynomials	Row combinations
x_0x_1	$\mathbf{g}_4$
$(x_1 + 1)x_0 = x_0x_1 + x_0$	$\mathbf{g}_4 + \mathbf{g}_6$
$x_1(x_0 + 1) = x_0x_1 + x_1$	$\mathbf{g}_4 + \mathbf{g}_5$
$(x_1 + 1)(x_0 + 1) = x_0x_1 + x_1 + x_0 + 1$	$\mathbf{g}_4 + \mathbf{g}_5 + \mathbf{g}_6 + \mathbf{g}_7$

Note that all polynomials are the sum of monomials in set $\mathcal{I}$ in decreasing order, and the monomial $f = x_0x_1$ is shared in all, which corresponds to the leading row $\mathbf{g}_4$ where $4 = (0\,0\,1)_2 = \text{ind}(\bar{f})$. Taking $\mathbf{g}_4$ as a codeword and a coset leader, we can find the rest of the codewords in the coset C_4 by row combinations. Forming the index set $\mathcal{K}_4 = \{5, 6\}$, which has $2^{|\mathcal{K}_4|} = 2^2 = 4$ subsets equivalent to the total number of $w_{\min}$-weight codewords, the codewords of weight $w_{\min}$ are formed by (5.21) in Chapter 5 as listed in the table above, where $\mathbf{g}_7$ in the bottom row is a balancing row and is located in the index set $\mathcal{M}(\mathcal{J})$ when $\mathcal{J} = \{5, 6\}$.

7.8 Structure of Codewords with Larger Weights

For minimum weight codewords, we considered the polynomials within the orbit of maximum degree monomials. However, for larger weight codewords, we need to consider the sum of polynomials in multiple orbits. We present two weight structures of polar codes (and in general decreasing monomial codes, which include Reed–Muller codes as well) called Type I and Type II codewords that characterize the structure of weights less than $2w_{\min}$ in terms of linear forms y_i in (7.9), resulting from the transformation of maximum degree monomials (degree r). The two classifications are as follows.

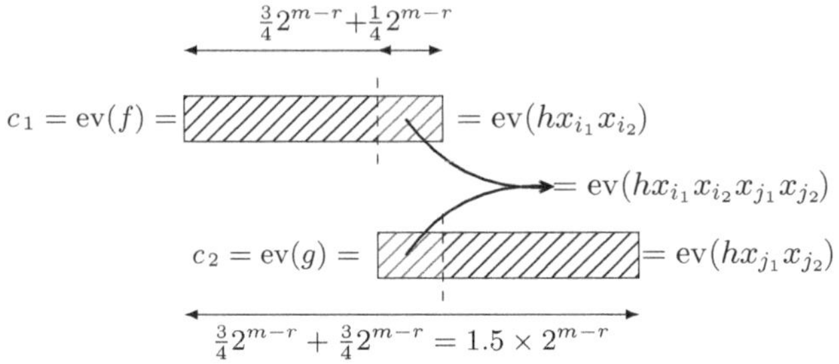

Figure 7.6 The structure of $1.5w_{\min}$-weight codewords: two codewords $\boldsymbol{c}_1, \boldsymbol{c}_2$ are added where the overlapping bits, constituting a quarter of codewords, are $\mathrm{ev}(hx_{i_1}x_{i_2}x_{j_1}x_{j_2})$. The resulting codeword is $\mathrm{ev}(f) + \mathrm{ev}(g) = \mathrm{ev}(f+g) = \mathrm{ev}(h(x_{i_1}x_{i_2} + x_{j_1}x_{j_2}))$.

Let $r < m$ and polynomial P be of $\deg(P) \leq r$ with $0 < \mathrm{w}(\mathrm{ev}(P)) < 2^{m+1-r}$. Then

- Type I: For $m \geq r + \mu, r \geq \mu \geq 3$
  $P = y_1 \dots y_{r-\mu}(y_{r-\mu+1} \cdots y_r + y_{r+1} \cdots y_{r+\mu})$, or alternatively
  $P \in \mathrm{LTA}(m,2) \cdot f + \mathrm{LTA}(m,2) \cdot g$ with $f, g \in \mathcal{M}_m$ having $\deg(f) = \deg(g) = r$.
- Type II: For $m - r + 2 \geq 2\mu \geq 2$
  $P = y_1 \dots y_{r-2}(y_{r-1}y_r + \cdots + y_{r+2\mu-3}y_{r+2\mu-2})$, or alternatively
  $P \in \sum_{i=1}^{\mu} \mathrm{LTA}(m,2) \cdot f_i$ with $f_i \in \mathcal{M}_m$ satisfying $\deg(f_i) = r$.

Depending on μ, which is the number of terms summed in Type II, the resulting weight will be

$$w\,(\mathrm{ev}(P)) = 2^{m+1-r} - 2^{m+1-r-\mu} = (2 - 1/2^{\mu-1})w_{\min}. \tag{7.16}$$

and for $\mu \geq 2$, the weight will be bounded to

$$1.5 \leq (2 - 1/2^{\mu-1}) < 2. \tag{7.17}$$

The particular case of $\mu = 2$, which corresponds to codewords of weight $1.5w_{\min}$ is the evaluation of the following polynomial

$$P \in \mathrm{LTA}(m,2)_h \cdot h \cdot \left(\mathrm{LTA}(m,2)_f \cdot \frac{f}{h} + \mathrm{LTA}(m,2)_g \cdot \frac{g}{h}\right), \tag{7.18}$$

where $f, g \in \mathcal{I}_r, h = \gcd(f,g)$, $\deg(h) = r - 2$. Figure 7.6 shows the structure of such codewords.

By finding all such monomial pairs $f, g \in \mathcal{I}_r$ satisfying $h = \gcd(f,g) \in \mathcal{I}_{r-2}$, all codewords of weight $1.5w_{\min}$ are characterized as

$$W_{1.5w_{\min}} = \bigcup_{\substack{f,g \in \mathcal{I}_r \\ h = \gcd(f,g) \in \mathcal{I}_{r-2}}} \mathrm{LTA}(m,2)_h \cdot h \cdot \left(\mathrm{LTA}(m,2)_f \cdot \frac{f}{h} + \mathrm{LTA}(m,2)_g \cdot \frac{g}{h}\right). \tag{7.19}$$

To count the number of $1.5w_{\min}$ codewords, we need to find the cardinality of the sum of two orbits. We can use the following property:

$$\left|\mathrm{LTA}(m,2)_h \cdot h \cdot \left(\mathrm{LTA}(m,2)_f \cdot \frac{f}{h} + \mathrm{LTA}(m,2)_g \cdot \frac{g}{h}\right)\right|$$
$$= \left|\mathrm{LTA}(m,2)_h \cdot h\right| \times \left|\mathrm{LTA}(m,2)_f \cdot \frac{f}{h} + \mathrm{LTA}(m,2)_g \cdot \frac{g}{h}\right|. \quad (7.20)$$

In (7.14), we learned how to find the cardinality of the orbit of a monomial f, $|\mathrm{LTA}(m,2) \cdot f|$. We also need to find the cardinality of the sum of two orbits. It turns out that the cardinality of the sum of two orbits does not necessarily equal the multiplication of the cardinalities, but it gives an upper bound as follows:

$$|\mathrm{LTA}(m,2) \cdot f + \mathrm{LTA}(m,2) \cdot g| \leq |\mathrm{LTA}(m,2) \cdot f| \times |\mathrm{LTA}(m,2) \cdot g|. \quad (7.21)$$

Example 7.25 Let $f = x_0x_2x_4$ and $g = x_1x_2x_3$. Then, any polynomial $P \in \mathrm{LTA}(m,2) \cdot f$ and $Q \in \mathrm{LTA}(m,2) \cdot g$ can be written as

$$P = (x_0 + \varepsilon_0)(x_2 + b_{2,1}x_1 + \varepsilon_2)(x_4 + b_{4,3}x_3 + b_{4,1}x_1 + \varepsilon_4),$$
$$Q = (x_1 + a_{1,0}x_0 + \gamma_0)(x_2 + a_{2,0}x_0 + \gamma_2)(x_3 + a_{3,0}x_0 + \gamma_3),$$

where the scalars $a_{i,j}, b_{i,j}, \varepsilon_i, \gamma_i \in \mathbb{F}_2$. The cardinality of the orbits are $|\mathrm{LTA}(m,2) \cdot f| = 2^{3+(2+1+0)} = 2^6$ and $|\mathrm{LTA}(m,2) \cdot g| = 2^{3+(1+1+1)} = 2^6$. Hence, we have

$$|\mathrm{LTA}(m,2) \cdot f + \mathrm{LTA}(m,2) \cdot g| \leq 2^{12}.$$

In this example, the upper bound is not achieved due to the existence of *collisions*.

Definition 7.5 Collision: Any pair of distinct polynomials $P, P^* \in \mathrm{LTA}(m,2) \cdot f$ and $Q, Q^* \in \mathrm{LTA}(m,2) \cdot g$ with $P \neq P^*, Q \neq Q^*$ such that they produce the same sum, that is, $P + Q = P^* + Q^*$, results in a collision.

Take f, g as in Example 7.25 and $P = x_0x_2(x_4 + x_3), Q = x_1x_2x_3$ and $P^* = x_0x_2x_4, Q^* = (x_1 + x_0)x_2x_3$. Both P, P^* are in orbit of f and Q, Q^* in orbit of g. Notice that we have produced a collision, i.e. $P + Q = P^* + Q^*$ by inserting in each monomial a variable from the other monomial. Whether there is a collision or not depends on the structure of the two monomials f, g. That is why we need to define the concept of degree of collisions.

Definition 7.6 Let $f = x_{i_1}x_{i_2}$ and $g = x_{j_1}x_{j_2}$ with $\gcd(f,g) = 1$ and $i_2 > j_2$. The *degree of collision* of f and g is

$$\alpha_{f,g} = \begin{cases} 0 & i_2 > i_1 > j_2 > j_1 \\ 1 & i_2 > j_2 > i_1 > j_1 \\ 2 & i_2 > j_2 > j_1 > i_1 \end{cases}.$$

Notice that $\alpha_{f,g}$ is equal to the inversion number of the indices of (i_2, i_1, j_2, j_1). Indeed, $\alpha_{f,g} = 0$ corresponds to positions $(1, 2, 3, 4)$, $\alpha_{f,g} = 1$ to $(1, 3, 2, 4)$, and $\alpha_{f,g} = 2$ to $(1, 3, 4, 2)$. For any given pair (P, Q) with $P \in \mathrm{LTA}(m, 2) \cdot f$ and $Q \in \mathrm{LTA}(m, 2) \cdot g$, the quantity $\alpha_{f,g}$ will allow us to count how many collisions we get for any fixed pair of polynomials. Indeed, as we will demonstrate, this parameter represents the number of independent linear equations required to satisfy collisions for a given pair of monomials f, g.

Example 7.26 Let $f = x_{i_1}x_{i_2}, g = x_{j_1}x_{j_2}$.

- $i_2 > j_2 > j_1 > i_1$ with $f = x_2x_6, g = x_3x_5$, which implies $\alpha = 2$. Let $P = (x_2 + x_1)(x_6 + x_4 + 1) \in \mathrm{LTA}(m, 2) \cdot f$ and $Q = (x_3 + x_2 + x_0 + 1)(x_5 + x_4 + x_2) \in \mathrm{LTA}(m, 2) \cdot g$. We can create two non-trivial distinct pairs P^*, Q^*
  - $P^* = (x_2 + x_1)(x_6 + x_4 + x_3 + x_2 + x_0)$ and $Q^* = (x_3 + x_2 + x_0 + 1)(x_5 + x_4 + x_1)$
  - $P^* = (x_2 + x_1)(x_6 + x_5 + x_3 + x_2 + x_1 + x_0)$ and $Q^* = (x_3 + x_1 + x_0 + 1)(x_5 + x_4 + x_1)$
- $i_2 > j_2 > i_1 > j_1$ with $f = x_2x_4, g = x_0x_3$. We obtain that $f + g = x_2(x_4 + x_0) + x_0(x_3 + x_2)$.

The parameter $\alpha_{f,g}$ will assist us in determining the number of polynomials that have been overcounted. To be more precise, our counting argument is based on the inclusion–exclusion principle. First, we start by letting all free variables coming from $\mathrm{LTA}(m, 2) \cdot f$ and $\mathrm{LTA}(m, 2) \cdot g$, and then we remove those variables that create collisions. Now, we are in a position to move from (7.21) to the exact cardinality of the sum of two orbits as follows: Let $\mathcal{I} \subseteq \mathcal{M}_m$ be a decreasing monomial set and $f = x_{i_1}x_{i_2}$ and $g = x_{j_1}x_{j_2}$ with $\gcd(f, g) = 1$. Then

$$|\mathrm{LTA}(m, 2) \cdot f + \mathrm{LTA}(m, 2) \cdot g| = \frac{|\mathrm{LTA}(m, 2) \cdot f| \times |\mathrm{LTA}(m, 2) \cdot g|}{2^{\alpha_{f,g}}}. \quad (7.22)$$

Figure 7.7 illustrates the algebraic structure of $1.5w_{\min}$-weight codewords reflected in (7.22). Now, considering (7.19) and (7.22) for all monomial pairs $f, g \in \mathcal{I}_r$, we can count all $1.5w_{\min}$-weight codewords using the following formula:

$$|W_{1.5w_{\min}}| = \sum_{\substack{f,g \in \mathcal{I}_r \\ h = \gcd(f,g) \in \mathcal{I}_{r-2}}} \frac{2^{r+2+|\lambda_h|+|\lambda_f(\frac{f}{h})|+|\lambda_g(\frac{g}{h})|}}{2^{\alpha_{\frac{f}{h},\frac{g}{h}}}}, \quad (7.23)$$

where $\frac{f}{h} = x_{i_1}x_{i_2}$, $\frac{g}{h} = x_{j_1}x_{j_2}$ with $i_2 > j_2$.

Therefore, the $w_{\min}$ and $1.5w_{\min}$ codewords can be described using $\mathcal{I}_r$, where they fall within an orbit $\mathrm{LTA}(m, 2) \cdot f$ (for the case of $w_{\min}$) or a sum of two orbits $\mathrm{LTA}(m, 2) \cdot f + \mathrm{LTA}(m, 2) \cdot g$ (for the case of $1.5w_{\min}$) with $f, g \in \mathcal{I}_r$ and $r = \max_{f \in \mathcal{I}} \deg(f)$. Continuing along the same path, any Type II codeword of a

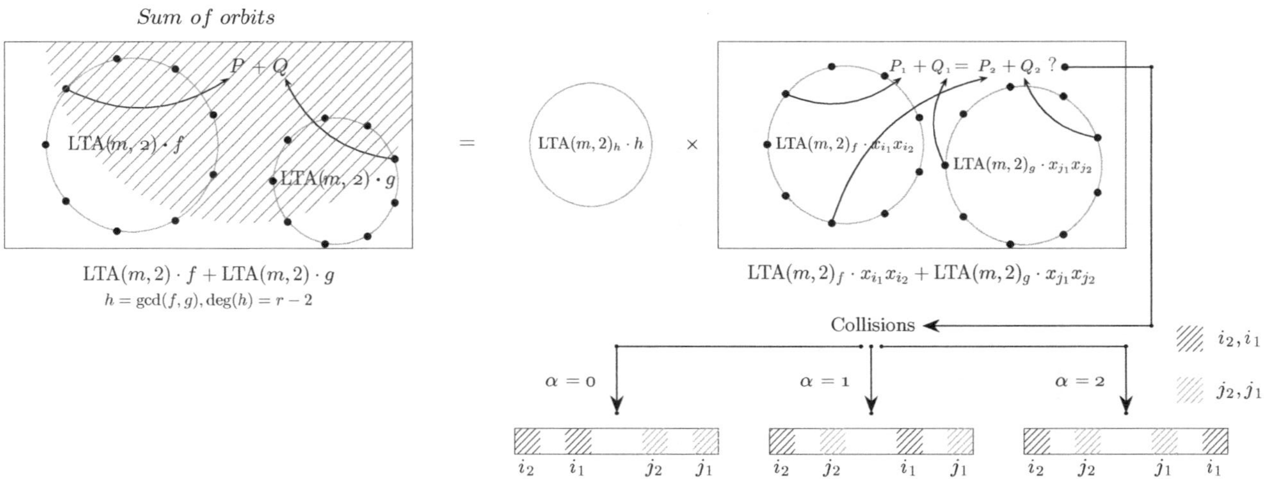

Figure 7.7 The set $\text{LTA}(m,2) \cdot f + \text{LTA}(m,2) \cdot g$ contains polynomials that provide $1.5w_{\text{min}}$ codewords (gray pattern within the left-hand box) as well as other weight, e.g. $2w_{\text{min}}$. The algebraic structure of $1.5w_{\text{min}}$-weight codewords is illustrated in the right part, i.e. any such codeword can be written as the evaluation of a polynomial, which can be decomposed into a product of two polynomials; one from $\text{LTA}(m,2)_h \cdot h$ and the other a sum $\text{LTA}(m,2)_f \cdot x_{i_1}x_{i_2} + \text{LTA}(m,2)_g \cdot x_{j_1}x_{j_2}$.

decreasing monomial code represented by the polynomial P resides within the sum of μ orbits of monomials from $\mathcal{I}_r$ as

$$P \in \text{LTA}(m,2)_h \cdot h \cdot \sum_{i=1}^{\mu} \text{LTA}(m,2)_{f_i} \cdot \frac{f_i}{h}, \tag{7.24}$$

where $\forall i, f_i \in \mathcal{I}_r$, $h = \gcd(f_i, f_j)$, for all $i,j \in [1,\mu]$ $(i \neq j)$ and $\deg(h) = r - 2$.

Straightforward, given a decreasing monomial code defined by the monomial set $\mathcal{I}$, one can enumerate the possible weights in the interval $[w_{\min}, 2w_{\min})$.

Example 7.27 Let us take the $\mathcal{R}(2,7)$ code. We can find the codeword weights using $2^{m+1-r} - 2^{m+1-r-\mu}$ for $m = 7, r = 2$, and $\mu \in \{2,3\}$ that gives the weights $\{32, 48, 56, 64\}$, corresponding to $w_{\min} = 2^{m-r} = 2^5, 1.5w_{\min} = 2^6 - 2^4, 1.75w_{\min} = 2^6 - 2^8$, and $2w_{\min} = 2^6$.

Considering $m - r + 2 = 7$ for Type II codewords, observe that we can have $\mu \in \{2,3\}$ corresponding to the number of terms in $x_0x_1 + x_2x_3$ and $x_0x_1 + x_2x_3 + x_4x_5$.

As additional examples, for $m = 7, r = 3$, let $\mathcal{I}_r = \{x_0x_1x_2, x_0x_1x_3, x_0x_1x_4\}$. In this case, there are no pairs or triplets of monomials, corresponding to $\mu \in \{2,3\}$, with $\deg(\gcd(f_1, f_2)) = 1$ or $\deg(\gcd(f_1, f_2, f_3)) = 1$.

Now, let us change the set to $\mathcal{I}_r = \{x_0x_1x_2, x_0x_1x_3, x_0x_1x_4, x_1x_2x_3, x_0x_1x_5, x_1x_2x_4, x_0x_2x_3\}$. Here, we can find $(x_0x_1x_5, x_0x_2x_3)$ as a valid pair for $\mu = 2$.

Exercises

1 For $m = 7$, find all 19 monomials in the generating set $\mathcal{I}$ of a decreasing monomial code represented by $\mathcal{I} = \{f : f \preceq x_1x_6, f \preceq x_0x_5, f \preceq x_6\}$.

2 In what ways can the concept of decreasing monomials be extended to analyze the weight enumerators of polar and Reed–Muller codes beyond their minimum weight codewords?

3 Discuss how understanding the structure of decreasing monomials in polar codes can lead to more efficient decoding algorithms for these codes.

4 Decoder Linearity: For any received LLR vector $\mathbf{L}$ and codeword $\mathbf{x} \in C$, show that the linear property holds for SC decoding:

$$\text{SC}\left(\mathbf{L} \cdot (-1)^{\mathbf{x}}\right) = \text{SC}(\mathbf{L}) \oplus \mathbf{x},$$

5 Show that the stage-shuffling of the SC factor graph is equivalent to a bit-index permutation of both the codeword vector $\mathbf{c}$ and the message vector $\mathbf{u}$ (interleaved with the frozen bits).

Bibliographical Notes

Bardet et al. [2016a] showed that polar codes, together with Reed–Muller codes, belong to a larger family of codes named decreasing monomial codes. Polar and Reed–Muller codes both have a polynomial structure, but the decreasing property also brings about new algebraic properties. Dragoi [2017] and Bardet et al. [2016b] discovered new structural properties, such as duals of decreasing monomial codes that are still decreasing monomial codes, the permutation group of decreasing monomial codes admits as a subgroup, the lower triangular affine group (LTA$(m, 2)$). Through a deeper understanding of the algebraic structure of polar codes, practical uses have been discovered. Furthermore, the permutation group revealed the complete structure of the minimum weight codewords of decreasing monomial codes and therefore, by extension, of polar codes. They also introduced a partial order of monomials, which was further elaborated and expanded in the literature, such as in papers by Wang and Drăgoi [2023] and Yao et al. [2024]. Li et al. [2021], Geiselhart et al. [2021a,b], Pillet et al. [2021, 2022], and Ivanov and Urbanke [2022] further studied the permutation group that has been used in the parallelized decoding of polar codes, called the automorphism (or permutation) ensemble decoding. The closed-form enumeration of polar codewords has been well studied in the literature, from minimum-weight ($w_{\min}$) codewords by Bardet et al. [2016b] to codewords of $1.5w_{\min}$ by Drăgoi et al. [2024], less than $2w_{\min}$ by Ye et al. [2024], and complete weight distribution for low-/high-rate polar codes by Rowshan et al. [2024]. As a new development, the introduction of a partial order based on the contribution of monomials to the weight distribution by Rowshan et al. [2024] could be instrumental in code design.

Bibliography

Magali Bardet, Julia Chaulet, Vlad Dragoi, Ayoub Otmani, and Jean-Pierre Tillich. Cryptanalysis of the McEliece public key cryptosystem based on polar codes. In *Post-Quantum Cryptography*, volume 9606, pages 118–143. 2016a.

Magali Bardet, Vlad Dragoi, Ayoub Otmani, and Jean-Pierre Tillich. Algebraic properties of polar codes from a new polynomial formalism. In *2016 IEEE International Symposium on Information Theory (ISIT)*, pages 230–234, 2016b.

Vlad Florin Dragoi. *An algebraic approach for the resolution of algorithmic problems raised by cryptography and coding theory*. PhD thesis, Normandie Université, 2017.

Vlad-Florin Drăgoi, Mohammad Rowshan, and Jinhong Yuan. On the closed-form weight enumeration of polar codes: 1.5d-weight codewords. *IEEE Transactions on Communications*, 72(10):5972–5987, 2024.

Marvin Geiselhart, Ahmed Elkelesh, Moustafa Ebada, Sebastian Cammerer, and Stephan ten Brink. Automorphism ensemble decoding of Reed–Muller codes. *IEEE Transactions on Communications*, 69(10):6424–6438, 2021a.

Marvin Geiselhart, Ahmed Elkelesh, Moustafa Ebada, Sebastian Cammerer, and Stephan ten Brink. On the automorphism group of polar codes. In *2021 IEEE International Symposium on Information Theory (ISIT)*, pages 1230–1235. IEEE, 2021b.

Kirill Ivanov, and Rüdiger Urbanke. Polar codes do not have many affine automorphisms. In *2022 IEEE International Symposium on Information Theory (ISIT)*, pages 2374–2378. IEEE, June 2022.

Yuan Li, Huazi Zhang, Rong Li, Jun Wang, Wen Tong, Guiying Yan, and Zhiming Ma. The complete affine automorphism group of polar codes. In *2021 IEEE Global Communications Conference (GLOBECOM)*, pages 01–06. IEEE, 2021.

Charles Pillet, Valerio Bioglio, and Ingmar Land. Polar codes for automorphism ensemble decoding. In *IEEE Information Theory Workshop (ITW)*, pages 1–6. IEEE, October 2021.

Charles Pillet, Valerio Bioglio, and Ingmar Land. Classification of automorphisms for the decoding of polar codes. In *ICC 2022 - IEEE International Conference on Communications*, pages 110–115. IEEE, May 2022.

Mohammad Rowshan, Vlad-Florin Drăgoi, and Jinhong Yuan. Weight structure of low/high-rate polar codes and its applications. In *2024 IEEE International Symposium on Information Theory (ISIT)*, pages 2945–2950. IEEE, 2024.

Hsin-Po Wang and Vlad-Florin Drăgoi. Fast methods for ranking synthetic BECs. In *2023 IEEE International Symposium on Information Theory (ISIT)*, pages 1562–1567. IEEE, 2023.

Liuquan Yao, Zhichao Liu, Yuan Li, Huazi Zhang, Jun Wang, Guiying Yan, and Zhi-Ming Ma. New partial orders of polar codes for BMSC. In *2024 IEEE International Symposium on Information Theory (ISIT)*, pages 2192–2197. IEEE, 2024.

Zicheng Ye, Yuan Li, Huazi Zhang, Jun Wang, Guiying Yan, and Zhiming Ma. On the distribution of weights less than $2w_{\min}$ in polar codes. *IEEE Transactions on Communications*, 72(10):5988–6000, 2024.

8

List Decoding of Polar Codes

"Sometimes the most interesting solution is not the first one you find, but the set of possible solutions."

— Peter Shor

Although the successive cancellation (SC) decoder, discussed in Chapter 4, offers low complexity, it falls short in terms of error correction performance at practical block lengths. This limitation arises because the SC decoding algorithm is inherently greedy, making locally optimal decisions based on the maximum likelihood (ML) rule for each symbol. Although these decisions are locally optimal, they are globally suboptimal, as they only consider one path through the decoding tree.

To address the weaknesses of SC decoding, this chapter explores enhanced algorithms that follow multiple paths in the decoding tree. These decoders, known as successive cancelation list (SCL) decoders, retain the recursive structure of SC decoding while improving its performance by considering several distinct solutions. We delve into both parallel and serial implementations of SCL decoders, highlighting their fundamental structures and trade-offs.

Through this chapter, readers will gain an understanding of the fundamental principles behind SC list decoders and learn how to exploit the structural properties of polar codes to enhance decoding efficiency.

8.1 SC List Decoding

The SC list (SCL) decoding algorithm scans the binary tree (in the case of binary codes) by following more than one path at a time. That is, a number of paths are followed in parallel. Figure 8.1 illustrates a tree where two paths are traversed simultaneously and the rest are prunned. This algorithm is a variant of *breadth-first search* algorithm that is used for searching a tree data structure. In Section 8.2.2,

Polar Codes: From Theory to Practice, First Edition. Mohammad Rowshan and Emanuele Viterbo.

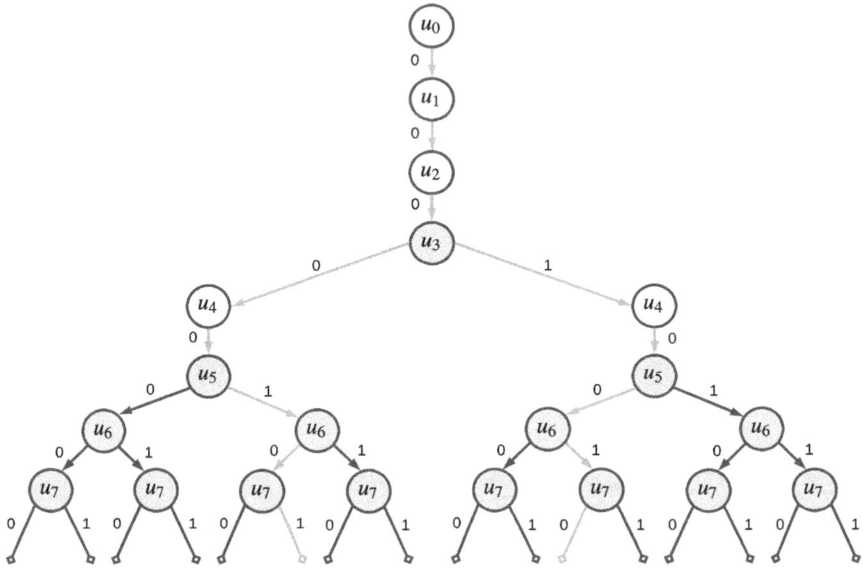

Figure 8.1 Tree search on code tree: Two closest paths to the received vector shown by light gray arrows are followed ($L = 2$).

we will learn about an iterative decoding algorithm that follows one path at a time at each iteration if the decoding attempt fails in the first iteration. That algorithm is also a multipath algorithm, but the path search is performed sequentially.

Since we deal with multiple paths, we need a metric for the paths to distinguish them based on their likelihood to be the correct solution to the decoding problem. Unlike SC decoding, which decides the value of u_i in step i and therefore follows a particular path at each decision step, SCL decoding considers both possible values $u_i = 0$ and $u_i = 1$ and finds a path through the decoding tree that has the highest probability of being the transmitted sequence u_0^{N-1}. For optimal decoding, the probability to *maximize* would be

$$P(\hat{u}_0^{N-1}|y_0^{N-1}) = \prod_{j=0}^{N-1} P\left(\hat{u}_j|\hat{u}_0^{j-1}, y_0^{N-1}\right). \tag{8.1}$$

However, we cannot traverse the entire decoding tree, that is considering all paths by SCL decoding under resource constraints. Therefore, the solution obtained from the decoding can be suboptimal. In SCL decoding, the probability of partial path l representing the sequence $\hat{u}_0^{i-1} = (\hat{u}_0, \hat{u}_1, \ldots, \hat{u}_{i-1})$ is computed as

$$P\left(\hat{u}_0^j[l]|y_0^{N-1}\right) = \prod_{j=0}^{j} P\left(\hat{u}_j[l]|\hat{u}_0^{j-1}[l], y_0^{N-1}\right). \tag{8.2}$$

In practice, it is more convenient and practical to deal with the logarithm of (8.1) and (8.2). Since $\log(x) < 0$ for $x < 1$, we change the sign to have a positive metric. Note that this converts the maximization of (8.1) to *minimization*. Therefore, we get the following logarithmic *path metric*

$$PM_l^{(i-1)} = -\log P\left(\hat{u}_0^{i-1}[l]|y_0^{N-1}\right)$$
$$= -\sum_{j=0}^{i-1} \log P\left(\hat{u}_j[l]|\hat{u}_0^{j-1}[l], y_0^{N-1}\right), \tag{8.3}$$

for the sequence $\hat{u}_0^{i-1}$ at position l.

Now, let the sequence $\hat{u}_0^i$ be obtained by appending $\hat{u}_i$ to $\hat{u}_0^{i-1}$ on the path l. The path metric for this extended sequence is

$$PM_l^{(i)} = -\sum_{j=0}^{i} \log P\left(\hat{u}_j[l]|\hat{u}_0^{j-1}[l], y_0^{N-1}\right) \tag{8.4}$$
$$= PM_l^{(i-1)} + \mu_l^{(i)}, \tag{8.5}$$

where $\mu_l^{(i)} = -\log P(\hat{u}_i[l]|\hat{u}_0^{i-1}[l], y_0^{N-1})$ denotes the *branch metric*, and $PM_l^{(-1)} = 0$. Hence, the path metric along a path ending at bit i is obtained by adding the path metric ending at bit $i-1$ to the branch metric at bit i. Throughout this chapter, we assume that the index $1 \leq l \leq L$ indicates the position of a path in a list sorted according to the path metric, that is $PM_1^{(i)} \leq PM_2^{(i)} \leq \cdots \leq PM_L^{(i)}$, and we let l_c denote the index of the correct path in the list.

To simplify the arithmetic operation, we can rewrite μ_i in terms of log-likelihoods λ as

$$\mu_l^{(i)} = -\log P(\hat{u}_i[l]|\hat{u}_0^{i-1}[l], y_0^{N-1})$$
$$= -\log\left(\frac{e^{(1-\hat{u}_i[l])\lambda_0^{(i)}[l]}}{e^{\lambda_0^i[l]}+1}\right) = \log\left(1 + e^{-(1-2\hat{u}_i[l])\lambda_0^{(i)}[l]}\right), \tag{8.6}$$

where the last equality holds only for the values of $\hat{u}_i[l] = 0$ and 1. For the value of $\hat{u}_i[l]$ that equals $h(\lambda_0^{(i)}[l])$, the term $e^{-(1-2\hat{u}_i)\lambda_0^{(i)}} = e^{-|\lambda_0^{(i)}|}$ is small, and hence $\log(1 + e^{-|\lambda_0^{(i)}|}) \approx 0$. Otherwise, we can approximate $\log(1 + e^{|\lambda_0^{(i)}|}) \approx |\lambda_0^{(i)}|$. Thus,

$$\mu_l^{(i)} = \mu_l^{(i)}(\lambda_0^{(i)}[l], \hat{u}_i[l]) \approx \begin{cases} 0 & \text{if } \hat{u}_t[l] = h(\lambda_0^{(i)}[l]) \\ |\lambda_0^{(i)}[l]| & \text{otherwise.} \end{cases} \tag{8.7}$$

As (8.7) shows, the path of the least likely bit value is penalized by $|\lambda_0^{(i)}[l]|$. This value is called *penalty* throughout this chapter. In each decoding step, the L paths with the smallest metrics are chosen among the $2L$ paths and stored in the ascending order from $PM_1^{(i)}$ to $PM_L^{(i)}$. After decoding the last bit, the path with the smallest path metric, that is $PM_1^{(N-1)}$ is selected as the estimated sequence.

Polar codes are traditionally constructed based on the reliability of the sub-channels in the polarized vector channel. Various methods were discussed in Chapter 6. This approach is well compatible with the SC decoder, which is a greedy search algorithm based on the maximum likelihood rule for each symbol. An incorrect estimation of one symbol results in a decoding failure. However, this is not the case for list decoding. That is, the correct path may remain in the list.

Algorithm 8.1: SC List Decoding of Polar Codes

input : N, y_0^{N-1}, Information set I, u_{I^c}, list size L

output: Estimated information bits $\hat{u}_I$

Ł $\leftarrow \{1\}$ `// a single path`

$PM_1^{(-1)} \leftarrow 0$

for $i \leftarrow 0$ **to** $N-1$ **do**

    Find $L_N^{(i)}[\ell]$, $\forall \ell \in$ Ł

    if $i \notin I$ **then**

        $\hat{u}_i[\ell] \leftarrow u_i[\ell]$, $\forall \ell \in$ Ł

        $PM_\ell^{(i)} \leftarrow PM_\ell^{(i-1)} + \mu_\ell^{(i)}$, $\forall \ell \in$ Ł `// cf. (8.6),(8.7),(8.4)`

    else

        for $\ell \leftarrow$ Ł **do**

            Copy the path ℓ to ℓ' and Ł $\leftarrow$ Ł $\cup \{\ell'\}$

            $\left(\hat{u}_i[\ell], PM_\ell^{(i)}\right) \leftarrow \left(0, PM_\ell^{(i-1)} + \mu_\ell^{(i)}\right)$

            $\left(\hat{u}_i[\ell'], PM_{\ell'}^{(i)}\right) \leftarrow \left(1, PM_{\ell'}^{(i-1)} + \mu_{\ell'}^{(i)}\right)$

        if |Ł| $> L$ **then**

            Ł $\leftarrow$ sort $PM_\ell^{(i)}, \forall \ell \in$ Ł in ascending order.

            Kill all paths $\ell \in$ Ł$[L+1, 2L]$

            Ł $\leftarrow$ Ł$[1, L]$

$\ell^* \leftarrow \arg\min_{\ell \in Ł} PM_\ell^{(N-1)}$

return $\hat{u}_I[\ell^*]$

Example 8.1 Suppose that the polar code (4, 2) with the frozen set $\mathcal{A}^c = \{0,\ 1\}$ is found using the density evolution (DE)/Gaussian approximation (GA) method. The vector $\mathbf{u} = [0,\ 0,\ 1,\ 1]$ is encoded to $\mathbf{x} = [0,\ 1,\ 0,\ 1]$. Passing through additive white Gaussian noise (AWGN) channels with $E_b/N_0 = 4$, the log-likelihood ratios (LLRs) of the received sequence is computed as $\mathbf{L}_8 = [8.4891, -1.5076,\ 2.2861, -4.7785]$. Figure 8.2 illustrates the decoding tree for this example under the SCL decoder with list size $L_{\max} = 2$. The decoding steps of the received $\mathbf{y}$ with the SCL decoder are listed below.

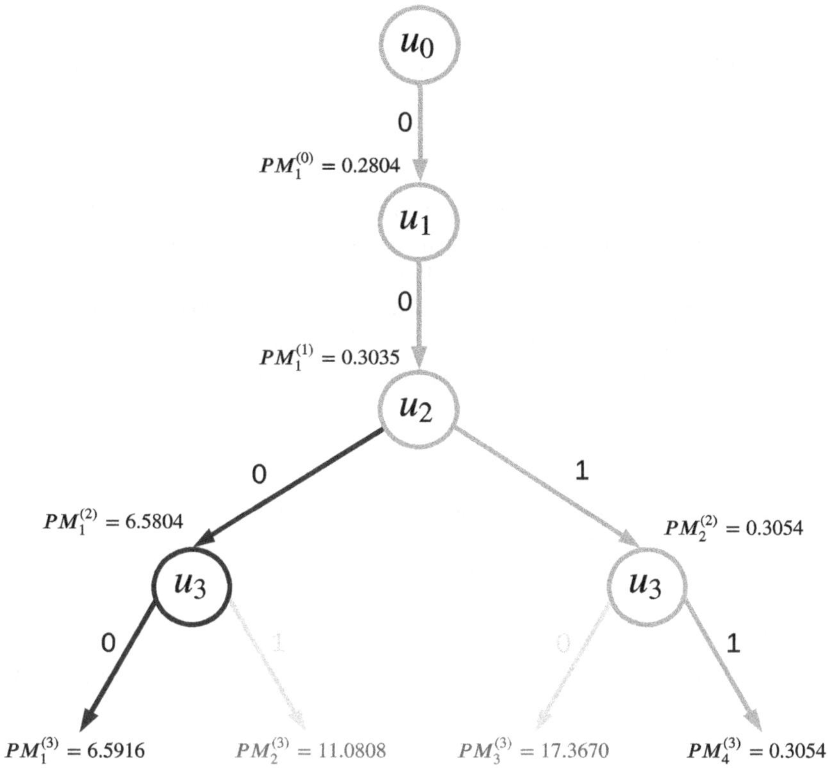

Figure 8.2 Decoding tree for SC list decoding with $L = 2$.

1) Initialize $PM_1^{(-1)} = 0$.
2) Find $L_4^{(0)} = 1.1279$.
3) "Extend" each path (here one path) and set $u_0 = 0$ as $0 \in \mathcal{A}^c$ and compute the branch metric $\mu_1^0 = 0.2804 = PM_1^{(0)}$ using (8.6).
4) Steps 1 and 2 are repeated for u_1 as $1 \in \mathcal{A}^c$.
5) Since $2 \in \mathcal{A}$ and therefore $u_2 \in \{0, 1\}$, each path (here, we have only one path) is "expanded" to two paths and the corresponding metrics $PM_l^{(2)}, l \in [1, 2]$ are computed. We reach the maximum number of paths $L_{\max} = 2$.
6) Step 5 is repeated for u_3 as $3 \in \mathcal{A}$. However, the number of paths is $4 > L_{\max}$, and we need to prune the unlikely paths with the highest metrics. Note that since this is the last bit, we may not need to do it.
7) Since we reach the last bit, we choose the fourth path with the smallest metric corresponding to the highest likelihood; that is, $\hat{\mathbf{u}} = [0,\ 0,\ 1,\ 1] = \mathbf{u}$.

8.1.1 Error Events in SCL Decoding

In SCL decoding, a decoding error event, denoted as $\mathcal{E}$, can be of two distinct types: $\mathcal{E}_1$ and $\mathcal{E}_2$. These two types of events represent mutually exclusive scenarios that contribute to the overall decoding failure.

- The Type I error event $\mathcal{E}_1$ occurs when the transmitted message (or its corresponding codeword) does not appear in the final list generated by the SCL decoder. The probability of this error event, $P_{SCL}\left(\mathcal{E}_1; L\right)$, depends primarily on the selection of the information index set $\mathcal{A}$, as discussed in Chapters 4 and 6. A poorly constructed code, where unreliable bit channels are used to transmit information bits, can increase the likelihood of this error. Another parameter affecting this error event is the list size L, which will be discussed in the following.
- Type II error event $\mathcal{E}_2$ occurs when a codeword other than the transmitted one (which also appears in the final candidate list) is closer to the received sequence, causing the decoder to make an incorrect decision. The probability of this error, $P_{SCL}\left(\mathcal{E}_2; L\right)$, is influenced by the minimum distance of the code and the number of codewords with minimum weight, both of which depend on the information index set $\mathcal{A}$. To improve $P_{SCL}\left(\mathcal{E}_2; L\right)$, we can use a genie (realized as cyclic redundancy check (CRC), see Section 8.1.3) to find the correct candidate in the list. Note that CRC-polar codes possess larger $d_{\min}$ or smaller $A_{d_{\min}}$. We refer the reader to the corresponding section in Chapter 6.

The overall error probability of the SCL decoder with list size L can be expressed as:

$$P_{SCL}(\mathcal{E}; L) = P_{SCL}\left(\mathcal{E}_1; L\right) + P_{SCL}\left(\mathcal{E}_2; L\right).$$

Note that when $L = 2^K$, the SCL decoder achieves maximum-likelihood (ML) decoding. An SCL decoder with a relatively large list size L has been numerically shown to achieve ML error correction performance. The probability of error of the ML decoder is dominated by the minimum distance $d_{\min}$ of the code and the number of minimum-weight codewords $A_{d_{\min}}$. For more information, refer to the discussion about the union bound in Chapter 2.

8.1.2 List Size in SCL Decoding

In SCL decoding, the size of the list plays a crucial role in determining the decoder's ability to retain the correct codeword in its list. Specifically, the probability of Type-I error, $P_{SCL}\left(\mathcal{E}_1; L\right)$, depends on whether the correct codeword remains in the list as decoding progresses. An approach to reducing this error probability is to carefully design the information sets to avoid consecutive less reliable information bits.

To better understand how to minimize this error event, we can analyze the decoding uncertainty using the concept of entropy. The uncertainty when decoding a particular bit u_m can be measured by the entropy of the set of information bits decoded up to u_m, denoted as $\mathbf{U}_{\mathcal{A}(i)}$, conditioned on the received signal $\mathbf{Y}$ and the frozen bits $\mathbf{U}_{\mathcal{A}^c(i)}$. This entropy is represented as

$$d_i(\mathbf{y}) \triangleq H\left(\mathbf{U}_{\mathcal{A}(i)} \mid \mathbf{Y} = \mathbf{y}, \mathbf{U}_{\mathcal{A}^c(i)}\right).$$

The corresponding random variable, D_m, describes the decoding uncertainty, and its expectation, $\overline{D}_i$, provides a measure of the overall uncertainty at step i:

$$\overline{D}_i \triangleq \mathbb{E}\left[d_i(\mathbf{y})\right] = H\left(\mathbf{U}_{\mathcal{A}(i)} \mid \mathbf{Y}, \mathbf{U}_{\mathcal{A}^c(i)}\right).$$

This conditional entropy $\overline{D}_i$ is critical in determining the required list size for the SCL decoder to achieve maximum likelihood decoding performance. To isolate the impact of decoding uncertainty introduced by a specific bit u_i, we define *incremental entropy* as the difference between the uncertainty at step i and that at step $i-1$:

$$\Delta_i \triangleq \overline{D}_i - \overline{D}_{i-1}.$$

The behavior of the incremental entropy is governed by whether u_i is an information bit or a frozen bit. The decoding uncertainty can increase or decrease based on this distinction. Hence, the incremental entropy by decoding u_i is:

$$\begin{cases} \Delta_i \geq 0, & \text{if } i \in \mathcal{A} \\ \Delta_i \leq 0, & \text{otherwise.} \end{cases}$$

This result indicates that if the information bits are chosen consecutively (i.e. $i, i+1 \in \mathcal{A}$), the decoder entropy increases continuously. The amount of this increase depends on the reliability of the sub-channels used for transmission. When information bits are transmitted over highly reliable sub-channels, the increase in entropy is negligible. However, in practice, particularly at finite block lengths, not all information bits can be transmitted over highly polarized sub-channels. When less reliable sub-channels are used, consecutive information bits lead to a continuous increase in decoder entropy, requiring a larger list size to retain the correct codeword.

When the list size is small, careful design of the information sets is crucial. An effective method to control the error probability is to insert frozen bits between consecutive information bits. This approach prevents the entropy from increasing too quickly, as frozen bits decrease the decoder uncertainty $\left(\Delta_i < 0\right.$ for $\left.i \in \mathcal{A}^c\right)$. Nevertheless, while inserting frozen bits between information bits can reduce $P_{SCL}\left(\mathcal{E}_1; L\right)$, it can increase the likelihood of generating minimum-weight codewords, potentially increasing $P_{SCL}\left(\mathcal{E}_2; L\right)$.

An alternative approach is to vary the list size throughout the decoding process. The stepwise change of list size can happen by steps of a single bit or segment-wise considering the uncertainty in the segments of bits. A practical metric that indicates an increase or decrease in uncertainty is the *path metric range* on the decoding bit i, defined as:

$$\rho_i = PM_i^{(L)} - PM_i^{(1)}.$$

This metric drops significantly over low-reliability bit coordinates. Observe that the LLR on low-reliability bits is small; hence, paths with large metrics remaining from the previous decoding step are pruned to make room for the extended paths based on the paths with small metrics remaining from the previous decoding step. Therefore, we can define a differential metric as:

$$\delta_i = \rho_i - \rho_{i-1}.$$

Note that we have large $\delta_i < 0$ on low-reliability coordinates, which reduces the path metric range ρ_i. Therefore, where ρ_i is small, the list size should be relatively large, and as ρ_i increases, the decoder can operate with a relatively smaller list size without significantly affecting error correction performance.

8.1.3 List Decoding of CRC-Polar Codes

It can be observed that when the SCL decoding fails, the correct decoding solution (equivalent to a path on the decoding tree, indicated by l_c) may still be in the list but not in the position of the most likely path, that is $1 < l_c \leq L$. Note that the most likely path is located in the first position according to the convention defined as $PM_1^{(i)} \leq PM_2^{(i)} \leq \cdots \leq PM_L^{(i)}$. In this case, as discussed in Chapter 4, adding a CRC to the message can assist the decoder in detecting the correct path among the L paths in the final list. To this end, the K information bits are systematically encoded, that is, an r-bit CRC vector is appended to the information vector $\mathbf{d}$. Following the notations employed in Chapter 3 for cyclic codes, let $\mathbf{d}(x)$ and $\boldsymbol{v}(x)$ denote the information bits in polynomial form and the coded message, respectively. If we have the CRC generator polynomial $\mathbf{g}(x)$, then the outer code is obtained by

$$\boldsymbol{v}(x) = x^r\mathbf{d}(x) + R_{\mathbf{g}(x)}[x^r\mathbf{d}(x)], \tag{8.8}$$

where $R_{\mathbf{g}(x)}$ denotes the remainder of dividing the argument, here $x^r\mathbf{d}(x)$, by $\mathbf{g}(x)$, and $x^r\mathbf{d}(x)$ is used for zero padding the message d by r zero-symbols. Needless to mention that $\mathbf{d}(x) = d_{K-1}x^{K-1} + \cdots + d_1x^1 + d_0$ is equivalent to $\mathbf{d} = [d_{K-1} \cdots d_1\, d_0]$. The outer code $\boldsymbol{v}$ is then polar coded into a vector $\mathbf{u}$ with length N in the similar fashion, as discussed in Chapter 4.

In the receiver, when list decoding ends, the syndrome of each candidate vector $\hat{\boldsymbol{v}}$ associated with the paths in the list is checked, starting from the path with the best metric to the worst one, using the same CRC generator polynomial $\mathbf{g}(x)$ that we use for outer encoding. The first candidate with syndrome $\mathbf{s}(x) = R_{\mathbf{g}(x)}[\hat{\mathbf{v}}(x)] = \mathbf{0}$ is assumed to be the correct path. The estimation of the transmitted message $\hat{\mathbf{d}}$ is extracted from the vector $\hat{\mathbf{v}}$ by discarding the appended r zero bits.

Note that this concatenation increases the polar code rate to $(K + r)/N$. We use the notation $P(N, K + r)$ to represent a polar code of length N with K information bits concatenated with r-bit CRC. The list decoding of CRC-polar codes is sometimes called *CRC-aided list decoding*.

8.1.4 List Decoding of PAC Codes

The concatenation of a polar code with a rate-1 convolutional code via generator polynomial $\mathbf{p} = [p_0\, p_1\, \cdots\, p_m], m > 0$, which is equivalent to convolutional pre-transformation of zero-interleaved information bits, is employed to improve the performance of polar codes under list decoding. As discussed in Chapter 6, this concatenation can improve the error coefficient of the code, that is, the number of minimum weight codewords, resulting in a superior performance over polar codes. The pre-transformation was discussed in Chapter 5. For outer decoding, the decision LLR as the output of the SC-based decoding is inversely transformed to the transmitted zero-interleaved information vector $\hat{\mathbf{v}}$ as follows: Similarly to the list decoding of polar codes, we consider two options of 0 and 1 for every $v_i, i \in \mathcal{A}$ and do the convolutional transform:

$$\hat{u}_i = \begin{cases} \hat{v}_i \oplus \bigoplus_{j=1}^{m} p_j \cdot \hat{v}_{i-j} & \text{if } i \in \mathcal{A} \\ \bigoplus_{j=1}^{m} p_j \cdot \hat{v}_{i-j} & \text{otherwise.} \end{cases} \tag{8.9}$$

Recall that $p_0 = 1$ by convention. In other words, when $i \notin \mathcal{A}$, we have $v_i[l] = 0$ resulting in $\hat{u}_i[l] = \oplus_{j=1}^{m} p_j \cdot \hat{v}_{i-j}$ and the path l is not split, since we have only one

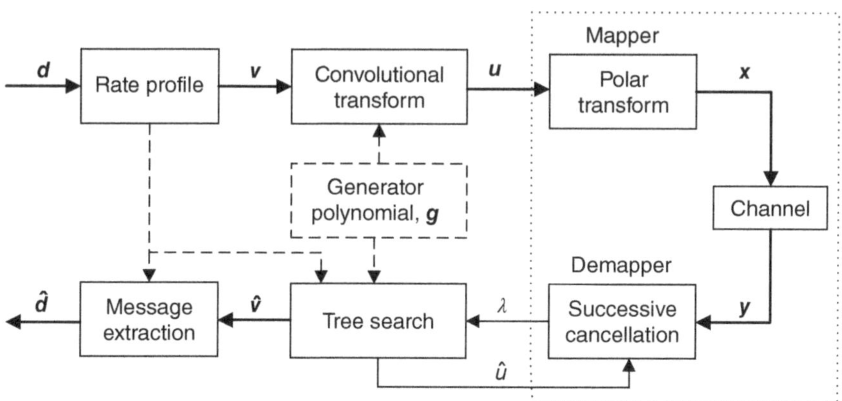

Figure 8.3 PAC coding scheme.

choice for the estimation of v_i. Whereas at $i \in \mathcal{A}$, the path l is split to $\hat{v}_i[l] = 0$ and $\hat{v}_i[L+l] = 1$, or vice versa, resulting in $\hat{u}_i[l] = \bigoplus_{j=1}^{m} p_j \cdot \hat{v}_{i-j}$ and $\hat{u}_i[l] = \hat{v}_i + \bigoplus_{j=1}^{m} p_j \cdot \hat{v}_{i-j}$ in $\mathbb{F}_2$. Once we have $\hat{u}_i$ for the path l, the branch metric is calculated similarly to the list decoding of polar codes by (8.7). Algorithm 8.2 demonstrates the decoding process, which is slightly different from Algorithm 8.1. Figure 8.3 illustrates the entire encoding and decoding process in which the inner encoding as a super-channel (see concatenated codes in Chapter 2) is placed in a dotted box.

Algorithm 8.2: List Decoding of PAC Codes

---

input : channel LLRs $\lambda_n^{0,N-1}$, $\mathcal{A}$, L, $\mathbf{p}$
output: recovered message bits $\hat{\mathbf{d}}$

$\mathcal{L} \leftarrow \{1\}$ // a single path in the list
$[\lambda, \beta] \leftarrow [\lambda_n^{0,N-1} + \{0\}, \{0\}]$ // Initialization of LLRs λ and partial sums β
for $i \leftarrow 0$ **to** $N-1$ **do**
  if $i \notin \mathcal{A}$ **then**
    for $l \leftarrow 1$ **to** $|\mathcal{L}|$ **do**
      $\lambda_0^i[l] \leftarrow$ updateLLRs($l, i, \lambda[l], \beta[l]$) // like SCLD
      $\hat{v}_i[l] \leftarrow 0$
      $[\hat{u}_i[l]$, cState$[l]] \leftarrow$ conv1bTrans(v_i, cState$[l]$, $\mathbf{p}$) // See Algorithm 6.2
      $PM_l^{(i)} \leftarrow PM_l^{(i-1)} + \mu_l^{(i)}, \forall l \in \mathcal{L}$ // cf. (8.6),(8.7),(8.4)
      $\beta[l] \leftarrow$ updatePartialSums($\hat{u}_i[l], \beta[l]$) // like SCLD
  else
    for $l \leftarrow 1$ **to** $|\mathcal{L}|$ **do**
      $\mathcal{L} \leftarrow$ duplicatePath($\mathcal{L}, l, i, \mathbf{p}$)
    if $|\mathcal{L}| > L$ **then**
      $\mathcal{L} \leftarrow$ prunePaths($\mathcal{L}$) // like SCLD
$\hat{\mathbf{d}} \leftarrow$ extractData($\hat{v}_1^N[0]$)
return $\hat{\mathbf{d}}$

8.2 Iterative SC-Based Decoding

The decoding of the received vector is not always successful. Suppose that we have a genie that tells us that we did not recover the message that was transmitted. In this case, the common practice is sending a request to the sender for retransmission. Another option is to attempt to decode the message again with a different strategy, hoping that we can correctly recover the transmitted message. In this section, we review different strategies for additional decoding attempts in an iterative manner for the SC and SCL decoding algorithms.

8.2.1 Adaptive List Decoding

When the list decoding fails, one can increase the number of surviving paths, that is increasing the list size L, in the hope that the correct path remains among the surviving paths in the next decoding attempts. This increase in the list size can be performed iteratively until the decoder finds a path with syndrome zero for the outer cyclic code, CRC, or until the decoder reaches the maximum list size $L_{\max}$. The maximum list size $L_{\max}$ is determined based on the available resources and/or the maximum admissible delay. On the other hand, knowing that the surviving paths represent the closest candidates in terms of Euclidean distance to the transmitted message, the maximum list size can be determined proportional to the number of minimum weight codewords and the error coefficient.

Denote each iteration or decoding attempt by t and the mean and maximum list size by L_{avg} and $L_{\max}$, respectively. One can begin the SC list decoding of a received sequence with $L = 1$, which is equivalent to SC decoding. Then, we can gradually increase it by a factor of two at each additional attempt. Hence, the list size at iteration t will be $L = 2^t \leq L_{\max}$. After decoding w received vectors, the average list size will be $L_{avg} = \left(\sum_{j=1}^{w} L_j\right)/w$, where L_j is the maximum list size used to decode the jth received vector and $L_j \leq L_{\max}$. Suppose that the total number of node operations in list decoding is $L \cdot N \log N$, then the average list size is proportional to the average computational complexity as $N \log N$ remains constant. Clearly, as the noise level reduces at high signal-to-noise ratio (SNR) regimes, fewer attempts are needed to decode the received vectors. Consequently, the average list size was significantly reduced. This is also the case even at low SNR regimes. The reason is that the random noise does not affect every received vector in a uniform way. Hence, the vectors less affected by noise will need fewer attempts, and most of the time they can be decoded correctly in the first attempt by $L = 1$. Observe that adaptive list decoding has an edge over single attempt decoding with a fixed list size of $L_{\max}$ in terms of average computational complexity. What about the time complexity? Suppose each iteration introduces one unit of delay; this delay is $2N - 2$ time steps for the standard SC-based decoding, as discussed in Chapter 5. Then, the average delay is $T_{avg} = \left(\sum_{j=1}^{w} T_j\right)/w$, where T_j is the number of attempts t for each received vector. Note that $T_{\max} = \log_2 L_{\max}$.

A practical variant of adaptive list decoding is a two-stage list decoding with a maximum of two attempts, which is simply a combination of SC decoding in the first stage and list decoding with a practical list size $L_{\max}$, for example,

$L_{\max} = 2^5 = 32$, in the second stage. This can be extended to multistage decoding, where the list size does not necessarily increase by a factor of two but can vary at each attempt. One can observe that this variant would have a smaller average delay T_{avg} at the cost of larger L_{avg} compared to the standard adaptive list decoding.

Algorithm 8.3: Adaptive SCL Decoding

input : received symbols y_0^{N-1}, maximum list size $L_{\max}$, CRC polynomial coefficients $\mathbf{g}_{CRC}$

output: recovered message bits $\hat{\mathbf{d}}$

$L \leftarrow 1$

$\hat{\mathbf{d}} \leftarrow \text{SCLDecoding}(y_0^{N-1}, \mathbf{g}_{CRC}, L)$ `// L = 1 for SC decoding.`

while CRCCheck$(\hat{\mathbf{d}}, \mathbf{g}_{CRC}) \neq \mathbf{0}$ AND $L < L_{\max}$ **do**

  $L \leftarrow 2 \times L$

  $\hat{\mathbf{d}} \leftarrow \text{SCLDecoding}(y_0^{N-1}, \mathbf{g}_{CRC}, L)$

return $\hat{\mathbf{d}}$

8.2.2 SC Decoding with Bit-Flipping

When SC decoding fails, wrong decisions have been made throughout the decoding process. The sequential nature of the SC and SCL decoders causes the propagation of errors from the first position of an incorrect decision. Recall from Chapter 4 that the hard decisions made on the value of the received signals are used to calculate the intermediate information required to decode the rest of the received signals. That is the reason for the error propagation.

If we knew the first position where a decision was made incorrectly, we could prevent the error propagation to the end of the decoding (in case of a single error) or to the next erroneous position not caused by the propagation. Recall again from Chapter 4 that sub-channels have different levels of reliability. The sub-channels with relatively low reliability are the positions prone to incorrect decisions. Therefore, the reliability measure of the sub-channels obtained in the unsuccessful decoding can be utilized to locate the erroneous position(s). In SC decoding, we can convert the decision based on the decision LLR α_i^0 to the potential position of the error(s). Since $|\alpha_i^0|$ can be considered a measure of reliability; the larger the value, the higher the reliability. Suppose that a single error at position j has occurred during the first decoding iteration and that the rest of the bits have been corrupted by propagation. We can sort the values $|\alpha_i^0|, i = 0, 1, \dots, N-1$ at the end of the first decoding iteration and flip the value of position(s) with small $|\alpha_i^0|$, starting from the smallest. In case of a single error, this approach can correct the error by checking the first few positions with the smallest value of $|\alpha_i^0|$.

However, in the case of multiple errors, finding the right combination of the positions to flip requires a massive number of attempts, resulting in a significant delay. Note that as the noise level reduces at high SNR regimes, the decoding failures are predominantly due to a single incorrect decision.

The key to correcting the error(s) in this decoding approach is to know the indices of sub-channels that contribute the most to the error event. The simple way to find the potential location of the error occurrence is to use the log-likelihood ratio to estimate the bits. The bit locations with the smaller magnitude of LLRs in the initial or previous decoding attempt are assumed to be the bits most prone to error in SC decoding. The index set of these sub-channels can be determined offline in a channel-independent manner. This set is called the *critical set* (CS). There exist some heuristic methods in the literature that are used to identify the first erroneous bit in the SC-flip (SCF) scheme (see the bibliographical notes). The sub-channels which are prone to the occurrence of the first error are equivalent to the first penalized bits of the correct path. Hence, these bits also contribute to the elimination of the correct path due to a variable number of penalties.

8.2.3 SCL Decoding with Diverse Path-Selection

In the SCL decoding, where the decoder deals with several candidates, the correct candidate is eliminated from the list throughout the decoding process due to single or multiple penalties (see the branch metric (8.7)) endured by the correct candidate. Recall the pruning rule at each step of the list decoding in which the decoder selects L best paths out of $2L$ paths in the expanded list. By changing the pruning rule and selecting the worst paths (in terms of metric), which is called *path-flipping* or *list-flipping*, or a selection of L best paths and L worst paths. In general, this operation is called *shifted-pruning* (SP), where we shift the pruning window by κ as illustrated in Figure 8.4. When $\kappa = L$, we choose the L worst paths and when $\kappa = L/2$, the selected paths are paths with indices $l = L/2 + k, k = 1, 2, \dots, L$.

8.2.4 SC/SCL Decoding with Perturbation

Perturbation refers to the introduction of small, controlled changes or "noise" to the received symbols or to the decoding process itself. This contrasts with the bit-flipping strategy, where the estimated/decoded bits are flipped, or the diverse path selection strategy, where the path pruning window is shifted to deviate from selecting paths with the highest likelihood during decoding. The similarity of this approach to the previous diversity-based iterative algorithms is that the decoding is repeated up to T attempts until the genie, here CRC, passes the estimated $\mathbf{d}$.

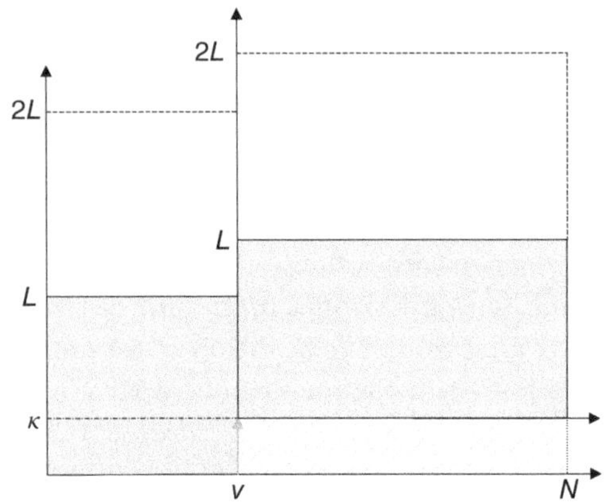

Figure 8.4 Shifting the pruning window by κ paths during the list pruning operation at bit $v \in \mathcal{V}$. Note that at coordinate v, the whole coordinate system is shifted up.

The main goal of perturbation is to prevent the decoder from getting stuck in local optima or trapping sets. In iterative decoding, the algorithm may sometimes converge to an incorrect codeword or fail to converge at all. Perturbation helps to "shake up" the decoding process and potentially guide it toward the correct solution. Perturbation introduces diversity by forcing the decoder to explore different parts of the solution space. This increased exploration can lead to better decoding decisions that might have otherwise been missed. The magnitude of the perturbation needs to be carefully chosen. Too small perturbation might not have a significant effect, while too large can degrade performance.

Let us $y'_i, i \in [0, N-1]$ denote the perturbed received symbols y_i as

$$y'_i = y_i + n'_i, \tag{8.10}$$

where $n'_i \sim \mathcal{N}(0, \sigma_p^2)$, and σ_p^2 is the perturbation noise power.

Algorithm 8.4 outlines the pseudocode for the iterative SC/SCL decoding algorithm, incorporating the addition of perturbations to the received symbols for every decoding attempt.

8.2.5 Partial Rewind of SC and SCL Decoding

In the iterative SC-based decoding algorithms, the decoding can be resumed from a position $j, 0 \leq j \leq N-2$ in each iteration or attempt to avoid redundant computations. This move from position $N-1$ to j is called *rewind*. If the resumption is not from the first bit, we call it *partial rewind*. Rewinding SC-based decoding seems trivial if we can access all $N \log_2 N$ values for intermediate LLRs and partial sums. However, as the block length increases, retaining all intermediate information becomes inefficient and impractical. We showed in Chapter 4 that full SC-based decoding requires a fraction of all intermediate information, the largest

Algorithm 8.4: Iterative SC/SCL Decoding with Perturbation

input : received symbols y_0^{N-1}, perturbation noise power σ_p^2, list size L, maximum decoding attempts T, CRC polynomial coefficients $\mathbf{g}_{CRC}$
output: recovered message bits $\hat{\mathbf{d}}$

$\hat{\mathbf{d}} \leftarrow \text{SCLDecoding}(y_0^{N-1}, \mathbf{g}_{CRC}, L)$ // $L = 1$ for SC decoding.
$t \leftarrow 1$
while CRCCheck$(\hat{\mathbf{d}}, \mathbf{g}_{CRC}) \neq \mathbf{0}$ AND $t < T$ **do**
  $y_0'^{N-1} \leftarrow y_0^{N-1} + n_0'^{N-1}$ // $n'_i \sim \mathcal{N}(0, \sigma_p^2)$
  $\hat{\mathbf{d}} \leftarrow \text{SCLDecoding}(y_0'^{N-1}, \mathbf{g}_{CRC}, L)$
  $t \leftarrow t + 1$
return $\hat{\mathbf{d}}$

$N - 1$ values of LLRs, and the partial sums. As we store a fraction of all the intermediate information in the memory-efficient SC-based decoding, we may be unable to rewind the SC process to the target bit index. The reason is that some of the stored intermediate information needed to decode the target symbol may have been overwritten. To recompute the lost information, we may need to rewind the process further. Before proposing the formal scheme for the rewinding process, we introduce a new operator for grouping the bit indices. This special grouping helps us find the bit index closest to the target index for rewinding. We also analytically prove that this approach gives access to the untouched intermediate information stored in the memory, which is essential to resuming the SC process.

Here, we use the properties of the SC algorithm discussed in Section 5.11.3 of Chapter 5 that can help us to rewind the process efficiently. We learned in Section 5.11 that we could save a significant amount of memory by knowing the maximum intermediate LLRs and the partial sums required to decode each bit. However, there is a drawback to this efficiency. Since we use a limited space for intermediate information instead of $N \log_2 N$ memory elements, we have to overwrite the used values throughout the SC decoding process, which we no longer need to proceed with decoding. In the normal decoding process, the overwriting operation does not cause any data corruption. However, if we need to move backward like in iterative decoding algorithms such as SCF and SP or in Fano decoding, we may no longer access the intermediate information as it may have been lost due to overwriting. In this section, based on the update properties we studied in Section 5.11 of Chapter 5, a scheme is proposed such that the rewind of the SC algorithm is performed efficiently with significantly fewer computations compared to restarting the algorithm.

Suppose the SC algorithm is decoding bit i and needs to rewind the SC process to bit $j, j < i$, with $i, j \in \mathcal{A}$. In the SCF scheme and SP scheme, we chose $i = 2^n - 1$; however, in Fano decoding, we may have $i \leq 2^n - 1, i \in \mathcal{A}$. Since the intermediate information required for decoding bit j may partially be overwritten, we may need to rewind further to a position denoted by $j_p < j$. From j_p, the SC algorithm proceeds with normal decoding up to position j. We shift the pruning window to this position or flip the bit u_j and then continue the normal SC-based decoding.

Now, the question is what the position j_p is? Assume $i \in \mathcal{Z}_{p'}$ and $j \in \mathcal{Z}_p$. Then,

$$j_p = \begin{cases} z_p & \text{if } z_p < z_{p'} \\ z_{p'} & \text{if } z_p = z_{p'}. \end{cases} \tag{8.11}$$

Example 8.2 Suppose $N = 2^5$ and we need to rewind the SC algorithm from position $i = 31 = (11111)_2$ to $j = 19 = (10011)_2$. We know that $i \in \mathcal{Z}_4$ and $j \in \mathcal{Z}_1$. Therefore, according to (8.11), $j_p = z_p = 16 = (10000)_2$.

Recursion for Case $z_p = z_{p'}$

For the case $z_p = z_{p'}$ in (8.11), we can choose a position k, $j_p < k \leq j$ to rewind, which is more efficient. To this end, let us $k \leftarrow j$ and $m \leftarrow n$, then while $\phi(k) \neq 0$:

- first, truncate the binary representation of $k = (k_{n-1} \dots k_1 k_0)_2$ by removing the bits from position $m - 1 - \phi(k)$ to the most significant bit (inclusive), that is position $m - 1$. Note that after truncation, we have a binary number of length $m = m - (\phi(k) + 1)$.
- second, find the new set $\mathcal{Z}_{p''}$ such that $k \in \mathcal{Z}_{p''}$ for $k = \sum_{t=m-(\phi(k)+1)}^{n-1} j_t \cdot 2^t - k$.
- then, $j_p = \sum_{t=m-(\phi(k)+1)}^{n-1} j_t \cdot 2^t + z_{p''}$.

We can continue the above procedure recursively to minimize $z_{p''} - j$. Note that in this recursion, k and m are replaced with new values at each iteration.

Example 8.3 Suppose $N = 2^5$ and we need to rewind the SC algorithm from position $i = 22 = (10110)_2$ to $j = 19 = (10011)_2$. We know that $i, j \in \mathcal{Z}_1$ and, therefore, $j_p = z_p = 16 = (10000)_2$. We truncate $j = (10011)_2$, as previously mentioned. We get $k = (011)_2$, $k \in \mathcal{Z}_0$, and $z_{p''} = 0$. Hence, the new j_p is $j_p = z_p + z_{p''} = 16$, the same as before.

Example 8.4 Suppose $N = 2^5$ and we need to rewind the SC algorithm from position $i = 22 = (10110)_2$ to $j = 20 = (10100)_2$. We know that $i, j \in \mathcal{Z}_1$ and, therefore, $j_p = z_p = 16 = (10000)_2$. However, if we truncate $j = (10100)_2$, as mentioned above, we get $t = (100)_2$, $t \in \mathcal{Z}_1$, and $z_{p''} = 4$. Hence, the new j_p is $j_p = z_p + z_{p''} = 16 + 4 = 20$.

One can observe that the recursion is not used in the schemes, where the rewind is performed from the last bit index. The reason is that the bit index $2^n - 1 \in \mathcal{Z}_n$, and this set have only one other element, which is $2^n - 2 = z_{n-1}$. On the other hand, if we need to rewind the SC process to a bit index smaller than z_{n-1}, the target bit index will fall into another set $\mathcal{Z}_p$ with different z_p. Hence, this may be used for Fano decoding, where the case $z_p = z_{p'}$ is possible. Note that we do not numerically evaluate this approach for Fano decoding, as we do not have any other approach to compare with. We can either use this approach or simply store all $N \log_2 N$ intermediate LLRs and partial sums and trade a significant complexity reduction with the memory efficiency.

Now, let us revisit the shifted pruning scheme. In this scheme (and in the SCF scheme), we may need to repeat the SC rewinding process up to T times. Therefore, we need to consider this. Assuming $t \in [1, T]$ indicates the current iteration, and $j(t)$ and $j_p(t)$ denote the j and j_p of iteration t, respectively, then j_p of the current iteration is obtained by considering $j_p(t-1)$ as follows:

$$j_p = \begin{cases} j_p(t-1) & \text{if } j_p(t) > j_p(t-1) \\ j_p(t) & \text{otherwise} \end{cases} \tag{8.12}$$

As (8.12) shows, if the destination position of the current iteration $j(t)$ is greater than the destination position of the previous iteration, the intermediate information is not valid. The reason is that some modification (bit-flipping or shifted-pruning) occurred at position $j(t-1)$ that affects not only the intermediate information but also the decoded data. In other words, we need to go to position $j(t-1)$, undo the modification and proceed with the decoding to position $j(t)$, and then perform the modification of the current iteration. Note that if both $j(t)$ and $j(t-1)$ are in the same $\mathcal{Z}_p$, then $j_p(t) = j_p(t-1)$, so there will be no difference. Figure 8.5 compares j and j_p for an example in which 5 iterations occur.

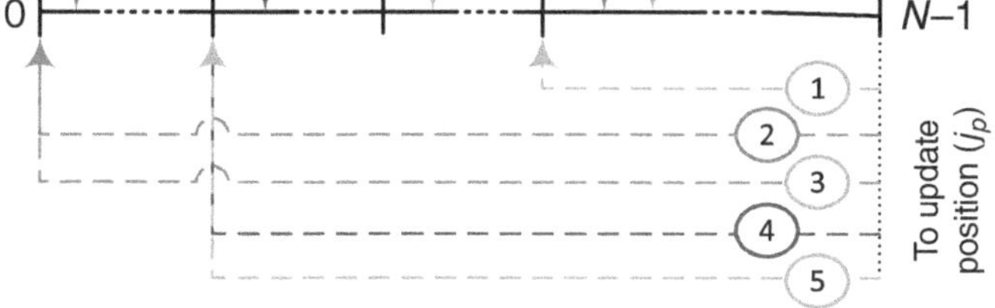

Figure 8.5 An illustrative example comparing the target position j and update position j_p.

Algorithm 8.5: SC-Flip Decoding with Partial Rewind

```
input : the received vector y_1^N, non-frozen set A
output: the recovered message bits û_1^N
[T, t, crcPass, z_{-1}] ← [20, 0, false, N − 1]
δ[1 : K + r, 1 : 2] ← {0}
do
    if t=0 then
        û_1^N, δ ← SCD(y_1^N, A, t, z_{-1})
        Sort δ with respect to |λ_0|, the 2nd column
    else
        û_1^N, z_{-1} ← SCD(y_1^N, A, t, z_{-1})
    if CRC(û_1^N) = success then
        crcPass = true
    if crcPass = false then
        t ← t + 1
while t < T AND crcPass = false
return û_1^N
subroutine SCD (y_1^N, A, t, z_{-1}):
    if t = 0 then
        z, j ← 0, 0
        for i ← z to N do
            λ_0 ← update_LLRs(i)      // updating the intermediate LLRs
            û_i ← h(λ_0)                        // hard decision on λ_0
            update_PSs(û_i)       // updating the intermediate partial
            sums
            δ[j, 1 : 2] ← [i, |λ_0|]
            j ← j + 1
        return û_1^N, δ
    else
        z ← RewindTo(δ[t, 1], z_{-1}, log_2(N))
        for i ← z to N do
            λ_0 ← update_LLRs(i)      // updating the intermediate LLRs
            û_i ← h(λ_0)                        // hard decision on λ_0
            if i = δ[t, 1] then
                û_i ← û_i ⊕ 1
            update_PSs(û_i)       // updating the intermediate partial
            sums
        return û_1^N, z
subroutine RewindTo (j, z_{-1}, n):
    z = 2^n − 2^{n−φ(j)}                     // Cf. (5.36) for φ(.)
    if z_{-1} < z then
        z = z_{-1}
    return z
```

Furthermore, when rewinding the list decoder from the last bit position, $N - 1$, to position j_p, some of the paths that existed at position j_p in the previous iteration could be eliminated in between and replaced with other paths. This potential replacement should be addressed when we have a list of paths/candidates, such as in the shifted-pruning scheme, not in the SCF scheme. To simplify the problem, we can limit the positions j_p to $j_p = 2^{n-1}$. Because all computations of intermediate LLRs from this position, 2^{n-1}, up to the last position, $2^n - 1$, are performed solely based on the channel LLRs and partial sums of stage $\psi(2^{n-1} - 1)$. Hence, we need to store the decoded data $\mathbf{u}[0 : 2^{n-1} - 1]$ and the path metric of all paths at position $2^{n-1} - 1$. Partial sums can also be stored or can be computed simply by $\mathbf{u}[0 : 2^{n-1} - 1]\mathbf{G}_{N/2}$.

Algorithm 8.5 illustrates the use of rewinding in the SCF algorithm at line 26 where the subroutine *RewindTo* returns the update position z, which is smaller than or equal to the target position $\delta[t, 1]$. As can be seen, depending on the iteration counter t, either the decoder starts decoding from bit $i = 0$ in lines 18—23 when $t = 0$ or from bit $i = z$ in lines 27–32 when $t > 0$. In the latter case, $\hat{u}_i$ is flipped at position $i = \delta[t, 1]$ in line 31.

Exercises

1 What do you expect if the path pruning operation is performed after multiple decoding steps? For instance, if currently decoding symbol i, the expansion is continued to symbol $i + 3$ and then pruning is performed. What are the advantages and disadvantages of this approach?
Hint: You should discuss the consequences of this approach in terms of expected performance (relative degradation or improvement) and required resources.

2 What would be the difference in the block error rate (BLER) performance of any polar code under adaptive list decoding with $L_{\max} = 8$, two-stage list decoding with $L = 1$ and $L = 8$, and single-attempt list decoding with $L = 8$?

3 What are the pros and cons of request for retransmission over multiple decoding attempts and vice versa?

4 Multiple attempts for decoding introduce a delay in the processing of the incoming signals from the channel. How can we tackle the queue imbalance in the input and output of the decoder?

Bibliographical Notes

As the performance of polar codes under SC decoding by Arıkan [2009] was not satisfactory, inspired by the list decoding algorithm used for Reed-Muller codes proposed by Dumer and Shabunov [2006], an elegant recursive multipath (list) decoding was proposed by Tal and Vardy [2011]. This was further improved by concatenating the polar codes with the CRC for detection of the correct path in the list. Note that the list decoding algorithm with list size M, called the M-Algorithm, was previously employed for convolutional codes by Anderson and Mohan [1984] as well. To reduce space complexity and memory usage, the LLR-based metric for list decoding was proposed by Balatsoukas-Stimming et al. [2014] and Yuan and Parhi [2014]. Given the importance of the list size in list decoding, Coşkun and Pfister [2022] analyzed the list size requirements in relation to code construction, employing a metric called differential entropy. In addition, techniques to adjust the size of the list during the decoding process have been explored by a number of researchers such as Chen et al. [2015], Fan et al. [2015], and Rowshan and Viterbo [2018]. The concatenation of polar codes with rate-1 convolutional codes, resulting in a class of codes called polarization adjusted convolutional (PAC) codes, was proposed by Arıkan [2019] for Fano decoding. This concatenation turned out to improve the error coefficient of the polar codes. SCL decoding and the List Viterbi decoding were adapted to PAC codes by Rowshan et al. [2021], Yao et al. [2021], and Rowshan and Viterbo [2021b], respectively. It is worth noting that the decoding errors were classified and numerically analyzed for list decoding, as discussed by Wilson and Altmayer [2021].

The idea of repeating the decoding process when decoding fails originated from the classical maximum likelihood decoding algorithms such as the one proposed by Chase [1972], where in each decoding attempt, a number of error patterns based on channel measurements are checked. The adaptive list decoding algorithm for CRC-Polar codes was employed by Li et al. [2012], where the list size increases in each decoding iteration until the decoding is successful. It is worth noting that a similar algorithm was earlier introduced by Gorji Zadeh and Soleymani [2005] for list decoding of convolutional codes. A different approach was proposed for correcting a single error using a likelihood test over proceeding frozen bits in the SC decoding over the binary erasure channel (BEC) by Parizi [2012] in his Master's thesis and later introduced by Afisiadis et al. [2014] for CRC-Polar codes based on the decision log-likelihood ratio (LLR) as a metric over the binary-input additive white Gaussian noise (BI-AWGN) channel. In another attempt, Zhang et al. [2017] proposed a static critical set (CS) of nonfrozen bits that collects the index of the first bit of the rate-1 constituent codes. It was numerically shown that over 99% of the first errors occur over the indices in the critical set. Later, the idea of bit-flipping was extended to SCL decoding by different path selections to avoid eliminating the correct path by Rowshan and Viterbo [2019,2021a,2022b] and Cheng et al. [2019].

Using CRC bits for block error detection, one can recover the correct path (transmitted data) in additional T attempts by making different decisions, that is flipping the initial estimate of the bit (s), in SC decoding [Afisiadis et al., 2014; Chandesris et al., 2018; Zhang et al., 2017], or effectively following different paths in the decoding tree in SC list decoding [Rowshan and Viterbo, 2019,2021a; Cheng et al., 2019; Zhang, 2023; Rowshan et al., 2019]. The earlier decoding is called SC-flip (SCF) decoding, and the latter is referred to as SCL decoding with shifted-pruning (SP), SCL-flip (SCLF) decoding, or other names such as SCL decoding with list-flipping and path flipping. Furthermore, there is an efficient way to start re-decoding in the additional iterations by Rowshan and Viterbo [2022a], that is, the re-decoding process is performed by partially rewinding the SC process, not necessarily from the first bit. This can significantly reduce the complexity of decoding in the SCF and SCLF decoding algorithms. Recently, a different approach called SCL-perturbation decoding has been developed [Wang et al., 2023], which is based on the idea of adding noise or perturbation to the received sequence in the literature, where small random perturbations are introduced to the received symbols before each SCL decoding attempt. Numerical results show that the perturbation yields a higher coding gain than flipping at a large block length with the same number of decoding attempts.

Bibliography

Orion Afisiadis, Alexios Balatsoukas-Stimming, and Andreas Burg. A low-complexity improved successive cancellation decoder for polar codes. In *2014 48th Asilomar Conference on Signals, Systems and Computers*, pages 2116–2120, 2014. doi: 10.1109/ACSSC.2014.7094848.

John B Anderson and Seshadri Mohan. Sequential coding algorithms: A survey and cost analysis. *IEEE Transactions on Communications*, 32(2):169–176, 1984. doi: 10.1109/TCOM.1984.1096023.

Erdal Arıkan. Channel polarization: A method for constructing capacity-achieving codes for symmetric binary-input memoryless channels. *IEEE Transactions on Information Theory*, 55(7):3051–3073, 2009.

Erdal Arıkan. From sequential decoding to channel polarization and back again. *arXiv preprint arXiv:1908.09594*, 2019.

Alexios Balatsoukas-Stimming, Mani Bastani Parizi, and Andreas Burg. LLR-based successive cancellation list decoding of polar codes. In *2014 IEEE International Conference on Acoustics, Speech and Signal Processing (ICASSP)*, pages 3903–3907, 2014. doi: 10.1109/ICASSP.2014.6854333.

David Chase. Class of algorithms for decoding block codes with channel measurement information. *IEEE Transactions on Information Theory*, 18(1):170–182, 1972.

Kai Chen, Bin Li, Hui Shen, Jie Jin, and David Tse. Reduce the complexity of list decoding of polar codes by tree-pruning. *IEEE Communications Letters*, 20(2):204–207, 2015.

Fengyi Cheng, Aijun Liu, Yingxian Zhang, and Jing Ren. Bit-flip algorithm for successive cancellation list decoder of polar codes. *IEEE Access*, 7:58346–58352, 2019. doi: 10.1109/ACCESS.2019.2914691.

Mustafa Cemil Coşkun and Henry D Pfister. An information-theoretic perspective on successive cancellation list decoding and polar code design. *IEEE Transactions on Information Theory*, 68(9):5779–5791, 2022.

Ilya Dumer and Kirill Shabunov. Soft-decision decoding of Reed-Muller codes: Recursive lists. *IEEE Transactions on Information Theory*, 52(3):1260–1266, 2006. doi: 10.1109/TIT.2005.864443.

YouZhe Fan, Ji Chen, ChenYang Xia, Chi-ying Tsui, Jie Jin, Hui Shen, and Bin Li. Low-latency list decoding of polar codes with double thresholding. In *2015 IEEE International Conference on Acoustics, Speech and Signal Processing (ICASSP)*, pages 1042–1046. IEEE, 2015.

Seyed Ali Gorji Zadeh and M Reza Soleymani. An adaptive M-algorithm convolutional decoder. In *VTC-2005-Fall. 2005 IEEE 62nd Vehicular Technology Conference, 2005*, volume 4, pages 2177–2181, 2005.

Bin Li, Hui Shen, and David Tse. An adaptive successive cancellation list decoder for polar codes with cyclic redundancy check. *IEEE Communications Letters*, 16(12):2044–2047, 2012.

Chandesris Ludovic, Savin Valentin, and Declercq David. Dynamic-SCFlip decoding of polar codes. *IEEE Transactions on Communications*, 66(6):2333–2345, 2018. IEEE.

Mani Bastani Parizi. *Polar codes: Finite length implementation, error correlations and multilevel modulation*. PhD thesis, Swiss Federal Institute of Technology Lausanne (EPFL), 2012.

Mohammad Rowshan and Emanuele Viterbo. Stepped list decoding for polar codes. In *2018 IEEE 10th International Symposium on Turbo Codes & Iterative Information Processing (ISTC)*, pages 1–5. IEEE, 2018.

Mohammad Rowshan and Emanuele Viterbo. Improved list decoding of polar codes by shifted-pruning. In *2019 IEEE Information Theory Workshop (ITW)*, pages 1–5, 2019. doi: 10.1109/ITW44776.2019.8989330.

Mohammad Rowshan and Emanuele Viterbo. Shifted pruning for path recovery in list decoding of polar codes. In *2021 IEEE 11th Annual Computing and Communication Workshop and Conference (CCWC)*, pages 1179–1184. IEEE, 2021a.

Mohammad Rowshan and Emanuele Viterbo. List Viterbi decoding of PAC codes. *IEEE Transactions on Vehicular Technology*, 70(3):2428–2435, 2021b. doi: 10.1109/TVT.2021.3059370.

Mohammad Rowshan and Emanuele Viterbo. Efficient partial rewind of successive cancellation-based decoders for polar codes. *IEEE Transactions on Communications*, 70(11):7160–7168, 2022a.

Mohammad Rowshan and Emanuele Viterbo. SC list-flip decoding of polar codes by shifted pruning: A general approach. *Entropy*, 24(9):1210, 2022b.

M Rowshan, E Viterbo, R Micheloni, and A Marelli. Repetition-assisted decoding of polar codes. *Electronics Letters*, 55(5):270–272, 2019.

Mohammad Rowshan, Andreas Burg, and Emanuele Viterbo. Polarization-adjusted convolutional (PAC) codes: Sequential decoding vs list decoding. *IEEE Transactions on Vehicular Technology*, 70(2):1434–1447, 2021. doi: 10.1109/TVT.2021.3052550.

Ido Tal and Alexander Vardy. List decoding of polar codes. In *IEEE International Symposium on Information Theory*, pages 1–5, St. Petersburg, Russia, July 2011.

Xianbin Wang, Huazi Zhang, Jiajie Tong, Jun Wang, Jianglei Ma, and Wen Tong. Perturbation-enhanced SCL decoder for polar codes. In *2023 IEEE Globecom Workshops (GC Wkshps)*, pages 1674–1679. IEEE, 2023.

Stephen G Wilson and Kumud S Altmayer. On error probability of CRC/polar codes with list decoding, 2021.

Hanwen Yao, Arman Fazeli, and Alexander Vardy. List decoding of Arıkan's PAC codes. *Entropy*, 23(7):841, 2021.

Bo Yuan and Keshab K Parhi. Successive cancellation list polar decoder using log-likelihood ratios. In *2014 48th Asilomar Conference on Signals, Systems and Computers*, pages 548–552, 2014. doi: 10.1109/ACSSC.2014.7094505.

Zhaoyang Zhang, Kangjian Qin, Liang Zhang, Huazi Zhang, and Guo Tai Chen. Progressive bit-flipping decoding of polar codes over layered critical sets. In *GLOBECOM 2017 - 2017 IEEE Global Communications Conference*, pages 1–6, 2017. doi: 10.1109/GLOCOM.2017.8254149.

9

Fast SC-Based Decoding

"The crux of the whole thing was, How do you design a good decoding algorithm?"

— Robert G. Gallager

The successive cancellation (SC) decoding algorithm discussed in Chapter 4, due to its recursive nature, requires computing $N \log_2 N$ nodes in the associated factor graph. However, as observed in Chapter 3 while decoding Reed–Muller codes, we can leverage the constituent codes of polar codes, eliminating the need to compute all $N \log_2 N$ nodes. This significantly reduces decoding complexity, giving rise to what is known as simplified or fast decoding.

This chapter explores various special nodes that correspond to constituent codes depending on the arrangement of frozen bits. We further generalize these nodes and demonstrate their application in SC list decoding, while noting that they can also enhance the efficiency of any SC-based decoding algorithm, including sequential decoding, which will be discussed in Chapter 10. Note that several other approaches and nodes further reduce complexity, albeit some of them come at the cost of slight performance degradation. We review some of these methods in the Bibliographical Notes section.

9.1 SC Decoding via Special Nodes

To generalize successive cancellation decoding, we provide a framework for decoding polar codes via their constituent codes. To this end, let $\mathbf{T}_n$ denote the full binary tree of depth $n, n > 0$. This binary tree has 2^n leaves at depth n. Now, take a node v at a depth $d_v < n$ on this tree. We refer to its parent node and the left and right child nodes by p_v, v^l, and v^r, respectively.

Polar Codes: From Theory to Practice, First Edition. Mohammad Rowshan and Emanuele Viterbo.

Let $\mathbf{V}_v$ denote the subtree rooted at node v. We index the leaves of the tree in a typical way in the set $[N]$. We define $\mathcal{I}_v \subseteq \mathcal{A}$ as the indices of all the information nodes that are descendants of the parent node v as follows where $l(u)$ gives the index of node u.

$$\mathcal{I}_v = \{i = l(u) \, : \, u \in \mathbf{V}_v, i \in \mathcal{A} \text{ and } u \text{ is a leaf node}. \tag{9.1}$$

Now, the constituent code corresponding to node v is $\mathbf{u}_v \boldsymbol{G}_{n-d_v}$, where $\mathbf{u}_v$ is obtained by interleaving the zeros and the information bits based on the index set $\mathcal{I}_v \cap \mathcal{A}$.

If $|\mathcal{I}_v \cap \mathcal{A}| = 2^{n-d_v}$, that is, all leaf nodes are associated with information bits, then we have a rate-1 constituent code at node v or we have a *Rate-1 node*. If $|\mathcal{I}_v \cap \mathcal{A}| = 0$, then we have a rate-0 constituent code at node v or the node is a *Rate-0 node*.

A constituent code at node v can be decoded locally since we constructed the code recursively. Assuming we employ the SC decoder at node v, then as Figure 9.2 illustrates, it calculates α^l via (Figure 9.1)

$$\alpha^l[i] = \alpha[2i] \boxplus \alpha[2i+1] \quad \text{for} \quad i = 0 \, : \, 2^{n-d_v-1} - 1, \tag{9.2}$$

and passes α_{v^l} to v^l. The binary operator $\boxplus$ combines two real numbers x and y to form $x \boxplus y := 2\,\text{atanh}(\tanh(x/2)\tanh(y/2))$. This operator has the useful property that

$$h(x \boxplus y) = h(x) \oplus h(y) \quad \text{if} \quad xy \neq 0, \tag{9.3}$$

where $\oplus$ denotes the modulo-two sum or addition in the field $\mathbb{F}_2$.

When the local decoder later receives β_{v^l} from node v^l, it calculates α_{v^r} via

$$\alpha_{v^r}[i] = (1 - 2\beta_{v^l}[i])\alpha_v[2i] + \alpha_v[2i+1] \quad \text{for} \quad i = 0 \, : \, 2^{n-d_v-1} - 1, \tag{9.4}$$

and passes α_{v^r} to v^r. When the local decoder eventually receives β_{v^r} from node v^r, it calculates β_v via

$$\beta_v[2i+1] = \beta_{v^r}[i] \quad \text{and} \quad \beta_v[2i] = \beta_{v^l}[i] \oplus \beta_{v^r}[i], \tag{9.5}$$

and passes β_v to p_v.

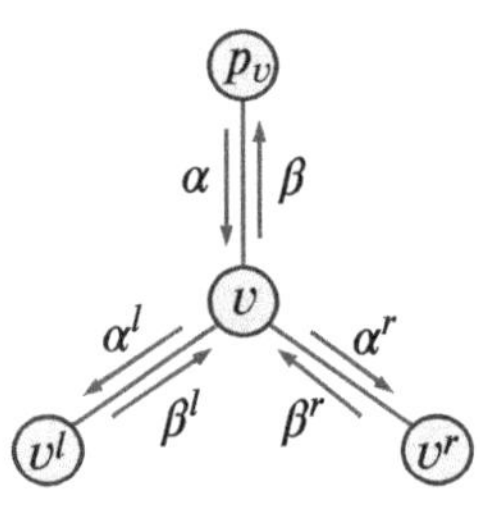

Figure 9.1 Inputs and outputs of a node v at depth d_v of the decoding tree.

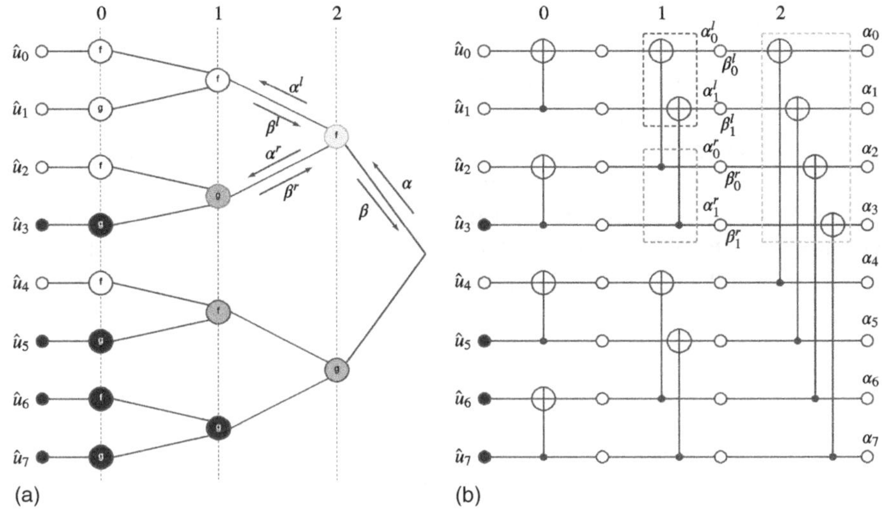

Figure 9.2 (a) The decoding tree for 8-bit code of rate 1/2 and (b) the corresponding factor graph.

Note that if node v is a leaf node, then the local decoder immediately sets β_v to

$$\beta_v = \begin{cases} h(\alpha_v) & \text{if } \mathcal{I}_v \in \mathcal{A}, \\ 0 & \text{otherwise}, \end{cases} \tag{9.6}$$

and passes it to parent node p_v. In the following, we review how we can simplify this process depending on the node rate. Figure 9.2 illustrates the decoding tree of a sample code of length 8 that we are going to simplify.

9.2 Rate-1 Node

When a Rate-1 node v is activated, it immediately calculates β_v via $\beta_v = h(\alpha_v)$ and all bits are immediately decoded using

$$\hat{\mathbf{u}}_v = \beta_v \boldsymbol{G}_{n-d_v}^{-1} = \beta_v \boldsymbol{G}_{n-d_v}, \tag{9.7}$$

where $\hat{\mathbf{u}}_v = [\hat{u}[\min \mathcal{I}_v], \ldots, \hat{u}[\max \mathcal{I}_v]]$. Recall from Chapter 5 that the inverse polar transform is $\boldsymbol{G}_{n-d_v}^{-1} = \boldsymbol{G}_{n-d_v}$.

9.3 Rate-0 Node

When a Rate-0 node v is activated, that is, $\mathbf{u}_v = \mathbf{0}$, we can immediately set $\beta_v = \mathbf{0}$ since $\mathbf{u}_v\ \boldsymbol{G}_{n-d_v} = \mathbf{0}\ \boldsymbol{G}_{n-d_v} = \mathbf{0}$.

9.4 Rate-C Node

One can observe that the two basic nodes, rate-0 and rate-1, are located deeper than the nodes with rates in between, denoted by rate-C nodes. Hence, to simplify the decoding further, we are interested in replacements for the rate-C nodes that are able to decode the rate-C constituent codes by calculating the output β_v directly from the input α_v—without the need to traverse the associated subtree. The following describes a few rate-C constituent codes. These nodes are decoded based on the following maximum-likelihood principle:

$$\beta_v = \arg\max_{\mathbf{x}_v \in \mathscr{C}_v} \sum_{j=0}^{2^{n-d_v}-1} (1 - 2\mathbf{x}_v[j])\alpha_v[j], \tag{9.8}$$

where $\mathscr{C}_v$ is the constituent code at node v and $1 - 2\mathbf{x}_v[i]$ is the bipolar modulation, that is $\{0, 1\} \longrightarrow \{1, -1\}$, of vector $\mathbf{x}_v$.

9.5 Repetition Nodes

When the subtree of a node v at depth d_v takes only one information bit at the rightmost leaf node, that is $\mathbf{u}_v = [0\,0\,\ldots\,0\,u_{\max \mathcal{I}_v}]$ and $|\mathcal{I}_v \cap \mathcal{A}| = 1$, the resulting constituent code is a *repetition* (REP) code. Observe that the single information bit $u_{\max \mathcal{I}_v}$ is repeated 2^{n-d_v} times after encoding: $\mathbf{x}_v = \mathbf{u}_v \mathbf{G}_{n-d_v} = u_{\max \mathcal{I}_v} \mathbf{g}_{\max \mathcal{I}_v} = u_{\max \mathcal{I}_v} \mathbf{1}$. Since all $\mathbf{x}_v[i]$ are identical, following (9.8), the repetition node is decoded under SC decoder by

$$\beta_v[i] = \arg\max_{\substack{\mathbf{x}_v \in \mathscr{C}_v, \\ \mathbf{x}_v \in \{\mathbf{0}, \mathbf{1}\}}} (1 - 2\mathbf{x}_v[i]) \sum_{j=0}^{2^{n-d_v}-1} \alpha_v[j] = h\left(\sum_{j=0}^{2^{n-d_v}-1} \alpha_v[j]\right), \tag{9.9}$$

Figure 9.3 demonstrates the formation of repetition node, labeled by REP, which follows the decoding tree shown in Figure 9.2.

9.5.1 General Repetition Node

In general, any node at depth d_v for which all its descendants are rate-0 nodes, except the rightmost one at depth $p < d_v$, that is, a generic rate-C node, can be decoded in a similar fashion. Figure 9.4 depicts the subtree of the general repetition nodes. Recall that

$$\alpha_i^r = g_t(\alpha) = \alpha_{i+2^t}^r + (1 - 2\beta_i^l)\alpha_i^r. \tag{9.10}$$

Since $\beta^l = 0$ and consequently $(1 - 2\beta_i^l) = 1$ at depth t to p, then we can compute α^r by a series of g functions based on the log-likelihood ratio (LLR) vector α of depth t. Hence, the LLRs vector at depth p is obtained by

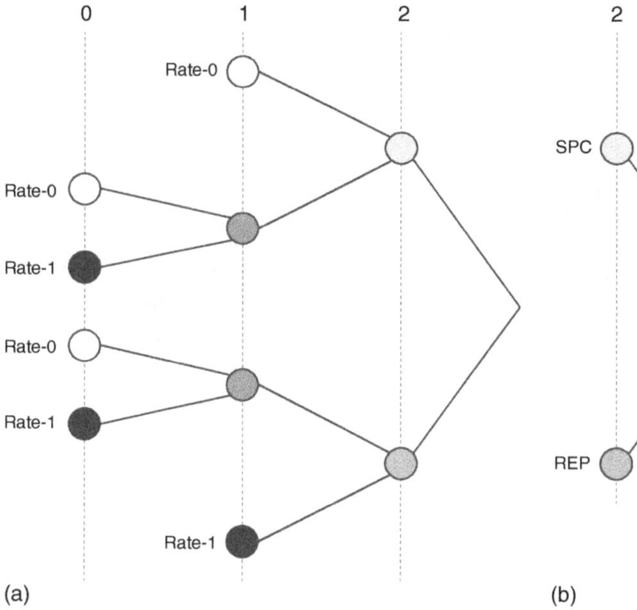

Figure 9.3 An example illustration of (a) rate-0 and rate-1 nodes (figure a), and (b) repetition (REP) and single parity check (SPC) node in the decoding tree for 8-bit code of rate 1/2.

$$\alpha_i^r = \sum_{j=0}^{2^p-1} \alpha_{i+j2^{t-p}}. \tag{9.11}$$

Now, the partial sum vector β^r of the rate-C descendant node at depth p is obtained based on α_i^r under SC decoding. Note that this node can further decompose into its constitute codes such as the aforementioned rate-1 code and parity check code which comes later. Then, since the partial sum vector β at depth t is calculated

Figure 9.4 A general repetition node.

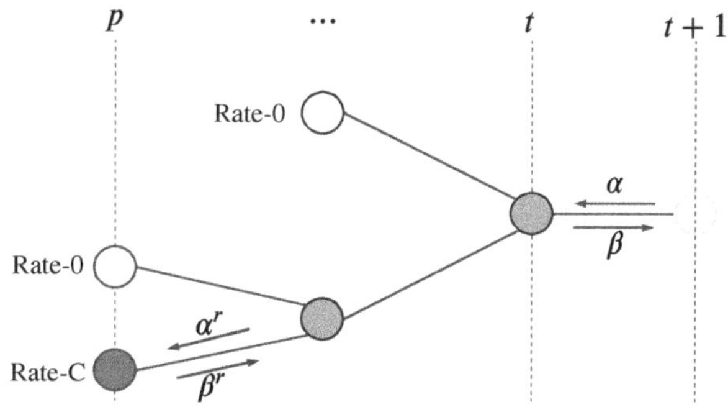

recursively using (9.5) with $\beta_i^l = 0$, the vector β^r is repeated 2^{t-p} times resulting in $\beta = [\underbrace{\beta^r \ \dots \ \beta^r}_{2^{t-p}}]$.

9.6 Parity Check Nodes

When the subtree of a node v at depth d_v takes only one frozen bit at the leftmost leaf node, that is $|\mathcal{I}_v \cap \mathcal{A}| = |\mathcal{I}_v| - 1$, the associated constituent code is called *single parity check* (SPC) code. Note that an odd number of errors can be detected by SPC because the parity checksum is not zero. Assuming the single frozen bit takes $u_0 = 0$, the code will have an even parity constraint because from the multiplication of $\mathbf{u}_v$ and the first column of $\boldsymbol{G}_{n-d_v}$, we have

$$\mathrm{x}_0 = \bigoplus_{i=1}^{2^{d_v}-1} u_i \oplus 0, \tag{9.12}$$

$$\bigoplus_{i=1}^{2^{d_v}-1} u_i \oplus \mathrm{x}_0 = 0. \tag{9.13}$$

That is, the value of the parity bit x_0 should make an even number of 1's on the left-hand side of (9.13). Recall that the addition and subtraction operations in $\mathbb{F}_2$ are identical.

The local decoding of this node is similar to the decoding of rate-1 nodes, except for considering the even parity constraint. Following the decoding process of rate-1 nodes, we get $\beta_v = h(\alpha_v)$. Then, we check whether the hard decision on the least reliable bit (LRB) position, $i_{LRB} = \arg\min_{0 \le j \le 2^{d_v}-1}(|\alpha_j|)$, satisfies the even parity constraint. That is, the dot product of β_v and the first column of $\boldsymbol{G}_{n-d_v}$ in $\mathbb{F}_2$, which is

$$\gamma = \bigoplus_{i=0}^{2^{d_v}-1} h(\alpha_i) \tag{9.14}$$

should result in $\gamma = 0$. This is in fact equivalent to $u_0 = 0$. If we get $\gamma \neq 0$, we need to flip the decision made on the value of the least reliable bit, $\beta_{i_{LRB}} = \beta_{i_{LRB}} \oplus \gamma$. This decoding scheme is called *Wagner decoding*.[1] In summary, the elements of β_v are obtained by

$$\beta_i = \begin{cases} h(\alpha_i) \oplus \gamma & \text{if } i = i_{LRB} \\ h(\alpha_i) & \text{otherwise.} \end{cases} \tag{9.15}$$

---

1 Wagner decoding is based on Wagner code, a codeword consisting of a sequence of $N - 1$ message bits followed by an additional bit used as a parity check. Observe that such a code is actually an SPC code.

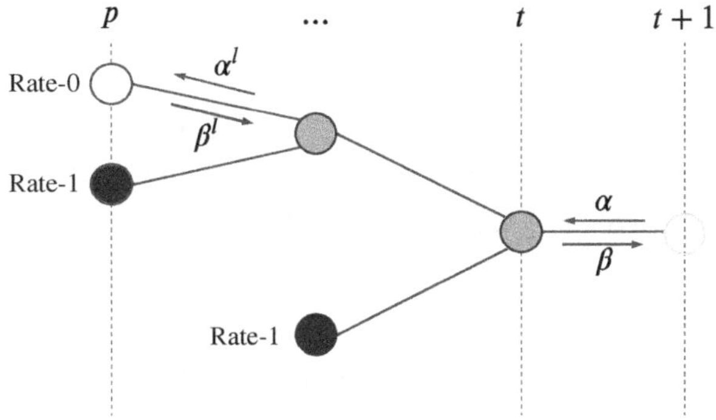

Figure 9.5 A general parity check node.

9.6.1 General Parity Check Node

In general, a node at depth t constitutes all rate-1 descendants, except for the left-most one at a depth $p < t$ which is rate-0, can be decoded as 2^p parallel single parity check codes due to having $N_p = 2^p$ independent parity check constraints by Wagner decoder. Figure 9.5 depicts the subtree of the general parity check nodes. The jth single parity check code ($j = 0, 1, \dots, N_p - 1$) should satisfy the following parity check constraint for $\gamma_j = 0$:

$$\gamma_j = \bigoplus_{i=0}^{2^{t-p}} h(\alpha_{iN_p+j}), \tag{9.16}$$

We divide the LLRs vector α of the general parity check node at depth t into N_p parts $\alpha^j, j = 0, \dots, N_p - 1$ such that $\alpha_i^j = \alpha_{iN_p+j}$ for $i = 0, \dots, 2^{t-p}$. For every subvector α^j, we find the index of the least reliable position as $i_{LRB}^j = \arg\max_i (\alpha_i^j)$, and the corresponding partial sum vector β^j is calculated by

$$\beta_i^j = \begin{cases} h(\alpha_i^j) \oplus \gamma_j & \text{if } i = i_{LRB}^j, \\ h(\alpha_i^j) & \text{otherwise.} \end{cases} \tag{9.17}$$

Finally, we obtain each element in the partial sum vector β of the general parity check node at depth t as

$$\beta_i = \beta_{\lfloor i/N_p \rfloor}^{i \bmod N_p}.$$

Example 9.1 Let us revisit Example 4.3 and illustrate the step-by-step fast successive cancellation decoding of the polar code $(8, 4, \{3, 5, 6, 7\})$.

We assume that the message data $\mathbf{d} = [1, 1, 0, 1]$ is sent. First, the message is rate-profiled into $\mathbf{u} = [0, 0, 0, 1, 0, 1, 0, 1]$ and encoded as $\mathbf{c} = [1, 1, 0, 0, 0, 0, 1, 1]$. With binary phase shift keying (BPSK) modulation, the codeword becomes $\mathbf{x} = [-1, -1, 1, 1, 1, 1, -1, -1]$.

Let $\mathbf{x}$ be transmitted over the additive white Gaussian noise (AWGN) channel with $E_b/N_0 = 2$ dB, resulting in the received sequence $\mathbf{y} = [-0.4167, -0.6357, 1.6604, -0.5944, 0.6982, 1.6727, -1.7881, -1.4540]$. The output values of the AWGN channel are obtained as explained in Chapter 1, Section 1.2.3.

As shown in Chapter 3, Section 3.1.1, the LLRs of the received symbols y_i are evaluated as $L_1^{(1)}(y_i) = \alpha(y_i) = \frac{2}{\sigma^2}y_i$, where $\sigma^2 = \frac{1}{2R\cdot 10^{\frac{\mathrm{SNR\,[dB]}}{10}}} = 0.6310$:

$$\alpha(\mathbf{y}) = [-1.321, -2.015, 5.263, -1.884, 2.213, 5.302, -5.668, -4.609].$$

Observe that according to rate-profile $\mathcal{A} = \{3, 5, 6, 7\}$, the left subtree at depth $d_v = 1$ takes only a single information bit u_3 and the rest of the bits $u_0 = u_1 = u_2 = 0$ are frozen. Thus, we have a repetition constituent code, where u_3 is repeated $2^{3-1} = 4$ times by encoding as

$$[0\ 0\ 0\ u_3]\mathbf{G}_{2^2} = u_3\mathbf{g}_3 = u_3[1\ 1\ 1\ 1] = [u_3\ u_3\ u_3\ u_3].$$

Now, we first find

$$\alpha_v[0] = \alpha(y_0) \boxplus \alpha(y_4) = -1.006,$$

Similarly, we can find $\alpha_v[1] = -1.979$, $\alpha_v[2] = -4.752$, and $\alpha_v[3] = 1.822$ as we did in Example 4.3. Since all bits are identical, that is $\beta[i] = u_3, i \in [0, 3]$, we can find it by

$$\beta_v[i] = h\left(\sum_{j=0}^{2^2-1} \alpha_v[j]\right) = h(-5.9150) = 1, \quad \text{for} \quad i \in [0, 3].$$

Now, to complete the decoding, we need to find the values of $u_i, i \in [0, 3]$ as

$$[u_0\ u_1\ u_2\ u_3] = [\beta_v[0]\ \beta_v[1]\ \beta_v[2]\ \beta_v[3]]\mathbf{G}_4^{-1} = [1\ 1\ 1\ 1]\mathbf{G}_4 = [0\ 0\ 0\ 1].$$

Figure 9.6 illustrates the decoding of the left subtree. The right subtree has a single frozen bit $u_4 = 0$; thus, we have a single parity check constituent code, or in other words, an SPC node. Let us denote this node by w. Then, we first find

$$\alpha_w[0] = (1 - 2\beta_v[0])\alpha(y_0) + \alpha(y_4) = 3.533.$$

Similarly, we can find $\alpha_w[1] = 7.318$, $\alpha_w[2] = -10.932$, and $\alpha_w[3] = -2.725$ and we find similar results as Example 4.3. Next, we find $\gamma = \bigoplus_{i=0}^{2^{d_v}-1} h(\alpha_i) = 0 \oplus 0 \oplus 1 \oplus 1 = 0$ and $i_{LRB} = 3$. Since the even parity constraint is satisfied, then we have

$$\beta_w[i] = h(\alpha_w[i]), \quad \text{for} \quad i \in [0, 3],$$

which results in $\beta_w[0] = 0$, $\beta_w[1] = 0$, $\beta_w[2] = 1$, and $\beta_w[3] = 1$. Similar to the left node, to complete the decoding, we need to find the values of $u_i, i \in [4, 7]$ as

$$[u_4\ u_5\ u_6\ u_7] = [\beta_w[0]\ \beta_w[1]\ \beta_w[2]\ \beta_w[3]]\mathbf{G}_4^{-1} = [0\ 0\ 1\ 1]\mathbf{G}_4 = [0\ 1\ 0\ 1].$$

Figure 9.7 illustrates the decoding of right subtree. Therefore, we could obtain the same decoding results as Example 4.3 with fewer operations.

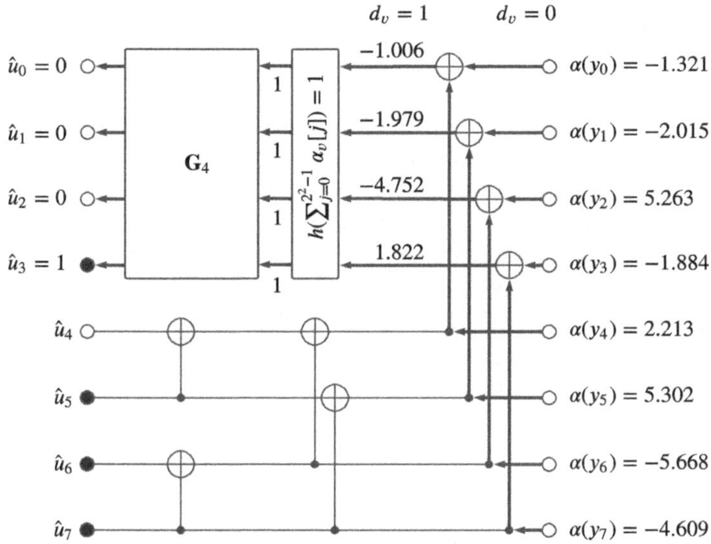

Figure 9.6 Decoding of the left subtree in Example 9.1.

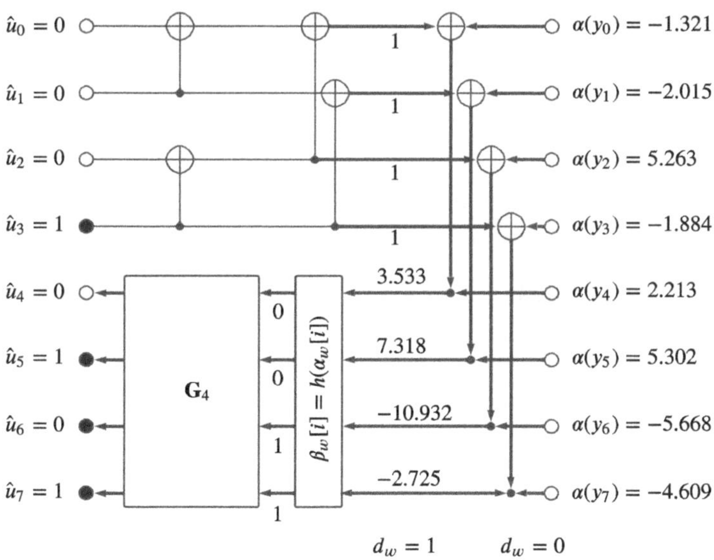

Figure 9.7 Decoding of the left subtree in Example 9.1.

9.7 Fast SC List Decodings

For list decoding, as discussed in Section 8.1, we need to consider both $\beta_i = 0$ and 1 for every $i \in \mathcal{A}$, calculate the metric for each path extension, and eventually prune the L least-likelihood paths. In this section, we discuss how to calculate the path metric using the partial sum vector β and the LLR vector α at node v.

Suppose we have β as the partial sum of node v regardless of the constituent code associated with the node. We calculate the path metric similarly to Section 8.1. The difference is that instead of estimation vector $\hat{\mathbf{u}}$, we use the estimation of vector β. For the case of Rate-0 node, we simply extend all the paths as $\beta = \mathbf{0}$ and we cannot have more options for each β_i but value 0. That is, assuming that we have the metric of path l from the last bit of the previous node as PM_i^l. The metric of this path at the rightmost bit of the Rate-0 node is

$$\mathrm{PM}_l^{(i+2^t-1)} = \mathrm{PM}_l^{(i)} + \sum_{j=0}^{2^t-1}(1+\exp^{-\alpha_j}). \tag{9.18}$$

Note that $(1-2\beta_j) = 1$ as $\beta_j = 0$ for every $j \in [0, 2^t-1]$. Hence, the exact branch metric in general is calculated by

$$\mu_l^{(i)} = (1+\exp^{-(1-2\beta_i)\alpha_i}). \tag{9.19}$$

Exercises

1 Show that for local SC decoding of a rate-1 node v, we have $\beta_v = h(\alpha_v)$?

2 Show that for local SC decoding of a rate-1 node v, we have $(\hat{u}[\min \mathcal{I}_v], \dots, \hat{u}[\max \mathcal{I}_v]) = \beta_v \boldsymbol{G}_{n-d_v}$?

Hint: Use induction approach.

3 Show that the parity check constraints at the general parity check nodes are independent.

Hint: Form the generator matrix for the subtree of the node, then show how each β^j returns 0 for the parity position of each j.

4 When there exist two frozen bits at the leftmost leaf nodes, that is $\mathbf{u}_v = [0\,0\,u_{\min \mathcal{I}_v} \cdots u_{\max \mathcal{I}_v}]$, how can we decode the node?

Hint: The even-indexed and odd-indexed bits in x_v constitute two separate single parity check codes with even parity as discussed earlier.

5 When there exist two information bits among the leaf nodes at the last two positions, $\mathbf{u}_v = [0\,0 \ldots 0\, u_{\max \mathcal{I}_v - 1}\, u_{\max \mathcal{I}_v}]$, how can we decode the node?

Hint: This results in the repetition of $[u_{\max \mathcal{I}_v} \oplus u_{\max \mathcal{I}_v - 1}\, u_{\max \mathcal{I}_v}]$ by 2^{n-d_v-1} times $\mathbf{u}_v \boldsymbol{G}_{n-d_v} = \left[\left([u_{\max \mathcal{I}_v - 1}\, u_{\max \mathcal{I}_v}] \cdot \boldsymbol{G}_2\right)_{\times n-d_v-1}\right]$. That is, the even-indexed and odd-indexed coded bits will be identical.

Bibliographical Notes

In order to reduce complexity, the SC decoder was simplified by Alamdar-Yazdi and Kschischang [2011] to decode the rate-1 constituent codes of polar codes locally without any negative impact on the block error rate. This simplification was extended to single parity constituent codes based on the Wagner rule and repetition codes by Sarkis et al. [2014]. Note that the Wagner rule was described by Silverman and Balser [1954] for correcting a single error in the single parity check codes.

The constituent-based decoding for successive cancellation list (SCL) decoding was employed by Sarkis et al. [2015] and improved by Hashemi et al. [2017]. Single parity check codes and repetition codes were extended to two and three parallel codes of these types by Hanif and Ardakani [2017] and adapted to SCL decoding by Hanif et al. [2018]. Condo et al. [2018] generalized decoding of the repetition and single parity check constituent codes/nodes. Furthermore, the sequence repetition (SR) node, introduced by Zheng et al. [2021], consists of a group of Rate-0 and REP nodes, along with a source node that can follow any pattern.

These special nodes can be utilized in other SC-based decoding methods, such as sequential decoding, as demonstrated by Ji et al. [2023], and in variants of polar codes, including polarization-adjusted convolutional (PAC) codes, as shown by Zhu et al. [2023].

Bibliography

Amin Alamdar-Yazdi and Frank R Kschischang. A simplified successive-cancellation decoder for polar codes. *IEEE Communications Letters*, 15(12):1378–1380, 2011. doi: 10.1109/LCOMM.2011.101811.111480.

Carlo Condo, Valerio Bioglio, and Ingmar Land. Generalized fast decoding of polar codes. In *2018 IEEE Global Communications Conference (GLOBECOM)*, pages 1–6, 2018. doi: 10.1109/GLOCOM.2018.8648105.

Muhammad Hanif and Masoud Ardakani. Fast successive-cancellation decoding of polar codes: Identification and decoding of new nodes. *IEEE Communications Letters*, 21(11):2360–2363, 2017. doi: 10.1109/LCOMM.2017.2740305.

Muhammad Hanif, Maryam H Ardakani, and Masoud Ardakani. Fast list decoding of polar codes: Decoders for additional nodes. In *2018 IEEE Wireless Communications and Networking Conference Workshops (WCNCW)*, pages 37–42, 2018. doi: 10.1109/WCNCW.2018.8369002.

Seyyed Ali Hashemi, Carlo Condo, and Warren J Gross. Fast and flexible successive-cancellation list decoders for polar codes. *IEEE Transactions on Signal Processing*, 65(21):5756–5769, 2017. doi: 10.1109/TSP.2017.2740204.

Houren Ji, Yifei Shen, Zaichen Zhang, Yongming Huang, Xiaohu You, and Chuan Zhang. Low-complexity fast Fano decoding for PAC codes. *IEEE Transactions on Vehicular Technology*, 72(12):15172–15184, 2023.

Gabi Sarkis, Pascal Giard, Alexander Vardy, Claude Thibeault, and Warren J Gross. Fast polar decoders: Algorithm and implementation. *IEEE Journal on Selected Areas in Communications*, 32(5):946–957, 2014.

Gabi Sarkis, Pascal Giard, Alexander Vardy, Claude Thibeault, and Warren J Gross. Fast list decoders for polar codes. *IEEE Journal on Selected Areas in Communications*, 34(2):318–328, 2015.

R A Silverman and Martin Balser. Coding for constant-data-rate systems-part I. A new error-correcting code. *Proceedings of the IRE*, 42(9):1428–1435, 1954.

Haotian Zheng, Seyyed Ali Hashemi, Alexios Balatsoukas-Stimming, Zizheng Cao, Ton Koonen, John M Cioffi, and Andrea Goldsmith. Threshold-based fast successive-cancellation decoding of polar codes. *IEEE Transactions on Communications*, 69(6):3541–3555, 2021.

Hongfei Zhu, Zhiwei Cao, Yuping Zhao, Dou Li, and Yanjun Yang. Fast list decoders for polarization-adjusted convolutional (PAC) codes. *IET Communications*, 17(7):842–851, 2023.

10

Alternative Decoding Algorithms

"Sometimes, if you explore alternatives, you discover unexpected improvements."

— Andrew Yao

While successive cancellation (SC) decoding and successive cancellation list (SCL) decoding are the most widely studied and implemented decoding techniques for polar codes, their inherent limitations in certain scenarios necessitate the exploration of alternative approaches. SC decoding, although computationally efficient, suffers from suboptimal error correction performance, especially at short to moderate block lengths. SCL decoding improves error performance by considering multiple paths in the decoding tree, but at the cost of increased latency and computational complexity. These trade-offs, combined with the growing demand for decoding algorithms tailored to diverse performance and application requirements, have spurred the development of advanced decoding strategies for polar codes.

This chapter delves into a variety of such advanced decoding algorithms that extend beyond the traditional SC and SCL techniques. Each algorithm offers unique advantages, targeting specific challenges in latency, complexity, or error performance. Among these, we explore belief propagation (BP) decoding, which leverages the graphical structure of polar codes to iteratively pass messages and achieve improved performance in certain configurations. We also cover sequential decoding, a tree-search-based approach that balances performance and complexity by prioritizing paths based on likelihood metrics. Furthermore, we examine sphere decoding, which focuses on finding the most likely codeword within a constrained search space, offering potential benefits in applications where computational resources are limited.

Beyond these specialized techniques, we discuss generic decoding algorithms, such as ordered statistics decoding (OSD) and guessing random additive noise

Polar Codes: From Theory to Practice, First Edition. Mohammad Rowshan and Emanuele Viterbo.

decoding (GRAND). These versatile methods are not specific to polar codes but can be adapted for their use, providing insights into alternative ways of decoding that may outperform conventional approaches in certain scenarios. OSD, for instance, relies on reordering and recombining decoding candidates to improve performance, while GRAND efficiently explores error patterns to recover transmitted messages.

By examining these advanced decoding strategies, this chapter aims to provide a comprehensive understanding of the broader decoding landscape for polar codes, highlighting the potential of these techniques to address various performance trade-offs and application needs.

10.1 Belief Propagation (BP) Decoding

In Chapter 3, we learned about the message-passing algorithm (MPA). We used this decoding algorithm to take into account the extrinsic information calculated iteratively by computing the marginals. Here, we use this decoder for polar codes. The main difference here is that the constraints are not based on the parity check matrix $\mathbf{H}$, but are based on the polar transform $\boldsymbol{G}_N$. Recall the membership function from Chapter 3, which is a type of predicate function that checks if the input meets some condition and returns true or false (alternatively 1 or 0) as the output. Here, the membership function evaluates to 1 if there exists a sequence $\mathbf{u}$ for which we get $\mathbf{u}\boldsymbol{G}_N = \mathbf{x}$. That is, for every $x_j, j = 0, 1, \dots, N-1$, we have the constraint $\sum_{i\in\mathcal{N}(x_j)} u_i = x_j$ in the binary field $\mathbb{F}_2$, or equivalently

$$x_j + \sum_{i\in\mathcal{N}(x_j)} u_i = 0, \tag{10.1}$$

where $\mathcal{N}(x_j)$ denotes the coordinates of non-zero elements in the jth column of $\boldsymbol{G}_N$, also known as neighbors of x_j. Hence, the membership function is

$$f\left(u_0, \dots, u_{N-1}, x_0, \dots, x_{N-1}\right) = \mathbb{1}_{\{\mathbf{x}\in\mathscr{C}\}} = \prod_{j=0}^{N-1} \mathbb{1}_{x_j + \sum_{i\in\mathcal{N}(x_j)} u_i = 0} = \begin{cases} 1 & \mathbf{u}\boldsymbol{G}_N = \mathbf{x}, \\ 0 & \text{otherwise.} \end{cases} \tag{10.2}$$

The sum $\sum_{i\in\mathcal{N}(x_j)} u_i$ in constraint $\sum_{i\in\mathcal{N}(x_j)} u_i = x_j$ can be progressively performed on u_i pairwise. Therefore, we can consider the basic Hadamard transform $\boldsymbol{G}_2$ and formulate the calculation of the messages for the message passing algorithm. Let $L_{s,i}$ and $R_{s,i}$ denote the messages corresponding to bit i at stage s passed to the left and right, respectively. The constraints corresponding to the function node for the left–right direction and the right–left direction are $u'_i \oplus u'_{i+2^{n-s}} = x'_i$ and $x'_i \oplus x'_{i+2^{n-s}} = u'_i$, where u' and x' are intermediate encoded bits except for stages $s = 0$ and n. Similar to MP algorithm for LDPC codes, a variable node passes a message

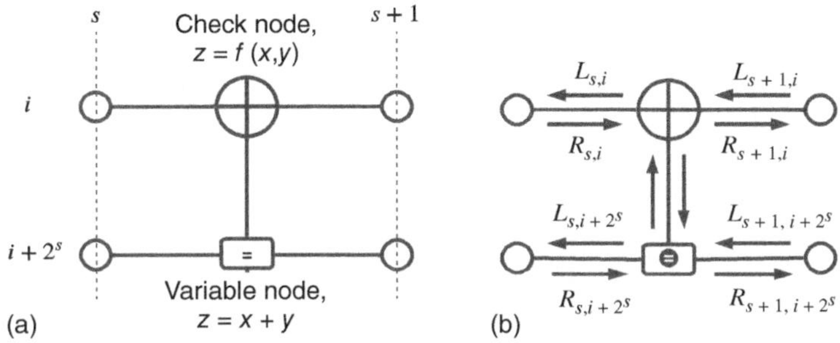

Figure 10.1 Basic computational block of the BP factor graph: (a) operations at the variable and check nodes. (b) Messages of the left (L) and right (R) propagation. The inputs x and y can be flowed into the corresponding node from any pair of edges.

to the function node which is the addition of $L_{s+1,i+2^{n-s}}$ and $R_{s,i+2^{n-s}}$ in the logarithm domain, where the messages are log-likelihood ratios (LLRs).

Figure 10.1 shows the basic computational block in the decoder. In an iterative decoding, we have message passing from left to right and right to left of this block. We need to calculate the messages out of the check node and variable node in both directions.

Note that if the messages are likelihood ratios (LR), the inputs to the variable node are multiplied and then passed to the check node. It is also worth noting that when the propagation direction is, for instance, toward the right, we only calculate the $R_{s,i}$ messages. However, we need $L_{s,i}$ at the variable node as well. This message is previously calculated in the left propagation direction.

Figure 10.2 illustrates two examples of the flow of messages and operations to calculate $L_{s,i}$ (a) and $L_{s+1,i+2^{n-s}}$ (b). Figure 10.3 shows the factor graph of the polar code (8,4) containing $\log_2 N + 1$ stages labeled as $s = 0, 1, \ldots, n$ and each stage includes N nodes labeled $i = 0, 1, \ldots, N-1$. From a decoding point of view, the right-most nodes labeled (n, i) are initially associated with the channel LLRs, and the left-most nodes are associated with the estimates of source bits $\hat{u}_i$. We label the left-propagating messages and right-propagating messages at iteration t by $L_{s,i}^{(t)}$ and

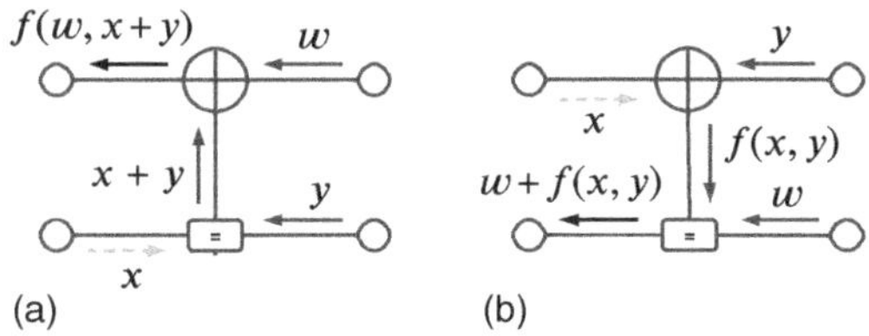

Figure 10.2 Examples showing the operations at the basic computational block of the BP factor graph. The dashed arrows show the messages from the immediate previous propagation in the opposite direction.

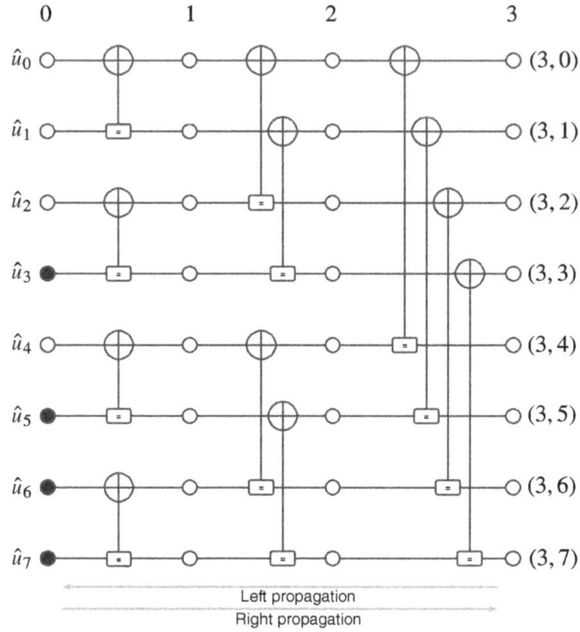

Figure 10.3 The factor graph of (8,4) polar code containing $\log_2 8 + 1 = 4$ stages, $s = 0, 1, 2, 3$.

$R_{s,i}^{(t)}$. Besides the channel LLRs denoted by $L_{n,i}^{(0)}$, we initialize the messages associated with left most nodes by

$$R_{0,i}^{(0)} = \begin{cases} +\infty & i \in \mathcal{A}^c, \\ 0 & \text{otherwise.} \end{cases} \tag{10.3}$$

We set every $i \in \mathcal{A}^c$ (frozen bit coordinates) to a large positive LLR value (indicated by $+\infty$) in agreement with the ML hard decision rule in (3.16) because we know that the demodulated value of u_i in bipolar mapping from $\{1, -1\} \longrightarrow \{0, 1\}$ is $u_i = 0$. On the other hand, we set $R_{0,i}^{(0)} = 0$ for $i \in \mathcal{A}$ due to the equal probability of taking 0 and 1 bit such u_i. All intermediate messages are set to $R_{s,i}^{(0)} = L_{s,i}^{(0)} = 0$ for $s = 2, \dots, n-1$.

During the right-propagation, the computational block shown in Figure 10.1 outputs

$$\begin{cases} R_{s+1,i}^{(t)} = f\left(R_{s,i}^{(t)}, R_{s,i+2^s}^{(t)} + L_{s+1,i+2^s}^{(t-1)}\right), \\ R_{s+1,i+2^s}^{(t)} = R_{s,i+2^s}^{(t)} + f\left(R_{s,i}^{(t)}, L_{s+1,i}^{(t-1)}\right). \end{cases} \tag{10.4}$$

while during the left propagation, the outputs are

$$\begin{cases} L_{s,i}^{(t)} = f\left(L_{s+1,i}^{(t)}, L_{s+1,i+2^s}^{(t)} + R_{s,i+2^s}^{(t)}\right), \\ L_{s,i+2^s}^{(t)} = L_{s+1,i+2^s}^{(t)} + f\left(L_{s+1,i}^{(t)}, R_{s,i}^{(t)}\right). \end{cases} \tag{10.5}$$

as illustrated in Figure 10.2. Note that a round-trip iteration starts from left-to-right and then right-to-left except for the first iteration denoted by $t = 0$, where it is just right-to-left and we use the initial LLR values for the left-to-right trip. Thus, we have the $L^{(t-1)}$ messages from the previous iteration $t - 1$ in (10.4). The function $f(.,.)$ was introduced in Chapter 3 by (3.20) as

$$f(x,y) = \log \frac{1 + e^{x+y}}{e^x + e^y} = 2\tanh^{-1}\left(\tanh\left(\frac{x}{2}\right) \cdot \tanh\left(\frac{y}{2}\right)\right), \tag{10.6}$$

and its approximation by (3.21) as

$$f(x,y) \approx \text{sign}(x) \cdot \text{sign}(y) \cdot \min(|x|, |y|). \tag{10.7}$$

After reaching a pre-set maximum iteration, that is when $t = T$, all $\hat{u}_i, i \in [0, N-1]$ are determined based on the hard-decision of the overall LLR values of nodes $(0, i)$, that is

$$\hat{u}_i = \text{h}\left(\text{LLR}_{0,i}^{(T)}\right) \triangleq \text{h}\left(L_{0,i}^{(T)} + R_{0,i}^{(T)}\right), \tag{10.8}$$

where $\text{h}(\cdot)$ is the hard decision function used in Chapter 4.

Similarly, the codeword $\hat{\mathbf{x}}$ can be estimated based on the hard decision of the overall LLR values of nodes (n, i), that is

$$\hat{x}_i = \text{h}\left(\text{LLR}_{n,i}^{(T)}\right) \triangleq \text{h}\left(L_{n,i}^{(T)} + R_{n,i}^{(T)}\right). \tag{10.9}$$

Knowing the relation between rate-profiled message $\mathbf{u}$ and codeword $\mathbf{x}$ in the polar encoding, $\mathbf{x} = \mathbf{u}\boldsymbol{G}_N$, we can terminate the message passing iterations earlier than the maximum T iterations if we get

$$\hat{\mathbf{x}} = \hat{\mathbf{u}}\boldsymbol{G}_N. \tag{10.10}$$

This is a commonly used criterion for *early termination* of BP decoding.

Example 10.1 Let us revisit Example 4.3 and use it to illustrate the BP decoding of the polar code $(8, 4, \{3, 5, 6, 7\})$ with maximum iterations $T = 2$ step by step.

We assume that the message data $\mathbf{d} = [1,\ 1,\ 0,\ 1]$ are going to be transmitted. First, the message is rate-profiled into $\mathbf{u} = [0,\ 0,\ 0,\ 1,\ 0,\ 1,\ 0,\ 1]$ and encoded as $\mathbf{c} = [1, 1, 0, 0, 0, 0, 1, 1]$. With BPSK modulation, the codeword becomes $\mathbf{x} = [-1, -1,\ 1,\ 1,\ 1,\ 1, -1, -1]$.

Let $\mathbf{x}$ be transmitted over the AWGN channel with $E_b/N_0 = 2$ dB, resulting in the received sequence $\mathbf{y} = [-0.4167, -0.6357,\ 1.6604, -0.5944,\ 0.6982,\ 1.6727,$

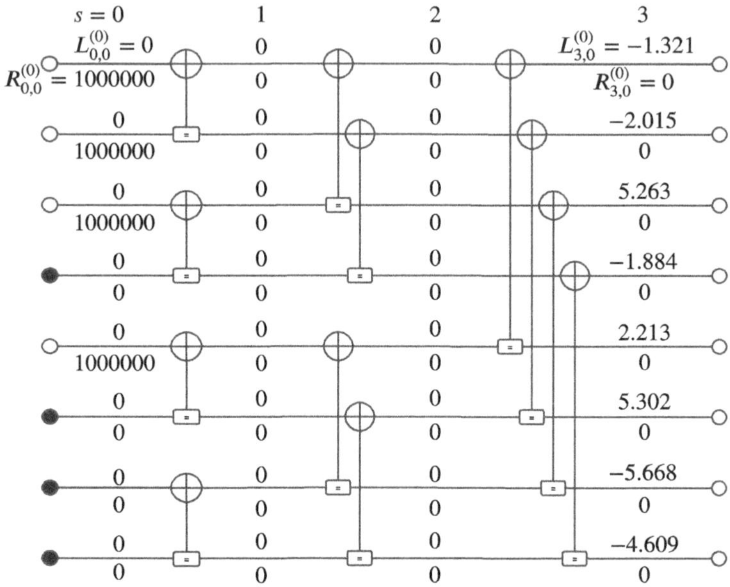

Figure 10.4 Initialization of the BP decoding algorithm in Example 10.1.

−1.7881, −1.4540]. To see how the output of the AWGN channel is obtained, see Chapter 1, Section 1.2.3.

As shown in Chapter 3, Section 3.1.1, the LLRs of the received symbols y_i are estimated by $L_{3,i}^{(0)}(y_i) = \frac{2}{\sigma^2}y_i$, where $\sigma^2 = \frac{1}{2R \cdot 10^{\left(\frac{\text{SNR [dB]}}{10}\right)}} = 0.6310$. Hence, the LLRs are

$$\mathbf{L}_3^{(0)} = \alpha(\mathbf{y}) = [-1.321, -2.015,\ 5.263, -1.884,\ 2.213,\ 5.302, -5.668, -4.609].$$

Furthermore, according to (10.3), we initialize $\mathbf{R}_{0,i}^{(0)} = 10^6$ for $i \in \mathcal{A}^c$ and set the rest to 0. Hence, we have

$$\mathbf{R}_0^{(0)} = \alpha(\mathbf{y}) = [1000000,\ 1000000,\ 1000000,\ 0,\ 1000000,\ 0,\ 0,\ 0].$$

The results are shown in Figure 10.4.

Following the update rules in (10.5) and (10.4), the messages are propagated to the left and then to the right for one round trip, as shown in Figure 10.5. The updated values are shown in boldface. The second iteration is also performed similarly, as shown in Figure 10.6.

Now, let us see whether the decoder has converged in the first iteration or not. We use the early termination criterion in (10.10). We estimate $\hat{\mathbf{x}}$ and $\hat{\mathbf{u}}$ according to (10.9) and (10.8), respectively, for all $i \in [0, 7]$ as follows:

$$L_{0,i}^{(1)} + R_{0,i}^{(1)} = [1000000.99,\ 999999.76,\ 999998.23, -2.93,\ 1000002.06, -6.39,\ 4.31, -4.61],$$

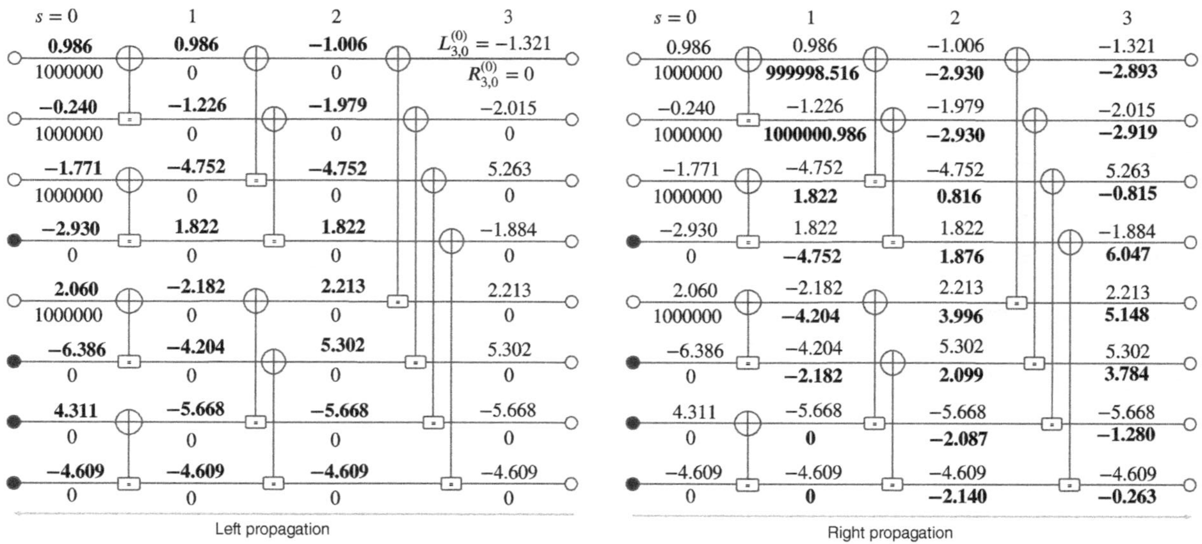

Figure 10.5 First iteration including left and right propagations—Example 10.1.

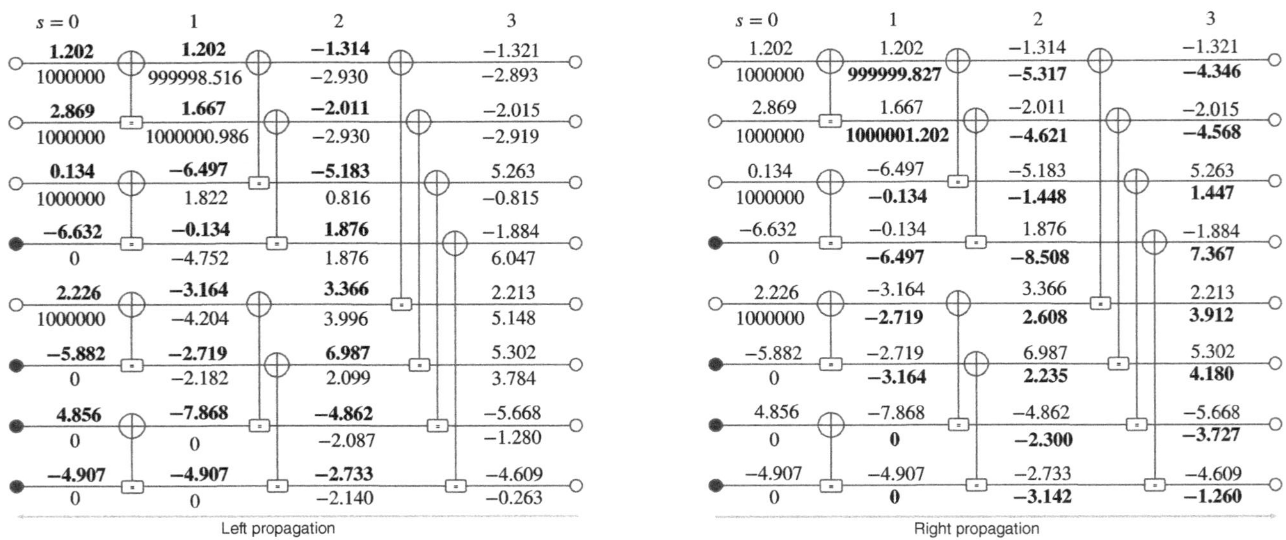

Figure 10.6 Second iteration including left and right propagations—Example 10.1.

$$L_{3,i}^{(1)} + R_{3,i}^{(1)} = [-4.21, -4.93,\ 4.45,\ 4.16,\ 7.36,\ 9.09, -6.95, -4.87].$$

Following the hard decision, we get

$$\hat{\mathbf{u}} = [0,\ 0,\ 0,\ 1,\ 0,\ 1,\ 0,\ 1],$$

$$\hat{\mathbf{x}} = [1,\ 1,\ 0,\ 0,\ 0,\ 0,\ 1,\ 1].$$

Observe that the equality in (10.10) is met. Therefore, we could stop the decoding process after the first iteration.

10.2 Sequential Decoding

Readers are first expected to read Section 3.5.1 in Chapter 3, where a general overview of sequential decoding algorithms such as Fano and stack decoding and their path metric used with these decoders are discussed. Here, we focus on Fano decoding. In the Fano algorithm, the decoder begins from the origin node of a binary tree, where its edge corresponds to u_0 and examines a sequence of adjacent nodes. At any step corresponding to the nonfrozen bits in the vector $\mathbf{u}$, it moves forward to one of the successor nodes or moves backward to the nonfrozen predecessor of the current node.

This tree traversal algorithm performs near the ML decoding; however, it has high time complexity and variable latency, which depends on the level of noise. This algorithm became more popular since it was used by Arıkan for the PAC code (128,64) with RM profile, demonstrating a performance achieving the performance of finite block-length codes. The superior performance resulted from the fact that the pre-transformation in PAC coding improves the weight distribution of polar codes, as discussed in Chapter 6.

Fano decoding traverses the binary decoding tree and moves forward to the successor nodes, and if the path is not sufficiently reliable (when the path metric passes the threshold), it moves backward to the predecessor nodes in search of the path with maximum likelihood. The following example shows the steps involved in the tree search.

Example 10.2 Figure 10.7 demonstrates the Fano decoding steps with $\Delta = 4$ for the polar code (4,3) with $\mathcal{A} = \{1, 2, 3\}$. Initially, we have $T = 0$; however, since the decoder cannot move forward, the threshold T is updated to $T = T - \Delta = -4$. The first bit is a frozen bit; hence, $u_0 = 0$. In Step 3, the edge corresponding to $u_1 = 1$ is selected due to higher likelihood, that is, $-2.9 > -3.5$. However, after examining the child nodes, the decoder cannot move forward via this path as the final path metrics of both extensions (-7.2 and -5.5) are smaller than the threshold -4.

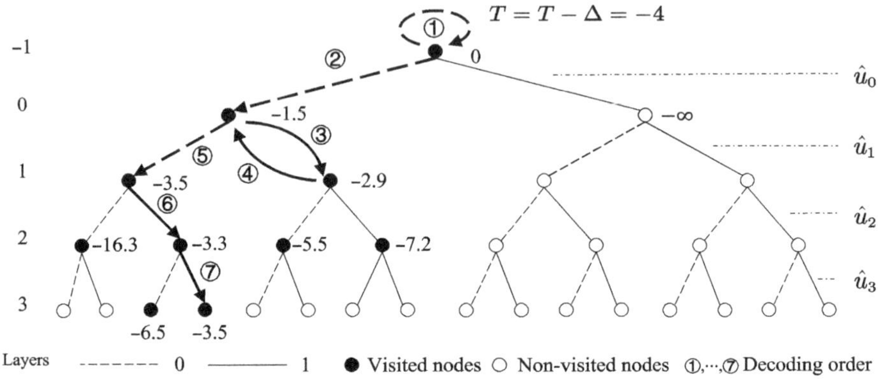

Figure 10.7 An example illustrating the tree traversal in Fano decoding.

Hence, the decoder moves backward in step 4 and selects the edge corresponding to $u_1 = 0$ in step 5. Observe that its metric −3.5 is larger than $T = -4$. The tree traversal follows smoothly by selecting the path extensions corresponding to $u_2 = 1$ and $u_3 = 1$ while they remain larger than $T = -4$. Note that the backward movements usually occur over low-reliability bits. Therefore, the estimated bit values are $\hat{u}_0^3 = [0\ 0\ 1\ 1]$.

During decoding, the path metric serves as a reliability indicator. Here we adapt the path metric used for the list decoding in Chapter 8 (see Equations (8.5) and (8.6)); however, we add a bias term b. The bias term based on characteristics of the code and the channel is generally chosen so that the metric does not overly favor longer paths, which could lead to excessive backtracking or getting stuck in less optimal paths. Hence, the path metric is as follows:

$$PM^{(i)} = \sum_{j=0}^{i} \log P\left(\hat{u}_j | \hat{u}_0^{j-1},\ y_0^{N-1}\right) + b \tag{10.11}$$

$$= PM^{(i-1)} + \mu^{(i)} + b, \tag{10.12}$$

where $\mu^{(i)} = \log P\left(\hat{u}_i | \hat{u}_0^{i-1}, y_0^{N-1}\right) = \log\left(1 + e^{-(1-2\hat{u}_i)\lambda_0^i[l]}\right)$. Furthermore, we initialize $PM^{(-1)} = 0$ and the bias term is suggested to be $b = -\log(1 - p_e^{(i)})$, where the error probability of bit i can be approximated using $L_N^{(i)}$ (see Chapter 6, Section 6.2.2) as

$$p_e^{(i)} \approx Q\left(\sqrt{L_N^{(i)}/2}\right), \quad 0 \le i \le N-1. \tag{10.13}$$

Note that the metric value is negative.

The decoding stops when it reaches a terminal node ($i = N - 1$). The search through the decoding tree is guided by a threshold T on the path metric (with an initial value of $T = 0$). If the metric becomes less than the threshold as shown in Example 10.2, the search is restored, and another path is followed. If no paths can be found with a metric above the threshold, the threshold value is lowered by the step-size Δ, and the process is continued. A node in the tree may be visited more than once in the forward direction but with a lower threshold each time.

10.3 Sphere Decoding

The core concept of the sphere decoder (SD) is to decrease the computational complexity of the maximum-likelihood decoder by limiting the search for valid $\mathbf{u}$'s over integer lattice points $\mathbf{1} - 2\mathbf{u}\boldsymbol{G}_N$ that fall within a hypersphere of radius r centered on the received noisy vector $\mathbf{y}$ in the N-dimensional space. In the first step of the algorithm, the sphere decoder performs a depth-first search with a very large initialized r_0. The radius r_0 may reduce later as the decoder explores the decoding tree looking for alternative candidates. By finding the candidate that allows the sphere to have a minimum radius, SD achieves the ML estimation for the transmitted sequence $\mathbf{u}$ with respect to the received vector $\mathbf{y}$.

We adopt the squared Euclidean distance to determine the likelihood of the followed path. Hence, for the estimated bits u_i^{N-1}, the branch metric $m_i\left(u_i^{N-1}\right)$ is

$$m_i(u_i^{N-1}) \triangleq \left(y_i - \left(1 - 2\bigoplus_{j=i}^{N-1} u_j g_{j,i}\right)\right)^2. \tag{10.14}$$

Based on the lower triangular form of $\mathbf{G}_N$, SD starts estimating the bit values from level $l = N-1$. The radius r_0 then works as a constraint on the path metrics, the left-hand side of Equation (10.15), for levels $l = N-1, \dots, 0$ so that for candidates within the sphere, we have

$$\sum_{i=l}^{N-1} m_i\left(u_i^{N-1}\right) \le r_0^2. \tag{10.15}$$

The first path candidate is found when it reaches $l = 0$ for the first time. The radius is then updated to the current path metric, that is, $r_0^2 = \sum_{i=0}^{N-1} m_i\left(u_i^{N-1}\right)$. Only one path is followed at a time during the decoding. To search for candidates with a smaller path metric, the level is moved backward (i.e. $l = l + 1$) to explore an alternative solution for u_{N-1}^l.

The branching condition in (10.15) is examined by the updated r_0 during the path expansion. Paths that do not meet the condition are located outside the sphere. Considered invalid candidates, they are pruned and will not be revisited. Recall that for $l \in \mathcal{A}^c$, u_l is set to 0. Hence, the path only splits when $l \in \mathcal{A}$.

To reduce computational complexity, path metrics with a lower bound provide a stricter branching condition. The bound $\Lambda_i = \min\limits_{\mathbf{u}\in\mathcal{U}}\left(m_i\left(u_i^{N-1}\right)\right)$ is given as the minimum branch metric for two possible cases at a certain estimated bit. Consequently, the bounded path metrics for branching evaluation can be computed as

$$\sum_{i=l}^{N-1} m_i\left(u_i^{N-1}\right) + \sum_{i=0}^{l-1} \Lambda_i \leq r_0^2. \tag{10.16}$$

The radius r_0 is updated once a new valid path is developed on the decoding tree, given that the new path has a metric smaller than the current r_0. Until all feasible paths have been attempted by shrinking the radius of the sphere, the tree search is completed. The survived path is the ML estimate demanded for **u** that has the minimum path metric.

10.4 Reliability-Based Generic Decoding

So far, we have focused on decoding algorithms based on the polar code structure, the SC process, or tree traversal. In this section, we describe two generic/universal decoding algorithms (see Chapter 3) that actually generate test error patterns (TEP) and find the codeword closest to the received sequence using the generator matrix or the parity check matrix.

10.4.1 Ordered Statistics Decoding (OSD)

Given a received sequence $\mathbf{y} = y_1^N$, the OSD decoder first reorders the $\mathbf{y}$ elements according to their absolute values $|y_i|, i \in [1, N]$, or their reliability, in descending order. The new sequence $\mathbf{r} = \lambda_1(\mathbf{y})$ is a permuted sequence, λ_1 is a permutation function corresponding to reordering, and $|r_1| > |r_2| > \cdots > |r_N|$. The OSD decoder then permutes the columns of the generator matrix $\boldsymbol{G}$ based on λ_1. that is, $\boldsymbol{G}' = \lambda_1(\boldsymbol{G}) = \left[\mathbf{g}_1', \mathbf{g}_2', \ldots, \mathbf{g}_N'\right]$. Clearly, the codeword $\mathbf{x}'$ obtained from $\lambda_1(\mathbf{d})\boldsymbol{G}'$ is the permuted version of the codeword $\mathbf{x}$ resulted from $\mathbf{d}\boldsymbol{G}$; that is,

$$\mathbf{x}' = \lambda_1(\mathbf{d})\boldsymbol{G}' = \lambda_1(\mathbf{d})\lambda_1(\boldsymbol{G}) = \lambda_1(\mathbf{d}\boldsymbol{G}) = \lambda_1\left(x_1^N\right).$$

Recall from Chapter 5 that in polar coding, the generator matrix is obtained by retaining the rows of the polar transform $\boldsymbol{G}_N$ with the indices in $\mathcal{A}$ and removing the rows corresponding to the frozen coordinates in $\mathcal{A}^c$.

The decoder then finds the first K independent columns, starting from the first column of $\boldsymbol{G}'$ corresponding to the coordinate of the element with the highest reliability in $\mathbf{z}$, while the remaining columns remain as in $\boldsymbol{G}'$. This gives the second permutation function λ_2; that is,

$$\boldsymbol{G}'' = \lambda_2\left(\boldsymbol{G}'\right) = \lambda_2\left(\lambda_1(\boldsymbol{G})\right).$$

Furthermore, by permuting the elements of $\mathbf{r}$ according to λ_2, we obtain $\mathbf{v} = \lambda_2(\mathbf{r}) = v_1^N$. These K columns are used to generate the systematic form of a new $K \times N$ matrix G''_{sys} by performing elementary row operations on it. Recall from Chapter 2 that in systematic form, the first K columns of $\boldsymbol{G}''_{sys}$ form an identity matrix. Note that $|v_1| > |v_2| > \cdots > |v_K|$ and $|v_{K+1}| > |v_{K+2}| > \cdots > |v_N|$.

Assuming $\theta(\boldsymbol{v})$ returns the hard-decision demodulation of the received vector $\boldsymbol{v}$, then we have $a_1^N = \theta(\boldsymbol{v})$ and a_1^K as its first K elements called the *most reliable independent* (MRI) symbols. The parity bits and hence the entire codeword are obtained by $c_1^N = a_1^K \boldsymbol{G}''_{sys}$. If the error(s) are located within the parity bits a_{K+1}^N, they will be corrected after encoding. We call this stage order-0 OSD. Note that to recover the original codeword, we need to perform the de-permutations as

$$\mathbf{x} = \lambda_1^{-1}\left(\lambda_2^{-1}\left(c_1^N\right)\right).$$

The aim is to find the closest modulated codeword $\mu(\mathbf{c})$ to $\mathbf{v}$, where $\mu(\cdot)$ is the modulation function. To measure the closeness, we can use the squared Euclidean distance as a metric:

$$d_E^2(\boldsymbol{v}, \boldsymbol{c}) = \sum_{i=1}^{n} \left(v_i - \mu\left(c_i\right)\right)^2. \tag{10.17}$$

More candidate codewords can be generated by

$$\boldsymbol{c} = \left(\theta\left(v_1^k\right) \oplus z_1^k\right) \boldsymbol{G}''_{sys},$$

where z_1^k is an error vector to perform the flipping and its Hamming weight $\mathrm{w}\left(z_1^k\right) = l$. Therefore, we can proceed and try to flip at most $j \in [1, l]$ bits of the MRI symbols in the l-order OSD. After testing all $\sum_{j=1}^{l} \binom{K}{j}$ possible error patterns in the form of $\mathbf{z}$ vectors and retaining the $\mathbf{c}$ with the smallest $d_E^2(\boldsymbol{v}, \boldsymbol{c})$, denoted by $\mathbf{c}_{\text{best}}$, the decoder declares the codeword $\hat{\mathbf{x}} = \lambda_1^{-1}\left(\lambda_2^{-1}\left(\mathbf{c}_{\text{best}}\right)\right)$ as a solution.

Observe that the total number of test error patterns is a constant $\sum_{j=1}^{l} \binom{K}{j}$, and it depends on the reprocessing order l and code dimension K, regardless of the channel condition. Fortunately, we have a criterion to terminate the attempts early, called the *early termination* (ET) mechanism. The early termination criterion is based on the sufficient condition for optimality where we stop the reprocessing of the order-l OSD at order $j \leq l$ and declare the closest/best-found codeword $\mathbf{c}_{\text{best}}$ given that *correlation discrepancy* metric

$$\lambda\left(\boldsymbol{v}, \mathbf{c}_{\text{best}}\right) = \sum_{i \in D_1\left(\mathbf{c}_{\text{best}}\right)} |v_i|, \tag{10.18}$$

is smaller than the bound on optimality Λ at the end of reprocessing order-l, that is,

$$\lambda\left(\boldsymbol{v}, \mathbf{c}_{\text{best}}\right) \leq \Lambda\left(\mathbf{c}_{\text{best}}, d_{\min}, l\right), \tag{10.19}$$

where

$$\Lambda\left(\mathbf{c}_{\text{best}}, d_{\min}, l\right) = \sum_{j=0}^{l} \left|v_{k-j}\right| + \sum_{j \in D_0^{(\delta')}\left(\mathbf{c}_{\text{best}}\right)} \left|v_j\right|, \tag{10.20}$$

$$\delta' = \max\left\{0, d_{\min} - \left|D_1\left(\mathbf{c}_{\text{best}}\right)\right| - (l+1)\right\}, \tag{10.21}$$

$$D_1(\mathbf{c}) \triangleq \left\{i : c_i \neq z_i, 1 \leq i \leq n\right\}, \tag{10.22}$$

$$D_0(\mathbf{c}) \triangleq [1, n] \backslash D_1(\mathbf{c}), \tag{10.23}$$

and $D_0^{(j)}(\mathbf{c})$ gives the set of first j indices in the set $D_0(\mathbf{c})$ that is sorted based on reliability in ascending order.

Algorithm 10.1 lists a pseudocode for the order-l OSD with early termination mechanism for a code with the generator matrix $\boldsymbol{G}$ and a BPSK-modulated received sequence $\mathbf{y}$ from an AWGN channel.

Algorithm 10.1: Order-l Ordered Statistics Decoding (OSD) with Early Termination

---

```
Input: Received vector y, generator matrix G of the code, order of decoding l
Output: Estimated codeword x̂
Find the permutation function λ_1 by sorting |y_1|, |y_2|, ..., |y_N| in decreasing order: r = λ_1(y);
Permute columns of G accordingly: G' ← λ_1(G);
Find the first K independent columns of G' from left: G'' ← λ_2(G'), v = λ_2(λ_1(y));
Perform Gaussian elimination on G'' to obtain the systematic form G''_sys = [I_K|P];
Find a_1^K = [a_1, a_2, ..., a_K], where a_i = 0 if v_i > 0, and a_i = 1 otherwise;
c_best ← a_1^K G''_sys;
d_best ← ∞;
ℓ ← 0;
while ℓ ≤ l and λ(v, c_best) > Λ(c_best, d_min, ℓ) do
    for j : 1 → j ≤ (K choose ℓ) do
        Generate TEPs z_1^K of weight ℓ, w(z_1^K) = ℓ. The number of test patterns is (K choose ℓ).;
        Compute the candidate codewords c = (a_1^K ⊕ z_1^K) G_sys'';
        if d_E^2(c, v) < d_best then
            d_best ← d_E^2(c, v);
            c_best ← c;
    ℓ ← ℓ + 1;
x̂ ← λ_1^{-1}(λ_2^{-1}(c_best));
return x̂;
```

Example 10.3 Let us take the polar code $(8, 4)$ with $\mathcal{A} = \{3, 5, 6, 7\}$ and $d_{\min} = 4$. As discussed in Chapter 5, the generator matrix $\boldsymbol{G}$ of this code is obtained by removing the rows of $\boldsymbol{G}_N$ with the indices in $\mathcal{A}^c$. To transmit the message vector

$$\mathbf{d} = [0\ 0\ 1\ 1],$$

we first encode it to

$$\mathbf{x} = [0\ 1\ 0\ 1\ 0\ 1\ 0\ 1],$$

and then modulate it by bipolar signaling (BPSK modulation) as

$$\mu(\mathbf{x}) = [1 - 1\ \ 1 - 1\ \ 1 - 1\ \ 1 - 1].$$

The receiver gets the following sequence after passing through an AWGN channel:

$$\mathbf{y} = [0.67 - 0.46\ \ 0.43 - 1.19\ \ 2.12\ \ 0.17\ \ 0.72 - 1.76].$$

Observe that the sign of the sixth element has been flipped, which means an error occurred. We want to decode the received sequence using order-2 OSD.

Following the steps discussed in this section, we obtain the following systematic form of permuted generator matrix.

$$\mathbf{G}''_{sys} = \begin{bmatrix} 1 & 0 & 0 & 0 & 1 & 1 & 0 & 1 \\ 0 & 1 & 0 & 0 & 1 & 0 & 1 & 1 \\ 0 & 0 & 1 & 0 & 1 & 1 & 1 & 0 \\ 0 & 0 & 0 & 1 & 0 & 1 & 1 & 1 \end{bmatrix}$$

We permute the received sequence as well and we get the following:

$$\mathbf{y}'' = [2.12 - 1.76 - 1.19\ \ 0.72\ \ 0.67 - 0.46\ \ 0.43\ \ 0.17].$$

Observe the erroneous element is the least reliable bit and plays the role of parity in the systematic form. As Table 10.1 shows, we could correct the error by order-0 OSD and we do not need to proceed further for two reasons: (1) the erroneous parity bit in $\boldsymbol{c}$ is corrected by encoding and (2) the early termination criterion is satisfied: $\lambda\left(\boldsymbol{c}_{\text{best}}\right) < \Lambda\left(\boldsymbol{c}_{\text{best}}, d_{\min}, \ell\right)$.

In the last step, $\boldsymbol{c}$ should be de-permuted as

$$[0\ 1\ 1\ 0\ 0\ 1\ 0\ 1] \longrightarrow [0\ 1\ 0\ 1\ 0\ 1\ 0\ 1].$$

Table 10.1 summarizes this single-step OSD decoding.

Table 10.1 OSD decoding with early termination, for Example 10.3, successful in a single step.

#	ℓ	z_1^4	a_1^4	c	d_E^2	c_{best}	$\lambda\left(c_{\text{best}}\right)$	$\Lambda\left(c_{\text{best}}, d_{\min}, \ell\right)$
1	0	[0 0 0 0]	[0 1 1 0]	[0 1 1 0 0 1 0 1]	4.0511	✓	0.1743	1.6133

Table 10.2 OSD Decoding with early termination, for Example 10.3, when the most likely solution is not the correct solution.

#	ℓ	z_1^4	a_1^4	c	d_E^2	c_{best}	$\lambda(c_{\text{best}})$	$\Lambda(c_{\text{best}}, d_{\min}, \ell)$
1	0	[0 0 0 0]	[0 0 1 0]	[0 0 1 0 1 1 1 0]	7.7791	✓	1.1063	0.8981
2	1	[1 0 0 0]	[1 0 1 0]	[1 0 1 0 0 0 1 1]	16.093			
3	1	[0 1 0 0]	[0 1 1 0]	[0 1 1 0 0 1 0 1]	11.1023			
4	1	[0 0 1 0]	[0 0 0 0]	[0 0 0 0 0 0 0 0]	9.9273			
5	1	[0 0 0 1]	[0 0 1 1]	[0 0 1 1 1 0 0 1]	11.4755		1.1063	1.9095

Now, to show you how an erroneous information bit is corrected by OSD, we introduced another error in the previous $\mathbf{y}$ vector and we get

$$\mathbf{y} = [0.67 - 0.46\ \ 0.43 - 1.19\ 2.12\ \ 0.17\ \ 0.72\ \ 1.76],$$

and the corresponding permuted vector:

$$\mathbf{y}'' = [2.12\ \ 1.76 - 1.19\ 0.72\ \ 0.67 - 0.46\ \ 0.43\ \ 0.17].$$

As you can see, the second element in $\mathbf{y}''$ is in error. Therefore, we expect to correct it by $z_1^4 = [0\ 1\ 0\ 0]$. Table 10.2 illustrates the process. We expect to get minimum d_E^2 at step 3.

As can be seen, OSD fails in this case because the most likely solution with minimum d_E^2 is not the correct codeword. Note that the transmitted symbol -1 was heavily affected by noise and changed to 1.76. This increases the distance with the correct codeword. This is an example case where the correct codeword fails. Needless to mention that $\mathbf{c}$ should be de-permuted as

$$[0\ 0\ 1\ 0\ 1\ 1\ 1\ 0] \longrightarrow [1\ 1\ 1\ 1\ 0\ 0\ 1\ 0].$$

10.4.2 Guessing Random Additive Noise Decoding (GRAND)

GRAND is a relatively new algorithm in the category of universal decoders. The key differences from OSD are as follows:

- The length of test error vectors $\hat{\mathbf{e}} = \lambda_1(\mathbf{z})$ is N, unlike OSD where the scope of guess is limited to code dimension, as the errors in parity bits are corrected via re-encoding. This can potentially increase the time complexity of decoding. Note that if we employ the systematic form of the parity check matrix $\mathbf{H} = [\mathbf{I}\ \mathbf{P}]$, although the scope of guessing can reduce to $n - k$ coordinates as $\mathbf{H}\hat{\mathbf{e}}^T = \hat{\mathbf{e}}_I + \mathbf{P}\hat{\mathbf{e}}_P^T$ where $\hat{\mathbf{e}} = [\hat{\mathbf{e}}_I\ \hat{\mathbf{e}}_P]$, the guessing order will not follow the (near) ML order and it becomes similar to OSD. Note that to obtain a unique $\hat{\mathbf{e}}_P$, $\mathbf{P}$ must have full column rank, that is, $\text{rank}(\mathbf{P}) = n - k$. This ensures that $\mathbf{P}$ has no null space, and equation $\mathbf{P}\hat{\mathbf{e}}_P^T = \mathbf{H}\hat{\mathbf{e}}^T - \hat{\mathbf{e}}_I$ has a unique solution for e_P^T. However, no additional constraints are required to obtain $\hat{\mathbf{e}}_I$ because $\hat{\mathbf{e}}_I$ directly corresponds to $\mathbf{H}\hat{\mathbf{e}}^T - \mathbf{P}\hat{\mathbf{e}}_P^T$, and the calculation is deterministic.

- The order of testing error patterns is in ML order in a variant of GRAND called soft GRAND, or in near ML order, such as in a variant called order reliability bit (ORB) GRAND. In the former, a stack is needed to store the TEPs that do not follow the ML order to be used later, and hence a likelihood metric is used for sorting generated TEPs and keeping them in an updated order. However, the TEPs in ORB-GRAND are tested after generation, with no likelihood calculations and comparisons are required, although this comes at the cost of a slight discrepancy with ML order.
- Unlike OSD, the resulting sequences $\theta\left(y_1^N\right) \oplus \hat{e}_1^N$ are not necessarily codewords; that is, $\mathbf{H} \cdot \left(\theta\left(y_1^N\right) \oplus \hat{e}_1^N\right)^T \neq \mathbf{0}$. Hence, we continue querying for the error vector $\mathbf{e}$ until we get $\mathbf{H} \cdot \left(\theta\left(y_1^N\right) \oplus \hat{e}_1^N\right)^T = \mathbf{0}$.
- Theoretically, to achieve (near) ML performance with GRAND, the maximum number of queries, called *abandonment threshold* denoted by b, should be $b \geq 2^{n-k}$. Therefore, as the code dimension K decreases, the required threshold b increases. This means GRAND is not suitable for medium and low code rates, as it cannot provide acceptable error performance with reasonable complexity. However, the complexity of OSD is proportional to the code dimension.

The rest of this section focuses on ORB-GRAND, as it is a relatively simple algorithm that works with basic arithmetic operations and does not require any metric calculations or comparisons. The test error patterns $\mathbf{z}$ in ORB-GRAND are generated by integer partitions of *logistic weights* $w_L = 1, 2, \dots, n(n+1)/2$.

The logistic weight w_L of a length-N binary vector $\mathbf{z}$ is defined as

$$w_L(\mathbf{z}) = \sum_{i=1}^{N} z_i \cdot i, \tag{10.24}$$

where $z_i \in \mathbb{F}_2$ is the ith element of the error pattern $\hat{\mathbf{e}}$ permuted in the ascending order of the received symbols' reliability $|y_i|, i \in [1, N]$. That is, the test error pattern is $\hat{\mathbf{e}} = \lambda_1(\mathbf{z})$, where $\lambda_1(\cdot)$ is the vector-wise permutation function that maps the binary vector $\mathbf{z}$ to the error pattern $\hat{\mathbf{e}}$. For element-wise mapping of this permutation, we will use $\pi(\cdot)$ for mapping the index of any elements in $\mathbf{z}$ to the corresponding element in $\hat{\mathbf{e}}$ and $\pi^{-1}(\cdot)$ for the reverse mapping. For the sake of simplicity, we refer to $w_L(\mathbf{z})$ by w_L.

To obtain all binary vectors $\mathbf{z}$ corresponding to a certain w_L, there is a simple approach. All coordinates j in $\mathbf{z}$, where $z_j = 1$ for a certain w_L, can be obtained from *integer partitions* of w_L with distinct parts and no parts being larger than the code length N. Let us define the integer partitions of w_L mathematically as follows.

Definition 10.1 The integer partitions of w_L are the elements of any subset $\mathcal{P} \subset [1, w_L]$ such that

$$w_L = \sum_{j \in \mathcal{P} \subset [1, w_L]} j. \tag{10.25}$$

Then, the binary vector $\mathbf{z}$ corresponding to any $\mathcal{P}$ consists of the elements $z_j = 1, j \in \mathcal{P}$ and $z_j = 0, j \notin \mathcal{P}$.

In Definition 10.1, we abused the notion of integer partitions and considered a single part/partition as well to cover all the error patterns obtained from every w_L. Observe that for every w_L, there exists at least one $\mathcal{P}$ with a single element w_L. For example, for $w_L = 1, 2$, we have a single $\mathcal{P} = \{w_L\}$. As w_L gets larger, the number of subsets $\mathcal{P} \subset [1, w_L]$ increases.

Example 10.4 Suppose we have the received sequence $\mathbf{y} = [0.5, -1.2, 0.8, 1.8, -1, -0.2, 0.7, -0.9]$.

We can get the following permutation based on $|y_i|, i \in [1, 8]$ in the ascending order:

$$\pi : [1, 2, 3, 4, 5, 6, 7, 8] \to [6, 1, 7, 3, 8, 5, 2, 4].$$

Assuming we have attempted all the error patterns generated based on $w_L = 1, 2, 3, 4, 5$ so far. Then, we need to find the error patterns based on $w_L = 6$. The integer partitions of $w_L = 6$ are $\mathcal{P} = \{6\}, \{1, 5\}, \{2, 4\}$, and $\{1, 2, 3\}$ that satisfy $w_L = \sum_{j \in \mathcal{P}} j, \mathcal{P} \subset [1, 6]$. We call every element in $\mathcal{P}$ as a *part*. Then, the $\mathbf{z}$ vector and the corresponding error patterns after the vector permutation λ_1 are

$$\mathbf{z} = [0\ 0\ 0\ 0\ 0\ 1\ 0\ 0] \to \hat{\mathbf{e}} = [0\ 0\ 0\ 0\ 1\ 0\ 0\ 0],$$

$$\mathbf{z} = [1\ 0\ 0\ 0\ 1\ 0\ 0\ 0] \to \hat{\mathbf{e}} = [0\ 0\ 0\ 0\ 0\ 1\ 0\ 1],$$

$$\mathbf{z} = [0\ 1\ 0\ 1\ 0\ 0\ 0\ 0] \to \hat{\mathbf{e}} = [1\ 0\ 1\ 0\ 0\ 0\ 0\ 0],$$

$$\mathbf{z} = [1\ 1\ 1\ 0\ 0\ 0\ 0\ 0] \to \hat{\mathbf{e}} = [1\ 0\ 0\ 0\ 0\ 1\ 1\ 0].$$

These test error patterns can be checked in an arbitrary order using

$$\mathbf{H} \cdot \left(\theta\left(y_1^N\right) \oplus \lambda_1(\mathbf{z})\right)^T = \mathbf{0}.$$

Algorithm 10.2: ORB-GRAND with Abandonment

Input: Received vector $\mathbf{y}$, parity check matrix $\mathbf{H}$ of the code, abandonment threshold b
Output: Estimated codeword $\hat{\mathbf{x}}$

Find the permutation function λ_1 by sorting $|y_1|, |y_2|, \dots, |y_N|$ in decreasing order: $\mathbf{r} = \lambda_1(\mathbf{y})$;
$w_L \leftarrow 0$;
$\ell \leftarrow 1$;
checkPassed $\leftarrow$ *false*;
while $\ell \leq b$ and checkPassed $=$ *false* **do**
  $\mathcal{P} \leftarrow$ Integer partitions of w_L;
  for $\mathcal{Q} \leftarrow \mathcal{P}$ **do**
    $\ell \leftarrow \ell + 1$;
    $\mathbf{z} \leftarrow$ Take an all-zero vector and flip the zeros at coordinates corresponding to integers in $\mathcal{Q}$;
    if $\mathbf{H} \cdot \left(\theta(\mathbf{y}) \oplus \lambda_1(\mathbf{z})\right)^T = \mathbf{0}$ **then**
      checkPassed $\leftarrow$ *true*;
      break;
  $w_L \leftarrow w_L + 1$;
$\hat{\mathbf{x}} \leftarrow \theta(\mathbf{y}) \oplus \lambda_1(\mathbf{z})$;
return $\hat{\mathbf{x}}$

Algorithm 10.2 lists a pseudocode for ORB-GRAND with abandonment threshold b for a code with the party check matrix $\mathbf{H}$ and a BPSK-modulated received sequence $\mathbf{y}$ from an AWGN channel.

Example 10.5 Let us take the polar code (8, 4) with $\mathcal{A} = \{3, 5, 6, 7\}$ and $d_{\min} = 4$ in Example 10.3. As discussed in Chapter 5, the transposed parity check matrix $\mathbf{H}$ of this code is obtained by removing the columns of $\boldsymbol{G}_N$ with indices in $\mathcal{A}$. This results in

$$\mathbf{H} = \begin{bmatrix} 1 & 1 & 1 & 1 & 1 & 1 & 1 & 1 \\ 0 & 1 & 0 & 1 & 0 & 1 & 0 & 1 \\ 0 & 0 & 1 & 1 & 0 & 0 & 1 & 1 \\ 0 & 0 & 0 & 0 & 1 & 1 & 1 & 1 \end{bmatrix}.$$

Assuming the codeword

$$\mathbf{x} = [0\ 1\ 0\ 1\ 0\ 1\ 0\ 1],$$

is modulated by bipolar signaling (BPSK modulation) as

$$\mu(\mathbf{x}) = [1\ -1\ \ 1\ -1\ \ 1\ -1\ \ 1\ -1].$$

and transmitted through an AWGN channel. The receiver gets the following sequence where the sixth bit is in error as the sign has flipped:

$$\mathbf{y} = [0.67\ -0.46\ \ 0.43\ -1.19\ \ 2.12\ \ 0.17\ \ 0.72\ -1.76].$$

Table 10.3 GRAND results, for Example 10.5, after first attempt with $w_L = 1$.

#	w_L	Integer parts	z	ê	$\mathbf{H} \cdot (\theta(\mathbf{y}) \oplus \lambda_1(\mathbf{z}))^T$
1	1	{1}	[1 0 0 0 0 0 0 0]	[0 0 0 0 0 1 0 0]	[1 1 1 1]

To guess the noise effect, or the error vector $\hat{\mathbf{e}}$, we first find the permutation function in the ascending order of the reliability of received symbols as follows:

$$\pi : [1\ 2\ 3\ 4\ 5\ 6\ 7\ 8] \rightarrow [6\ 3\ 2\ 1\ 7\ 4\ 8\ 5].$$

The hard decision of the $\mathbf{y}$ is $\theta(\mathbf{y}) = [0\ 1\ 0\ 1\ 0\ 0\ 0\ 1]$, which is equivalent to $w_L = 0$. The receiver realizes that this sequence is not a codeword as the syndrome is non-zero:

$$\mathbf{H} \cdot \theta(\mathbf{y})^T = [1\ 1\ 0\ 1].$$

Table 10.3 shows that the error pattern is found in the first attempt where we have $w_L = 1$.

Now, assuming another error at the eighth bit has also been introduced due to noise. Note that the magnitude of the eighth symbol is relatively large.

$$\mathbf{y} = [0.67 - 0.46\ \ 0.43 - 1.19\ \ 2.12\ \ 0.17\ \ 0.72\ 1.76].$$

The demodulation of this sequence results in $\theta(\mathbf{y}) = [0\ 1\ 0\ 1\ 0\ 0\ 0\ 0]$ and the syndrome will be also non-zero as $\mathbf{H} \cdot \theta(\mathbf{y})^T = [0\ 0\ 1\ 0]$.

Table 10.4 illustrates the process of guessing the error vector by finding all possible integer partitions of $w_L = 1, 2, \ldots$. Observe that a valid codeword is found at 12th attempt where the syndrome is an all-zero vector. However, this does not give the actual error pattern. As discussed in Example 10.3, this is one of the cases that an ML decoder fails. Although the decoding is terminated at this step in practice, we continue processing other error patterns and as can be seen, the actual error pattern is found at the 20th attempt. Although ORB-GRAND may require more decoding attempts, the processes are simpler and ith does not require likelihood computation at every attempt for comparison purposes.

Exercises

1 What is the impact of cycles in the factor graph of polar codes on the performance of belief propagation decoding?
Hint: When cycles exist in the factor graph, the messages passed between the nodes can get "reused" before the algorithm converges, causing information from the same variable node to influence itself prematurely.

Table 10.4 GRAND results, for Example 10.5, when the most likely solution is not the correct solution.

#	w_L	Integer parts	z	ê	$\mathbf{H}\cdot(\theta(\mathbf{y})\oplus\lambda_1(\mathbf{z}))^T$
1	1	{1}	[1 0 0 0 0 0 0 0]	[0 0 0 0 0 1 0 0]	[1 1 1 1]
2	2	{2}	[0 1 0 0 0 0 0 0]	[0 0 1 0 0 0 0 0]	[1 0 0 0]
3	3	{3}	[0 0 1 0 0 0 0 0]	[0 1 0 0 0 0 0 0]	[1 1 1 0]
4	3	{1, 2}	[1 1 0 0 0 0 0 0]	[0 0 1 0 0 1 0 0]	[0 1 0 1]
5	4	{4}	[0 0 0 1 0 0 0 0]	[1 0 0 0 0 0 0 0]	[1 0 1 0]
6	4	{1, 3}	[1 0 1 0 0 0 0 0]	[0 1 0 0 0 1 0 0]	[0 0 1 1]
...	...	...	...	...	...
10	6	{6}	[0 0 0 0 0 1 0 0]	[0 0 0 1 0 0 0 0]	[1 1 0 0]
11	6	{1, 5}	[1 1 0 0 0 0 0 0]	[0 0 0 0 0 1 1 0]	[0 1 0 0]
12	6	{2, 4}	[0 1 0 1 0 0 0 0]	[1 0 1 0 0 0 0 0]	[**0 0 0 0**] ×
13	6	{1, 2, 3}	[1 1 1 0 0 0 0 0]	[0 1 1 0 0 1 0 0]	[1 0 0 1]
...	...	...	...	...	...
19	8	{8}	[0 0 0 0 0 0 0 1]	[0 0 0 1 0 0 0 0]	[1 0 1 1]
20	8	{1, 7}	[1 0 0 0 0 0 1 0]	[0 0 0 0 0 1 0 1]	[**0 0 0 0**] ✓

2 Why is belief propagation referred to as an "iterative" decoding method, and what are the typical stopping criteria used in practice?

3 What is the impact of the bias term on Fano decoding, and how can adjusting it influence the balance between error performance and computational complexity?

4 What are the advantages of using a Stack decoder over a Fano decoder when dealing with long codes or high-noise channels?

5 What role does reliability information play in the OSD and GRAND algorithms and how does it impact the selection of test error patterns?

Bibliographical Notes

Belief propagation (BP) decoding for polar codes has been extensively studied in recent years, with growing interest due to its ability to parallelize, as shown by Arıkan [2010] in a pipeline implementation. This decoder was based on the

Forney BP decoder for Reed-Muller (RM) codes using factor graph representations of the generator matrix $\boldsymbol{G}$. Early studies demonstrated that BP decoding could offer slightly better performance compared to SC, particularly when multiple iterations are employed, though it tends to underperform in comparison to more advanced techniques like SC-List decoding. Given the extensive literature on BP decoding of LDPC codes and the familiarity of more researchers with this algorithm, numerous papers in the literature discuss various techniques to improve BP decoding. These include concatenation with outer LDPC codes for low-reliability sub-channels by Guo et al. [2014], facilitating belief propagation by utilizing specific constituent codes by Xu et al. [2015], permutated BP list decoding by Elkelesh et al. [2018], early termination criteria by Yuan and Parhi [2014], other early termination schemes compromising performance for reduced complexity by Ren et al. [2015], a memory-efficient stage-combined belief propagation (BP) decoding by Sha et al. [2015], converting dense factor graph to a sparse bipartite factor graph [Cammerer et al., 2018], among others. Interested readers may also consider reading a survey paper on BP decoding by Arli and Gazi [2021].

Sequential decoding and sphere decoding are ML decoding algorithms that have been adapted to polar codes. Due to the high complexity of sequential decoding, interested readers can refer to several techniques available in the literature, such as a score function by Trifonov [2018], Jeong and Hong [2019], and decomposition of a polar code decoded into a number of outer codes by Trofimiuk et al. [2020]. Other variants of sequential decoding are also available, such as stack decoding by Niu and Chen [2012], Miloslavskaya and Trifonov [2014], and creeper decoding by Ivanov et al. [2022]. Recently, Fano decoding has also been adapted to PAC codes initially by Arıkan [2019] and then improved by Rowshan et al. [2020], Rowshan and Viterbo [2021], Moradi [2021], Ji et al. [2023], and a hardware architecture was designed for this decoder by Mozammel [2021] and proposed a sequential-like decoding algorithm that enhances the performance of modified G_N-coset codes by integrating a successive cancellation ordered search (SCOS) approach, which systematically prioritizes and explores decoding paths based on reliability metrics, achieving a significant reduction in computational complexity while maintaining near-optimal error correction capabilities for modern communication systems.

Sphere decoding was adapted to polar codes by Kahraman and Çelebi [2012], and then several complexity reduction techniques and approaches such as bounded metrics for SD, list SD (LSD), and cyclic redundancy check (CRC)-aided SD were proposed by Guo and i Fabregas [2015], Piao et al. [2018], Hashemi et al. [2015], Husmann et al. [2017], and Zhou et al. [2020] to explore the structure of the decoding tree in search of the (near) optimal path.

For a variant of polar codes named PAC codes, various decoding algorithms have been adapted, such as sequential decoding by Arıkan [2019], Rowshan et al.

[2020], Rowshan [2021], and Mozammel [2021], list decoding by Rowshan et al. [2021] and Yao et al. [2021], list Viterbi decoding by Rowshan and Viterbo [2021], belief propagation decoding by Zhang et al. [2022], GRAND by Rowshan and Yuan [2022, 2023, 2025] and Wang et al. [2023], stack decoding by Rowshan et al. [2021], Jiang et al. [2023], Wu et al. [2023], and Zhang et al. [2023], fast and simplified list decoding by Saber et al. [2024], Dai et al. [2024], and other decoding algorithms by Yu and Shi [2021] and Zhang [2023].

The brute-force search is a naive ML decoder which is impractical except for low-rate codes with short block-length. In general, ML decoding is prohibitively complex for most codes, as it was shown to be an NP-complete problem [Van Tilborg et al., 1978]. Ordered statistics decoding by Fossorier and Lin [1997] is a near-ML decoding algorithm. This algorithm permutes the columns of the generator matrix $\mathbf{G}$ with respect to the reliability of the received symbols $\mathbf{y}$ and performs elementary row operations on the K-independent columns extracted from the permuted generator matrix, resulting in the systematic form. Observe that these operations have to be performed on every block of received symbols as the reliability of the received symbols varies from one block to another. Then, the hard decision codeword plus an error sequence constructed based on the reliability of $\mathbf{y}$ is obtained and modulated to get $\hat{\mathbf{d}}$. The testing error patterns can have a Hamming weight of up to l, $0 \leq l \leq K$ in the l-order OSD. The Euclidean distance between the testing $\hat{\mathbf{m}}$ and reordered $\mathbf{y}$ based on the permutations, denoted by $\hat{\mathbf{c}}$, determines the most likely error sequence. Since OSD evaluates all error sequences with Hamming weight l, the complexity is fixed. Furthermore, the sufficient condition for optimality [Lin and Yan, 2014] can significantly reduce the number of queries.

Recently, Duffy et al. [2018] suggested a hard-decision scheme in which the error patterns were ordered from the most likely to the least likely ones based on a statistical channel model, and then the incorrect error patterns were sequentially removed, searching for the first error patterns corresponding to a valid codeword. This original idea, which was later called guessing random additive noise decoding (GRAND), was further developed into a soft decision scheme or sGRAND by Duffy and Médard [2019], where error patterns were generated based on the reliability of the symbols and sequential insertion and removal of the error patterns from an ordered stack until the first valid codeword was found. The sGRAND algorithm was shown by Duffy et al. [2019] to be a capacity-achieving algorithm and ML algorithm, although it cost a significant computational complexity because the error patterns in the stack needed to be sorted after the insertion of new patterns into the stack. This variant of GRAND was improved by Zheng and Ma [2024]. GRAND was further evolved by Duffy et al. [2022] using a simple metric that provides test error patterns in near-ML order. This variant was called ordered reliability bits (ORB) GRAND. This step was a significant boost for GRAND toward making it practical for high-rate and short codes. The approximate scheduling of the error sequences is based on a distinct integer partitioning of positive integers, which is significantly less complex. An attempt towards reducing the complexity of ORB-GRAND was made by Rowshan and Yuan [2022, 2025] utilising the code structure in the process of generating test error patterns.

Bibliography

Erdal Arıkan. Polar codes: A pipelined implementation. In *Proceedings of the 4th ISBC*, volume 2010, pages 11–14, 2010.

Erdal Arıkan. From sequential decoding to channel polarization and back again. *arXiv preprint arXiv:1908.09594*, 2019.

Ahmet Çağri Arli and Orhan Gazi. A survey on belief propagation decoding of polar codes. *China Communications*, 18(8):133–168, 2021.

Sebastian Cammerer, Moustafa Ebada, Ahmed Elkelesh, and Stephan ten Brink. Sparse graphs for belief propagation decoding of polar codes. In *2018 IEEE International Symposium on Information Theory (ISIT)*, pages 1465–1469. IEEE, 2018.

Jingxin Dai, Hang Yin, Yansong Lv, Yuhuan Wang, and Rui Lv. Fast list decoding of PAC codes with new nodes. *IEEE Communications Letters*, 28(3):449–453, 2024.

Ken R Duffy and Muriel Médard. Guessing random additive noise decoding with soft detection symbol reliability information-SGRAND. In *2019 IEEE International Symposium on Information Theory (ISIT)*, pages 480–484. IEEE, 2019.

Ken R Duffy, Jiange Li, and Muriel Médard. Guessing noise, not code-words. In *2018 IEEE International Symposium on Information Theory (ISIT)*, pages 671–675. IEEE, 2018.

Ken R Duffy, Jiange Li, and Muriel Médard. Capacity-achieving guessing random additive noise decoding. *IEEE Transactions on Information Theory*, 65(7):4023–4040, 2019.

Ken R Duffy, Wei An, and Muriel Médard. Ordered reliability bits guessing random additive noise decoding. *IEEE Transactions on Signal Processing*, 70:4528–4542, 2022.

Ahmed Elkelesh, Moustafa Ebada, Sebastian Cammerer, and Stephan Ten Brink. Belief propagation list decoding of polar codes. *IEEE Communications Letters*, 22(8):1536–1539, 2018.

M P C Fossorier and Shu Lin. Weight distribution for closest coset decoding of —u—u+v— constructed codes. *IEEE Transactions on Information Theory*, 43(3):1028–1030, 1997.

Jing Guo and Albert Guillén i Fabregas. Efficient sphere decoding of polar codes. In *2015 IEEE International Symposium on Information Theory (ISIT)*, pages 236–240. IEEE, 2015.

Jing Guo, Minghai Qin, Albert Guillen i Fabregas, and Paul H Siegel. Enhanced belief propagation decoding of polar codes through concatenation. In *2014 IEEE International Symposium on Information Theory*, pages 2987–2991. IEEE, 2014.

Seyyed Ali Hashemi, Carlo Condo, and Warren J Gross. List sphere decoding of polar codes. In *2015 49th Asilomar Conference on Signals, Systems and Computers*, pages 1346–1350. IEEE, 2015.

Chistopher Husmann, Panagiotis Chatzi Nikolaou, and Konstantinos Nikitopoulos. Reduced latency ML polar decoding via multiple sphere-decoding tree searches. *IEEE Transactions on Vehicular Technology*, 67(2):1835–1839, 2017.

Fedor Ivanov, Ilya Chetverikov, Alexey Kreshchuk, and Ilya Timokhin. Successive cancellation creeper decoding of polar codes. In *2022 IEEE International Multi-Conference on Engineering, Computer and Information Sciences (SIBIRCON)*, pages 60–64. IEEE, 2022.

Min-Oh Jeong and Song-Nam Hong. SC-Fano decoding of polar codes. *IEEE Access*, 7:81682–81690, 2019.

Houren Ji, Yifei Shen, Zaichen Zhang, Yongming Huang, Xiaohu You, and Chuan Zhang. Low-complexity fast Fano decoding for PAC codes. *IEEE Transactions on Vehicular Technology*, 72(12):15172–15184, 2023.

Zongsheng Jiang, Zhiliang Huang, Youyan Zhang, and Baohong Zhou. A soft-output stack decoding of polarization-adjusted convolutional codes. In *International Conference on Cryptography, Network Security, and Communication Technology (CNSCT 2023)*, volume 12641, pages 272–279. SPIE, 2023.

Sinan Kahraman and M Ertuğrul Çelebi. Code based efficient maximum-likelihood decoding of short polar codes. In *2012 IEEE International Symposium on Information Theory Proceedings*, pages 1967–1971. IEEE, 2012.

Jun Lin and Zhiyuan Yan. Efficient list decoder architecture for polar codes. In *Proceedings of IEEE International Symposium on Circuits and Systems (ISCAS)*, pages 1022–1025, June 2014.

Vera Miloslavskaya and Peter Trifonov. Sequential decoding of polar codes. *IEEE Communications Letters*, 18(7):1127–1130, 2014.

Mohsen Moradi. On sequential decoding metric function of polarization-adjusted convolutional (PAC) codes. *IEEE Transactions on Communications*, 69(12):7913–7922, 2021.

Amir Mozammel. Hardware implementation of Fano decoder for polarization-adjusted convolutional (PAC) codes. *IEEE Transactions on Circuits and Systems II: Express Briefs*, 69(3):1632–1636, 2021.

Kai Niu and Kai Chen. Stack decoding of polar codes. *Electronics Letters*, 48(12):695–697, 2012.

Jinnan Piao, Jincheng Dai, and Kai Niu. CRC-aided sphere decoding for short polar codes. *IEEE Communications Letters*, 23(2):210–213, 2018.

Yuanrui Ren, Chuan Zhang, Xing Liu, and Xiaohu You. Efficient early termination schemes for belief-propagation decoding of polar codes. In *2015 IEEE 11th International Conference on ASIC (ASICON)*, pages 1–4. IEEE, 2015.

Mohammad Rowshan. *Towards Enhanced Decoding of Polar Codes and PAC Codes*. PhD thesis, Monash University, 2021.

Mohammad Rowshan and Emanuele Viterbo. List Viterbi decoding of PAC codes. *IEEE Transactions on Vehicular Technology*, 70(3):2428–2435, 2021.

Mohammad Rowshan and Jinhong Yuan. Constrained error pattern generation for grand. In *2022 IEEE International Symposium on Information Theory (ISIT)*, pages 1767–1772. IEEE, 2022.

Mohammad Rowshan and Jinhong Yuan. Segmented GRAND: Combining sub-patterns in near-ML order. *arXiv preprint arXiv:2305.14892*, 2023.

Mohammad Rowshan and Jinhong Yuan. Segmented GRAND: Complexity reduction through sub-pattern combination. *IEEE Transactions on Communications*, 2025. doi: 10.1109/TCOMM.2025.3541094.

Mohammad Rowshan, Andreas Burg, and Emanuele Viterbo. Complexity-efficient Fano decoding of polarization-adjusted convolutional (PAC) codes. In *2020 International Symposium on Information Theory and Its Applications (ISITA)*, pages 200–204. IEEE, 2020.

Mohammad Rowshan, Andreas Burg, and Emanuele Viterbo. Polarization-adjusted convolutional (PAC) codes: Sequential decoding vs list decoding. *IEEE Transactions on Vehicular Technology*, 70(2):1434–1447, 2021.

Hamid Saber, Homayoon Hatami, and Jung Hyun Bae. Simplified successive cancellation list decoding of PAC codes. *arXiv preprint arXiv:2401.13922*, 2024.

Jin Sha, Xing Liu, Zhongfeng Wang, and Xiaoyang Zeng. A memory efficient belief propagation decoder for polar codes. *China Communications*, 12(5):34–41, 2015.

Peter Trifonov. A score function for sequential decoding of polar codes. In *2018 IEEE International Symposium on Information Theory (ISIT)*, pages 1470–1474. IEEE, 2018.

Grigorii Trofimiuk, Nikolai Iakuba, Stanislav Rets, Kirill Ivanov, and Peter Trifonov. Fast block sequential decoding of polar codes. *IEEE Transactions on Vehicular Technology*, 69(10):10988–10999, 2020.

H C A Van Tilborg, R J Mceliece, and E R Berlekamp. On the inherent intractability of certain coding problems. 1978.

Yueh Wang, Zhiping Shi, Ziyu Han, Zu'en Wei, and Kunyang Li. Improved ORB-grand for PAC codes. In *2023 8th International Conference on Communication, Image and Signal Processing (CCISP)*, pages 477–481, 2023.

Yuze Wu, Zhiliang Huang, Youyan Zhang, and Shuihong Zhou. Multi-stack decoding algorithm for polarization-adjusted convolutional codes. In *2023 3rd International Conference on Electronic Information Engineering and Computer Science (EIECS)*, pages 1194–1197. IEEE, 2023.

Jingwei Xu, Tiben Che, and Gwan Choi. XJ-BP: Express journey belief propagation decoding for polar codes. In *2015 IEEE Global Communications Conference (GLOBECOM)*, pages 1–6. IEEE, 2015.

Hanwen Yao, Arman Fazeli, and Alexander Vardy. List decoding of Arıkan's PAC codes. *Entropy*, 23(7):841, 2021.

Qingping Yu and Zhiping Shi. Threshold-based list decoding for PAC codes. In E*2021 IEEE 21st International Conference on Communication Technology (ICCT)*, pages 1296–1299. IEEE, 2021.

Bo Yuan and Keshab K Parhi. Early stopping criteria for energy-efficient low-latency belief-propagation polar code decoders. *IEEE Transactions on Signal Processing*, 62(24):6496–6506, 2014.

Wei Zhang. A novel SCL bit-flipping decoding of polarization-adjusted convolutional (PAC) codes. *arXiv preprint arXiv:2307.05871*, 2023.

Xianwen Zhang, Ming Jiang, Mingyang Zhu, Kailin Liu, and Chunming Zhao. CRC-aided adaptive BP decoding of PAC codes. *Entropy*, 24(8):1170, 2022.

Li Zhang, Haina Liu, and Yejun He. Improved stack decoding for PAC codes. In *2023 32nd Wireless and Optical Communications Conference (WOCC)*, pages 1–5, 2023.

Xiangping Zheng and Xiao Ma. A universal list decoding algorithm with application to decoding of polar codes. *IEEE Transactions on Information Theory*, 71(2):975–995, 2024.

Huayi Zhou, Warren J Gross, Zaichen Zhang, Xiaohu You, and Chuan Zhang. Efficient sphere polar decoding via synchronous determination. *IEEE Transactions on Vehicular Technology*, 69(6):6777–6781, 2020.

11

Rate-Compatible Polar Codes

"Price is what you pay. Value is what you get."

— Warren Buffett

Wireless channels have a time-varying nature. To address different error protection requirements depending on the channel condition, we need a flexible channel coding scheme that allows us to improve the error correction capability by changing the code rate. This is known as explicit *link adaptation*. Moreover, the availability of a feedback path on wireless channels enables us to use hybrid automatic repeat request (HARQ) protocols under any channel conditions. These protocols adjust the parity bits in response to a negative acknowledgment (NACK) from the receiver, indicating the decoding failure of the last transmission. This can be considered an implicit link adaptation. In terms of overall throughput, the implicit link adaptation is superior to the explicit one because the effective code rate is reduced only when needed by retransmission, not for every transmission. Note that HARQ schemes require additional redundancy, where the effective code rate is reduced, or retransmission of the entire codeword, where the accumulated received E_b/N_0 increases after each retransmission. The former scheme is called *incremental redundancy* (IR), and the latter is called *chase combining* (CC). Although the overall throughput of the implicit link adaptation is higher, the explicit link adaptation results in a shorter delay.

In the case of polar codes, the code length is constrained to a power of two by definition, that is, $N = 2^n$ for some $n \geq 1$. This constraint arises due to the construction of polar codes based on the Kronecker power of the basic kernel, $\boldsymbol{G}_2$. However, the error correction capability of short codes with small n is not desirable. To address this, one can choose a relatively large n and then adjust the length using puncturing or shortening.

Polar Codes: From Theory to Practice, First Edition. Mohammad Rowshan and Emanuele Viterbo.

This chapter provides an in-depth discussion on the concepts of code modification through puncturing and shortening, both for linear block codes in general and specifically for polar codes. We explore how these techniques allow polar codes to achieve flexible code lengths and rates, enabling their application in real-world systems like the 5G standard. Popular low-complexity approaches for puncturing and shortening are introduced, with references to more advanced methods provided in the bibliographical notes.

11.1 Modification of Block Codes

A new code can be constructed by modifying an existing code. The modification can span from adding additional parity bit(s), which is called extension, deleting parity check bit(s), which is called puncturing, deleting message coordinate(s), that is shortening; adding message coordinate(s), that is lengthening; adding new codewords, that is, augmenting; and removing some codewords, that is, expurgating. In this section, we limit the scope to the most popular modifications, define the punctured and shortened codes for any linear block codes, and study their impacts on the resulting codes.

11.1.1 Shortened Codes

Given a linear block code $\mathscr{C}(n,k,d)$ with information length k, block length n, and minimum Hamming distance d, this code is called the *mother code*, a new code $\mathscr{C}'$ can be created by keeping only the codewords $\mathbf{c} = (c_0, c_1, \dots, c_{n-1})$ with $c_j = 0$ of the given code and then deleting the jth position. Then, *shortened code* $\mathscr{C}'(n',k',d')$ is a subcode of $\mathscr{C}$ characterized by the parameters $n' = n-1$, $k' = k-1$, and $d' \geq d$. Observe that if none of the codewords of the subcode has the Hamming weight d, or in other words if all the minimum weight codewords in $\mathscr{C}$ have a non-zero value at position j, then the minimum Hamming distance of the subcode will be $d' > d$. In fact, shortening a code can result in an increase in the minimum Hamming distance of a code, which can improve the error correction capability of the subcode at the cost of a smaller code size, that is, $2^{k-1} < 2^k$.

11.1.2 Punctured Codes

The *punctured code* $\mathscr{C}'$ of $\mathscr{C}$ is obtained by eliminating jth symbol in the codeword, $c_j \in \{0,1\}$, irrespective of its value. This symbol in the systematic linear block code is assumed to be a parity check symbol. This operation is equivalent to removing a column from the generator matrix of the code. For a given mother code, puncturing in different columns may not give equivalent codes. The punctured code $\mathscr{C}'(n', k', d')$ is characterized by the parameters $n' = n - 1, k' = k$, and $d' \leq d$. The minimum distance d of the mother code reduces to $d' = d - 1$ if $\mathscr{C}$ has a minimum weight codeword with a nonzero value at position j.

Puncturing and shortening are often performed in multiple positions in an analogous manner. Observe that the code rate decreases $\left(\frac{k-1}{n-1} < \frac{k}{n}\right)$ by shortening and increases $\left(\frac{k}{n-1} > \frac{k}{n}\right)$ by puncturing the mother code. The advantage of puncturing is that while the effective code rate increases, the same decoder can be employed for decoding.

On the receiving side, since the decoder does not have any stochastic information about the punctured or shortened symbols, so their initial log-likelihood ratios (LLRs, defined as $\log \frac{P(y_j|u_j=0)}{P(y_j|u_j=1)}$, see Chapter 3) are set to:

- Zero for punctured symbols: The values of the punctured symbols could be 0 or 1, hence due to uncertainty, the LLRs of the punctured symbols are set on the boundary of positive values where $P(y_j|u_j = 0) > P(y_j|u_j = 1)$, and negative values where $P(y_j|u_j = 0) < P(y_j|u_j = 1)$.
- $+\infty$ or a large positive value for shortened symbols: Since the shortened symbols are known to be 0, by default, a positive LLR would lead to the desired relation of $P(y_j|u_j = 0) > P(y_j|u_j = 1)$.

Example 11.1 Let us take the extended Hamming code (7+1,4,3+1) with the generator matrix in systematic form as

$$\boldsymbol{G} = \begin{bmatrix} 1&0&0&0&1&1&0&1 \\ 0&1&0&0&1&0&1&1 \\ 0&0&1&0&0&1&1&1 \\ 0&0&0&1&1&1&1&0 \end{bmatrix}. \tag{11.1}$$

Observe that the extended Hamming code above is obtained by adding an overall parity check bit. The additional parity bit can be obtained by the additional column with elements that give zero when summing up the row elements. Puncturing operation by deleting a parity check (a column in $\boldsymbol{G}$ corresponding to a parity bit) can give us the original Hamming code (7,4,3) with generator matrix

$$\boldsymbol{G}_p = \begin{bmatrix} 1&0&0&0&1&1&0 \\ 0&1&0&0&1&0&1 \\ 0&0&1&0&0&1&1 \\ 0&0&0&1&1&1&1 \end{bmatrix}, \tag{11.2}$$

whereas the shortened code (7,3,4) can be obtained by deleting a message coordinate (a row in $\boldsymbol{G}$) resulting in the generator matrix

$$\boldsymbol{G}_s = \begin{bmatrix} 1 & 0 & 0 & 1 & 1 & 0 & 1 \\ 0 & 1 & 0 & 1 & 0 & 1 & 1 \\ 0 & 0 & 1 & 0 & 1 & 1 & 1 \end{bmatrix}. \tag{11.3}$$

11.2 Punctured Polar Codes

To get a punctured code (E, K) with length $E < N$ where $N = 2^{\lceil \log_2 E \rceil}$, a mother code (N, K) is used as a framework for encoding and decoding. In the mother code, we need to select the $N - E$ puncturing indices called the *puncturing pattern*. As the dimension of the code remains the same, these indices can be selected from the frozen set $\mathcal{A}^c$. Recall from Chapter 4 that during the encoding process or channel combining, the random information bits and known frozen bits are combined to form coded bits. Hence, puncturing a coded bit corresponding to a frozen index would result in access to fewer information about the information bits in the decoding process or channel splitting.

When a coded bit x_i is punctured, we have $I(W_i) = 0$, that is, the channel W_i becomes *incapable*. Without loss of generality, consider the case of $N = 2$ for polarized channels by $\boldsymbol{G}_2$ where we have

$$I(W_2^{(0)}) \leq \min\{I(W_0), I(W_1)\}, \tag{11.4}$$

$$I(W_2^{(1)}) \geq \max\{I(W_0), I(W_1)\}, \tag{11.5}$$

$$I(W_2^{(0)}) + I(W_2^{(1)}) = I(W_0) + I(W_1). \tag{11.6}$$

From (11.4), observe that sub-channel $W_2^{(0)}$ is incapable, that is $I(W_2^{(0)}) = 0$ if channel W_0 or channel W_1 or both are incapable. From (11.5) and (11.6), on the other hand, one can see that sub-channel $W_2^{(1)}$ is incapable, that is, $I(W_2^{(1)}) = 0$ if both channels W_0 and W_1 are incapable. Thus, polarization by $\boldsymbol{G}_2$ preserves the number of incapable channels $W_i, i = 0, 1$. This principle can be extended to higher-order polarization by $\boldsymbol{G}_2^{\otimes n}$. That is, the symmetric capacities of a number of split channels are degraded to zero when the same number of symbols corresponding to the split channels are not transmitted.

Definition 11.1 ***Incapable bits:*** The uncoded bits u_i in $\mathbf{u}$ that correspond to the zero-capacity sub-channels, $I(W_N^{(i)}) = 0$, due to puncturing are called *incapable bits*. The values of the incapable bits similar to frozen bits are set to zero, $u_i = 0$.

Let us conclude that any puncturing bit pattern with $\mathcal{R}$ punctured bits in $\mathbf{x}$ leads to exactly the same number of incapable bits in $\mathbf{u}$.

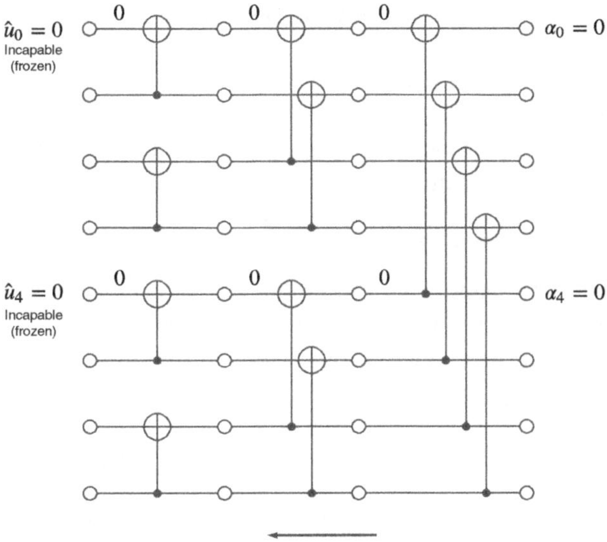

Figure 11.1 Bit-reversal puncturing pattern $\mathcal{R} = \{0, 1\}$ of polar code with $E = 6$ on the factor graph.

Example 11.2 Suppose that the coded bits x_0 and x_4 for a polar code of length 8 are punctured. Observe that the bit reverse of $4 = 100_2$ is $1 = 001_2$. When decoding, as Figure 11.1 illustrates, we set the LLRs of the punctured coordinates to zero. This results in the evolved LLRs being zero for these coordinates, regardless of the LLRs of other coordinates. The reason lies in the following general rules:

- The output of the f_- nodes when one of the inputs is 0; that is, according to (4.41), we have

$$f_-(0, \alpha) \triangleq \log\left(\frac{e^{0+\alpha}+1}{e^0+e^\alpha}\right) = \log\left(\frac{e^\alpha+1}{e^\alpha+1}\right) = \log(1) = 0 \text{ for any } \alpha, \quad (11.7)$$

- The output of the nodes f_+ when both inputs are 0; that is, according to (4.42), we have

$$f_+(0, 0, u) \triangleq (-1)^u \cdot 0 + 0 = 0 \text{ for any } u. \quad (11.8)$$

11.3 Shortened Polar Codes

Given a mother polar code (N, K), the shortened polar code is a subcode $(N - S, K - S)$ of the mother code, where $S < K$ bits of the codewords are restricted to 0, by default. A trivial way to get a guaranteed zero-value encoded bit in polar coding is to pass zero through the sub-channel that is independent of other sub-channels, that is the last sub-channel with index $N - 1 = (11 \cdots 11)_2$ which is the most reliable channel as well. The second sub-channel is chosen so that it depends only on the sub-channel $N - 1$. There are several candidates for this sub-channel. The two commonly used ones are: (1) the channel with index $N - 2 = (11 \cdots 10)_2$ and (2) its bit-reversal counterpart that has index $N/2 - 1 = (01 \cdots 11)_2$. This can be continued to find more sub-channels that are only dependent on the previous sub-channels with zero input.

Example 11.3 Suppose that the coded bits x_6 and x_7 for a polar code of length 8 are shortened. Figure 11.2a illustrates the last two sub-channels with a shortening pattern $\mathcal{R} = \{6, 7\}$ that are constrained to zero outputs. Alternatively, one can choose the shortening pattern $\mathcal{R} = \{3, 7\}$ which is the bit-reversal of the previous pattern. These patterns can be seen in the form of matrix multiplication due to the lower triangular shape of $\boldsymbol{G}_2^{\otimes n}$. As Figure 11.3 shows, we have $x_7 = u_7$. Hence, by setting $u_7 = 0$, we get $x_7 = 0$. The other shortened bit could be x_6 (a), which depends on u_7 through $x_6 = u_6 \oplus u_7$, or x_3 (b), which also depends on u_7 through $x_3 = u_3 \oplus u_7$. Hence, either x_3 or x_6 can get 0 if we set $u_3 = 0$ or $u_6 = 0$, respectively. From Figure 11.2b, one can observe that $I(W_8^{(6)}) = I(W_8^{(7)}) = 1$ virtually forms the one-capacity split channels $W_8^{(6)}$ and $W_8^{(7)}$. These channels are called *overcapable* channels. Observe that similar rules to those in Example 11.2 are as follows:

- The output of f_{-} nodes when both inputs are very large (∞; i.e. according to (4.43), we have

$$f_{-}(\infty, \infty) \approx \tilde{f}_{-}(\infty, \infty) \triangleq \text{sign}(\infty)\text{sign}(\infty) \min\{|\infty|, |\infty|\} = \infty, \tag{11.9}$$

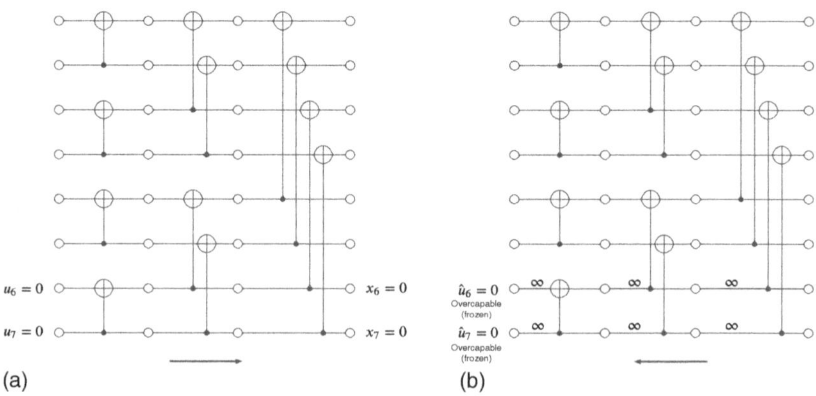

Figure 11.2 Encoding (a) and decoding (b) of polar code with shortening pattern $\mathcal{R} = \{6, 7\}$ on the factor graph.

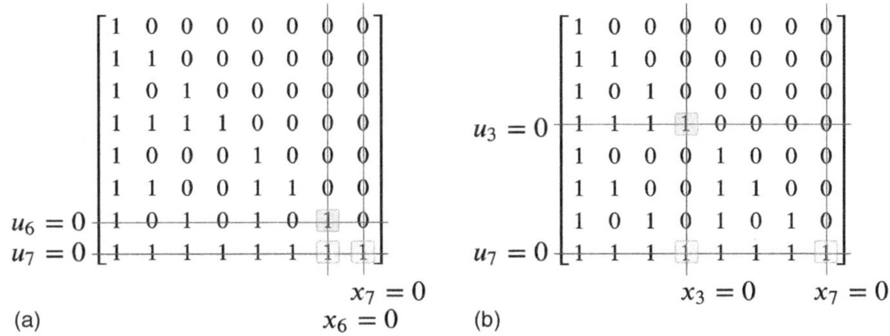

Figure 11.3 Natural (a) and bit-reversal (b) shortening patterns $\mathcal{R} = \{6,7\}$ and $\{3,7\}$ of polar code with $E = 6$ on matrix $\mathbf{G}_8$.

- The output of f_+ nodes when one of the inputs is ∞; that is, according to (4.42), we have

$$f_+(\infty, \alpha, u) \triangleq (-1)^u \cdot \alpha + \infty = \infty \text{ for any } u \text{ and } \alpha. \tag{11.10}$$

Definition 11.2 ***Overcapable bits:*** The uncoded bits u_i in $\mathbf{u}$ that correspond to the one-capacity sub-channels, $I(W_N^{(i)}) = 1$, due to shortening of known coded bits are called *overcapable bits*. The values of the overcapable bits similar to frozen bits are set to zero, $u_i = 0$.

Similar to punctured bits, it can be shown that any shortening bit pattern with $\mathcal{R}$ shortened bits in $\mathbf{x}$ leads to exactly the same number of overcapable bits in $\mathbf{u}$. Note that similar to puncturing, shortening influences channel polarization and changes the reliability of subchannels. Hence, the reliability of the rest of the sub-channels can be evaluated by considering the mean of LLRs of the shortened sub-channels in the density evolution methods equal to infinity or a large value.

11.4 Rate-Matching in 5G NR

As we learned, the length of polar codewords, represented by the bit sequence $d_0, d_1, \ldots, d_{N-1}$, is constrained to a power of two by definition, that is $N = 2^n$ for some $n \geq 1$. However, for a given number of information bits $A \leq 1706$ and the desired rate R depending on channel conditions, a codeword with length $E = A/R$, whose bits are denoted by $f_0, f_1, \ldots, f_{E-1}$, is needed in typical 5G applications. Adjusting the codeword length is achieved by classical techniques, namely, puncturing, shortening, and extending. These techniques are employed for the purpose of rate-matching, or to be accurate length-matching with respect to the desired code rate.

As a result of codeword modification, some of the bits may not be transmitted. A bit interleaving (or permutation) is needed to retain the more important bits from the error correction point of view. Bit interleaving is performed before puncturing, shortening, or extension primarily to enhance performance and ensure fairness in bit protection levels across the codeword. This is discussed later in this section.

11.4.1 The Mother Code and Rate-Matching Choice

The length of the mother code $N = 2^n$ is first determined. The minimum length of the mother code is considered $2^{n_{\min}}$, $n_{\min} = 5$, while the maximum length corresponds to $n_{\max} = 10$ for uplink and $n_{\max} = 9$ for downlink. Given the minimum code rate $R_{\min} = 1/8$, the mother code length $N = 2^n$ is obtained by

$$n = \max(\min(n_1, n_2, n_{\max}), n_{\min}), \tag{11.11}$$

where $n_2 = \lceil \log_2 \frac{K}{R_{\min}} \rceil$ and $n_1 = \lceil \log_2 E \rceil - 1$ if $E \leq (9/8) \cdot 2^{n_1}$, while the effective code rate is $K/E < 9/16$, otherwise $n_1 = \lceil \log_2 E \rceil$.

Example 11.4 For encoding $K = 42$ bits in the downlink at the desired rate $R = 3/8$, we need to transmit $E = K/R = 112$ bits after rate-matching, where $\lceil \log_2 E \rceil - 1 = 6$. The length of the mother code is obtained as

$$n = \max(\min(n_1, n_2, 9), 5) = 7. \tag{11.12}$$

11.4.2 Subblock Interleaving

Bit interleaving rearranges the order of bits in the codeword such that the effects of puncturing or shortening are distributed more evenly across the entire codeword. This helps in maintaining the balance of protection across all bits.

The sequence of coded bits $d_0, d_1, \ldots, d_{N-1}$ is partitioned into 32 subblocks. Clearly, when we have $N = 32$, every subblock takes a single bit. Suppose the output of the interleaver is denoted by $y_0, y_1, \cdots, y_{N-1}$. The interleaver is defined by the relation $j = J(i)$ between the index $i = 0, 1, \cdots, N-1$ of d_i and j of y_j as

$$j = J(i) = P\left(\left\lfloor \frac{32i}{N} \right\rfloor\right) \frac{N}{32} + i \bmod \frac{N}{32}, \tag{11.13}$$

where $P(\cdot)$ is the interleaving pattern based on Table 11.1. Figure 11.4 illustrates an interleaving example for $N = 32$.

Note that the de-interleaving should be performed before decoding in the receiver.

Table 11.1 Subblock interleaver pattern $P(i)$.

i	$P(i)$	i	$P(i)$	i	$P(i)$	i	$P(j)$	i	$P(i)$	i	$P(i)$	i	$P(i)$	i	$P(i)$
0	0	4	3	8	8	12	10	18	12	20	14	24	24	28	27
1	1	5	5	9	18	13	18	17	20	21	22	25	25	29	29
2	2	8	6	10	9	14	11	18	13	22	15	28	28	30	30
3	4	7	7	11	17	15	19	19	21	23	23	27	28	31	31

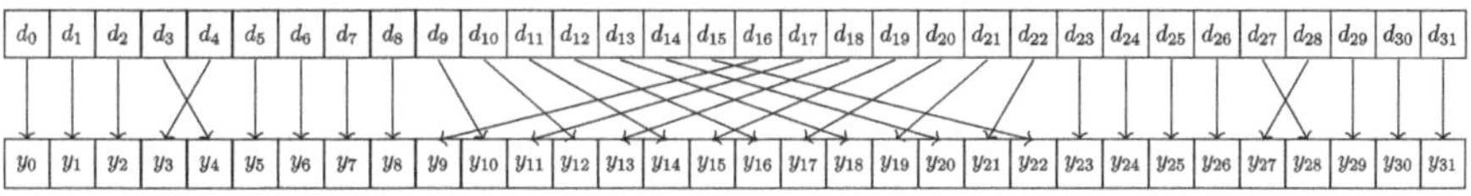

Figure 11.4 Subblock interleaver J for $N = 32$.

11.4.3 Bit-Selection

The length of the interleaved sequence $y_0, y_1, \dots, y_{N-1}$ needs to be adjusted depending on the desired output length E. Recall that $E = \lfloor G/C \rfloor$, where $C \in \{1, 2\}$ is the number of segments for coded bits, and G is the desired coded block. For adjusting the length by puncturing or shortening (when $E < N$) or repetition (when $E > N$), the sequence $y_0, y_1, \dots, y_{N-1}$ is processed as follows:

- Repetition: When $E > N$, every $e_k, k = 0, 1, \cdots, E-1$ is obtained by

  $$e_k = y_{k \bmod N}.$$

  That is, the sequence $y_0, y_1, \dots, y_{N-1}$ is followed by $y_0, y_1, \dots, y_{E-N-1}$, which means the first $E - N$ bits are repeated.
- Puncturing: When $E \leq N$ and $K/E \leq 7/16$, every $e_k, k = 0, 1, \dots, E-1$ is obtained by

  $$e_k = y_{k+N-E}.$$

- Shortening: When $E \leq N$ and $K/E > 7/16$, every $e_k, k = 0, 1, \dots, E-1$ is obtained by

  $$e_k = y_k.$$

Note that this process can be implemented while writing the sequence $y_0, y_1, \dots, y_{N-1}$ into a circular buffer of length N shown in Figure 11.5.

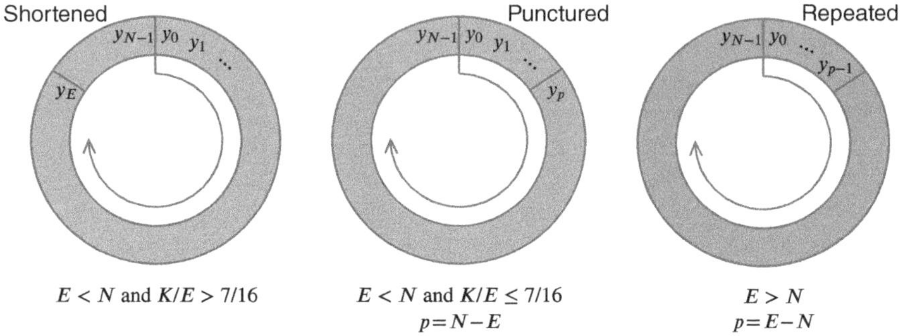

Figure 11.5 Implementation of length-matching by a circular buffer.

Exercises

1 What is the relation between bit-reversal order and natural order puncturing and shortening?

2 What is the fundamental difference between puncturing and shortening schemes in rate-compatible polar codes?

3 How do puncturing and shortening affect the reliability of bit channels in polar codes?

4 In a practical communication system, what criteria would you consider when choosing between a puncturing and a shortening scheme for rate adaptation?

5 How might extending (i.e. adding redundant bits to increase block length) offer alternative strategies for rate compatibility? What potential advantages and challenges could arise from employing extending schemes in comparison to puncturing or shortening?

6 How can puncturing and shortening schemes be effectively integrated into Hybrid ARQ protocols to achieve flexible rate adaptation? What are the key considerations when applying these techniques to balance retransmission efficiency and error correction performance in real-time communication systems?

Bibliographical Notes

Code lengths of polar codes are restricted to the powers of two, which can be a limitation in practice. To accommodate the diverse requirements of practical applications, puncturing, and shortening techniques have been introduced for rate-compatible polar codes, providing flexibility for the length of transmitted sequence.

Puncturing and shortening of polar codes employ puncturing and shortening patterns obtained from mother codes. Various puncturing patterns have been developed to improve the performance of punctured polar codes, including the Quasi-uniform puncturing (QUP) algorithm Niu et al. [2013] and bit-reversal scheme Bioglio et al. [2017].

In the formation of shortened polar codes, the shortening patterns are treated as fixed values and are assumed to be known by the decoder. Therefore, the shortened bits must be independent of the information sets of the polar codes. The generation of shortening patterns based on the column weights of the generator matrix and based on bit-reversal order were proposed by Wang and Liu [2014] and Bioglio et al. [2017], respectively. In addition, the properties of the puncturing patterns for polar codes were studied by Chandesris et al. [2017]. Among the variants of polar codes, puncturing and shortening techniques for systematic and non-systematic PAC codes have been investigated by Sun et al. [2021, 2023], Arıkan [2020], Gu et al. [2023], Zhang et al. [2023], Qiu and He [2023], and Jiang et al. [2023].

For further reading on different code modification techniques, the potential readers are referred to the following papers, out of many papers existing in the literature: El-Khamy et al. [2017], Hong et al. [2017], and Jang et al. [2019].

The third-generation partnership project (3GPP) has standardized the rate-compatible polar codes in the technical specification report [2022].

Bibliography

3rd Generation Partnership Project (3GPP). NR; multiplexing and channel coding. Technical Speification (TS) 38.212 v15.13.0 (Release 15), Sophia Antipolis, France, 2022.

Erdal Arıkan. Systematic encoding and shortening of PAC codes. *Entropy*, 22(11), 2020. ISSN: 1099-4300 URL https://www.mdpi.com/1099-4300/22/11/1301.

Valerio Bioglio, Frederic Gabry, and Ingmar Land. Low-complexity puncturing and shortening of polar codes. In *2017 IEEE Wireless Communications and Networking Conference Workshops (WCNCW)*, pages 1–6. IEEE, 2017.

Ludovic Chandesris, Valentin Savin, and David Declercq. On puncturing strategies for polar codes. In *2017 IEEE International Conference on Communications Workshops (ICC Workshops)*, pages 766–771. IEEE, 2017.

Mostafa El-Khamy, Hsien-Ping Lin, Jungwon Lee, and Inyup Kang. Circular buffer rate-matched polar codes. *IEEE Transactions on Communications*, 66(2):493–506, 2017.

Xinyi Gu, Mohammad Rowshan, and Jinhong Yuan. Rate-compatible shortened PAC codes. In *2023 IEEE/CIC International Conference on Communications in China (ICCC Workshops)*, pages 1–6. IEEE, 2023.

Song-Nam Hong, Dennis Hui, and Ivana Marić. Capacity-achieving rate-compatible polar codes. *IEEE Transactions on Information Theory*, 63(12):7620–7632, 2017.

Min Jang, Seok-Ki Ahn, Hongsil Jeong, Kyung-Joong Kim, Seho Myung, Sang-Hyo Kim, and Kyeongcheol Yang. Rate matching for polar codes based on binary domination. *IEEE Transactions on Communications*, 67(10):6668–6681, 2019.

Shirong Jiang, Jiahao Wang, Chunfang Xia, and Xiang Li. Construction of PAC codes with list-search and path-splitting critical sets. *arXiv preprint arXiv:2304.11554*, 2023.

Kai Niu, Kai Chen, and Jia-Ru Lin. Beyond turbo codes: Rate-compatible punctured polar codes. In *2013 IEEE International Conference on Communications (ICC)*, pages 3423–3427. IEEE, 2013.

Ziqi Qiu and Yejun He. Construction of shortened systematic PAC codes based on Monte-Carlo algorithm. In *2023 32nd Wireless and Optical Communications Conference (WOCC)*, pages 1–5. IEEE, 2023.

He Sun, Emanuele Viterbo, and Rongke Liu. Optimized rate-profiling for PAC codes, 2021.

Zhuang Sun, Dengsheng Lin, Yue Xiao, and Ming Xiao. An efficient construction of polarization-adjusted convolutional codes. In *2023 IEEE Globecom Workshops (GC Wkshps)*, pages 1243–1248, 2023.

Runxin Wang and Rongke Liu. A novel puncturing scheme for polar codes. *IEEE Communications Letters*, 18(12):2081–2084, 2014.

Xianwen Zhang, Ming Jiang, Mingyang Zhu, Chunming Zhao, and Lijie Hu. Rate-compatible puncturing and shortening of short PAC codes for 6G URLLC. In *2023 IEEE 24th International Workshop on Signal Processing Advances in Wireless Communications (SPAWC)*, pages 301–305. IEEE, 2023.

12

Polar-Coded Modulation

We know the past but cannot control it. We control the future but cannot know it.

— Claude Shannon

Channel coding introduces redundancy into source bits, enabling the receiver to correct a certain number of errors within a transmitted block. While this redundancy can either decrease the transmission rate over band-limited channels or necessitate increased bandwidth to maintain the desired rate, it offers a significant advantage: it reduces the transmit power required to achieve a target bit or block error rate. To address the disadvantage of channel coding, we can jointly optimize the underlying channel coding scheme and the modulation scheme. The aim of combining the coding and modulation schemes is to improve error correction performance through code diversity and time diversity. This technique is called the *coded-modulation* that effectively results in spectrally efficient channel coding. Spectrally efficient codes constructed using this technique are called the *block-coded modulation* (BCM) codes defined in Euclidean space. Hence, one of the aims of coded-modulation is to increase the minimum Euclidean distance. The modulation schemes with memory, such as trellis-coded modulation (TCM) are different from BCM. In this chapter, we discuss the major block-coded-modulation schemes and symbol labeling rules. In particular, we learn how we can combine the higher-order modulation schemes with polar coding or the so-called *polar-coded modulation* (PCM). There are two types of PCM: Bit-interleaved PCM (BIPCM) and multi-level PCM (MLPCM). We discuss each type in the following. But first, let us review higher-order modulation in digital communications.

Polar Codes: From Theory to Practice, First Edition. Mohammad Rowshan and Emanuele Viterbo.

12.1 Higher-Order Modulation

In digital modulation, discrete signals are modulated into analog carrier signals, therefore the modulation is considered a digital-to-analog conversion. The carrier signal changes by a finite number of modulation alphabet or symbols, M, which is a power of two, $M = 2^m$, where m is the number of bits packed in each symbol. This number is called *modulation order*. The simplest digital modulation is of order two, where there are only two symbols usually denoted by "−1" and "1," for instance, in the binary phase shift keying (BPSK). For a higher rate of data transmission, we have to use a larger modulation alphabet, namely, a larger M, although the detection of the symbols will be harder at the receiver due to the shorter distance between the modulation symbols. Modulations of order $M = 4$ and higher are called *higher order modulations*. A popular example is 2^m-ary *phase-shift keying modulation* (2^m-PSK) or 2^m-ary *quadrature amplitude modulation* (2^m-QAM). When the average signal energy is fixed, the larger M results in a larger number of signal points in the constellation space, which would tend to decrease the distance between the signal points. The reduction in the distance between the pairs of constellation points results in an increase in the misdetection rate. In this case, the key to tackling the difficulty in signal detection is to combine coding and modulation and transmit the constrained symbol sequences. Note that each symbol represents several bits. Therefore, any erroneous symbol may cause a burst error. The coding constraints imposed on symbols will help reduce the misdetection rate. Next, we look at the constellations and their properties.

12.1.1 Constellations and Distance Between Signal Points

A signal constellation S is a set of points that can be one dimensional on real line $\mathbb{R}$, or two-dimensional on a plane $\mathbb{R}^2$ (sometimes on a complex plane). The *amplitude-shift keying* (ASK) is a one-dimensional constellation. There is another form of one-dimensional signal modulation called *pulse-amplitude modulation* (PAM), where the encoded bits are mapped to varying amplitudes of a series of signal pulses. Demodulation is performed by detecting the carrier's amplitude level in every period. The points in the two-dimensional constellation are placed on a circle or on a square grid. The two most commonly used two-dimensional constellations are M-PSK and M-QAM.

***M*-PSK** The M points in the *phase-shift keying* (PSK) are lying on a circle with radius $\sqrt{E_s}$ and phases $i \cdot \Delta\phi, i = 0, 1, \ldots, M = 2^m$, where $\Delta\phi = 2\pi/M$. Note that the energy of the signal points, E_s, is sometimes normalized to 1 in the literature. When there are two points $\pm\sqrt{E_s}$ in the PSK constellation, it is called binary PSK (BPSK), and when $M = 4$, we have 4-PSK which is also called quaternary

PSK (QPSK). In M-PSK, the minimum squared Euclidean distance of the signal points is given as $d_0^2 = 4\sin^2\Delta\phi/2 = 2(1-\cos\Delta\phi)\sqrt{E_s}$. As we shall see later, if we partition the set of signal points into two subsets such that the minimum Euclidean distance between points in different subsets is maximized (e.g. see Figure 12.2 for 8-PSK), the minimum squared distance between points in subsets will increase to $d_1^2 = 2(1-\cos 2\Delta\phi)\sqrt{E_s}$. In general, the minimum squared distance for level l is

$$d_l^2 = 2\left(1-\cos(2^l\Delta\phi)\right)\sqrt{E_s},\quad l = 0, 1, \ldots, m-1,\quad \Delta\phi = \frac{2\pi}{2^m}.$$

***M*-QAM** When the constellation points are located on a grid, the constellation space is referred to as *quadrature amplitude modulation* (QAM). When we have $M = 32$ and 128, the constellation points are arranged in a cross shape instead of a rectangular shape to reduce the average signal energy.

The constellations used in the 5G standard are QPSK, 16QAM, 64QAM, and 256QAM. Figure 12.1 illustrates the constellations with eight points.

Let $s \in \mathcal{S}$ be a signal in a two-dimensional constellation space in $\mathbb{R}^2$ represented by a point $(\phi_1(s), \phi_2(s))$. The *squared Euclidean* (SE) distance between $s, s' \in \mathcal{S}$ is

$$d_E^2(s,s') = (\phi_1(\mathbf{s}) - \phi_1(\mathbf{s}'))^2 + (\phi_2(\mathbf{s}) - \phi_2(\mathbf{s}'))^2, \tag{12.1}$$

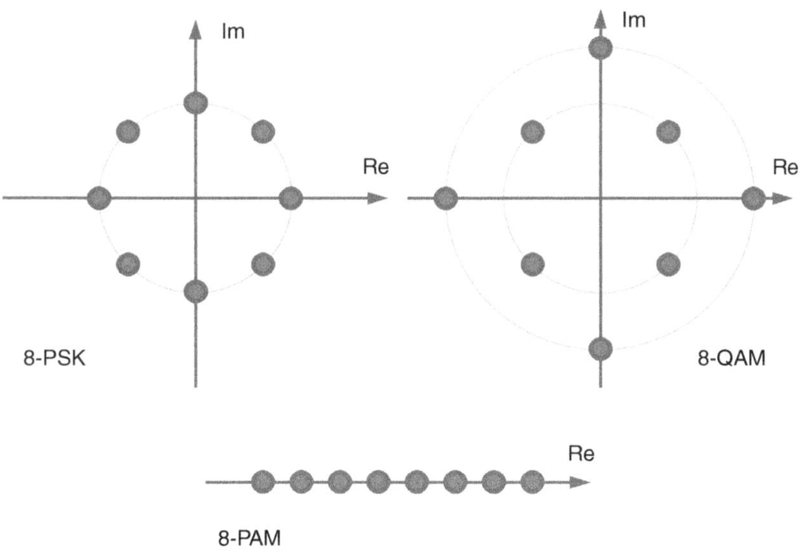

Figure 12.1 Constellations with 8 points: two-dimensional 8-PSK and 8-QAM modulations, and one-dimensional 8-PAM modulation with shape circular (4,4), that is 4 points on each circle. Note that 8-QAM can have different shapes such as rectangular and triangular.

and the squared Euclidean distance between two sequences of signal points $\mathbf{v} = [s_0, s_1, \ldots, s_{N/m-1}]$ and $\mathbf{v}' = [s'_0, s'_1, \ldots, s'_{N/m-1}]$ is

$$d_E^2(\mathbf{v}, \mathbf{v}') = \sum_{i=0}^{N/m-1} (\phi_1(s_i) - \phi_1(s'_i))^2 + (\phi_2({}_is) - \phi_2(s'_i))^2. \quad (12.2)$$

The *spectral efficiency* of a constellation $\mathcal{S}$ is defined as the number of uncoded source/information bits mapped to each symbol, that is, $R_c \cdot M$, where R_c is the code rate, at each signaling interval T_s that can be transmitted over a given bandwidth B [Hz]. Then the spectral efficiency is defined as follows.

$$\eta = \frac{R_c \cdot M/T_s}{B} \text{ (bits/s)/Hz.} \quad (12.3)$$

Now assuming we have a BCM code $\mathscr{C}$ over the constellation $\mathcal{S}$, the minimum squared Euclidean (MSE) distance is defined as

$$d_E^2(\mathscr{C}) = \min\{d_E^2(\mathbf{v}, \mathbf{v}'), \mathbf{v}, \mathbf{v}' \in \mathscr{C}, \mathbf{v} \neq \mathbf{v}'\}. \quad (12.4)$$

12.1.2 Binary Labeling of Signal Points

Each symbol in the constellation is labeled with a binary m-tuple, and we say that the symbols in this constellation have m bit levels. A binary labeling rule, denoted by $\mathcal{L} : \{0,1\}^m \rightarrow \mathcal{S}$, maps binary m-tuples bijectively to the 2^m input symbols $s \in \mathcal{S}$. The modulated constellation points $s_j, 0 \leq j \leq 2^m - 1$ are represented by binary labels $v_{m-1} \ldots v_1 v_0$, where $v_i \in \{0, 1\}$ is the binary label for the protection level i, which is also called *binary level*. The coded bits $v_0, v_1, \ldots, v_{m-1}$ are mapped to the 2^m-ary symbol by a bijective mapping function,

$$f : \mathbf{v} \rightarrow s = f(\mathbf{v}),$$

where $\mathbf{v} = (v_{m-1} \ldots v_1 v_0)$ and $s \in \mathcal{S}$. The popular labeling rules are gray labeling and set partitioning. In the following, we review the two major polar-coded modulation schemes. The 2^m-ary channel $W : \mathcal{S} \rightarrow \mathcal{Y}$ takes an input symbol from the constellation set $\mathcal{S}$, where $|\mathcal{S}| = 2^m$, and returns an output symbol from the alphabet $\mathcal{Y}$.

12.2 Multilevel-Coded Modulation

Multilevel coding (MLC) aims to protect each bit of a constellation point with a binary code $\mathscr{C}_i$ of different rates. This leads to m protection levels. This scheme was proposed by Imai and Hirakawa in 1976. Before describing the details of the scheme, we give an overview here. In a multilevel coding approach, the raw channel W is effectively decomposed into m binary sub-channels while preserving their mutual information. Let us denote this channel decomposition, which is called the sequential binary partition (SBP) as

$$\psi : W \to \left\{ B_{\psi}^{(1)}, B_{\psi}^{(2)}, \dots, B_{\psi}^{(m)} \right\},$$

where $B_{\psi}^{(j)} : \{0,1\} \to \mathcal{Y} \times \{0,1\}^{j-1}$ is the binary sub-channel corresponding to the bit level $j = 1, 2, \dots, m$. Therefore, an M-ary channel W is divided into m bit levels (or sub-channels) $B_{\psi}^{(i)} (0 \leq i < m)$ that are binary-input discrete memoryless channels (BI-DMCs) as long as W is a DMC. In SBP, each sub-channel $B_{\psi}^{(j)}$ has knowledge of both the channel output $y \in \mathcal{Y}$, and their previous bit levels obtained by decoding. Mutual information between channel input S and channel output $\mathcal{Y}$ of W, assuming equiprobable source symbols, is referred to as coded modulation capacity C_{cm}(W) and defined as

$$C_{cm}(\mathrm{W}) := I(X;Y) = \sum_{i=0}^{m-1} I\left(B_{\psi}^{(i)}\right). \tag{12.5}$$

This capacity does not depend on a specific labeling rule $\mathcal{L}$. Note that according to Arıkan's seminal work, as the block length of the component codes increases, the polar component codes approach the bit level capacities $I\left(B_{\psi}^{(i)}\right)$.

One drawback of the MLC approach in practical use is the necessity of employing several relatively short component codes with different rates for various bit levels. This requirement, combined with the need for code lengths to be powers of two in polar coding, adds to the disadvantages of the multilevel coding approach.

12.2.1 Labeling Signal Points in $\mathcal{S}$ by Partition Chain

Before discussing the component codes with different rates, we consider a Euclidean space $\mathcal{S}$ with 2^m signal points. A chain of binary partitions $\mathcal{S} \supset \mathcal{S}_1 \supset \cdots \supset \mathcal{S}_m$ for the signal space $\mathcal{S}$ is formed, where the subspace $\mathcal{S}_i$ consists of 2^{m-i} points for $i \in [1, m]$. Let δ_i^2 denote the minimum intra-set distance at the partition level i. The partition of $\mathcal{S}$ is carried out so that δ_i^2 increases monotonically by i. The partition chain gives a unique hierarchical labeling of the 2^m signal points in $\mathcal{S}$ by a string of length m denoted as $v_{m-1} \dots v_1 v_0$. Hence, we can define a bit-to-signal mapping $f(v_{m-1} \dots v_1 v_0)$. Let us denote the set of 2^{m-i} signal points with a common label $v_{i-1} \dots v_1 v_0$ by $\mathcal{Q}(v_{i-1} \dots v_1 v_0)$.

Figure 12.2 illustrates an example of set partitioning for 8-PSK. The set partitioning was first described by Ungerböck. At partition level 0, where the value of $v_0 \in \{0,1\}$ does not depend on any previously coded/mapped bit, the signal set is divided into two subsets such that the minimum Euclidean distance between points in different subsets is maximized: a subset of signals with $v_0 = 0$, denoted by $\mathcal{Q}(v_0 = 0)$, and the other subset where $v_0 = 1$, $\mathcal{Q}(v_0 = 1)$. At each subsequent level,

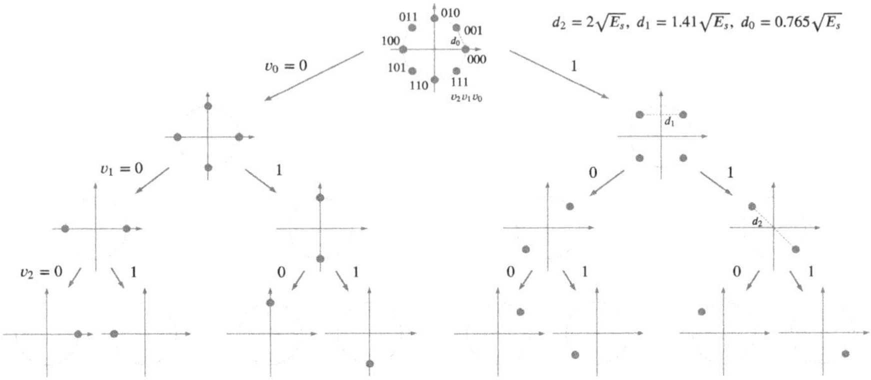

Figure 12.2 Set partitioning for 8-PSK constellation.

each subset is further divided into two smaller subsets. At the partition level i, all subsets $Q(v_{i-1} \dots v_0)$ are divided into two subsets $Q(0v_{i-1} \dots v_0)$ and $Q(1v_{i-1} \dots v_0)$, where $v_i = 0$ and $v_i = 1$, respectively. The partition of signal points continues to the level $m - 1$, where every pair of signal points is at the maximum Euclidean distance. In the partition tree, every subset at the partitioning level i is a path uniquely labeled by $(v_{i-1} \dots v_0)$ from the root.

Now, if we look at a block coded modulation (BCM) code as a concatenated code (see Chapter 2), we can consider the signal points in a constellation as codewords of the inner code in the concatenation. We can group this code into subcodes, each subcode includes some codewords. The grouping of the signal set is called *partitioning*. This partitioning is performed so that the minimum distance of a subcode is maximized. Each subcode can be partitioned into subcodes with a smaller number of codewords. Clearly, the minimum distance increases by further partitioning.

12.2.2 Choosing *m* Binary Component Codes

Now, let us look at the outer code. Given a block of K source bits, $\mathbf{d} = [d_0, d_1, \dots, d_{K-1}]$, $d_i \in \{0, 1\}$, the block is divided into m partitions:

$$\mathbf{d}^i = [d_0^i, d_1^i, \dots, d_{K_i-1}^i], \quad i \in [0, m-1] \tag{12.6}$$

of length K_i with

$$K = \sum_{i=0}^{m-1} K_i. \tag{12.7}$$

Each set of bits $\mathbf{d}^i$ is encoded by an individual binary encoder Enc_i that generates coded bits $\mathbf{x}^i = [x_0^i, x_1^i, \dots, x_{N-1}^i]$, $x_j^i \in \{0, 1\}$ of the component code $\mathscr{C}_i(N, K_i, d_i)$ of length N, dimension K_i, and minimum Hamming distance d_i. Each component code contributes a labeling bit for the signal points in the constellation. The resulting BCM code consists of 2^K signal sequences. Here, we assume that the lengths of the component codes are equal and that the choice of component codes is arbitrary.

12.2.3 Interleaving Component Codes and Mapping

In this section, we will see how the bits of component codes are mapped to signal points. Let us look at the case where there are two component codes. Suppose

$$\mathbf{x}^0 = [x_0^0, x_1^0, \dots, x_{N-1}^0], \ \ \mathbf{x}^0 \in \mathscr{C}_0,$$

$$\mathbf{x}^1 = [x_0^1, x_1^1, \dots, x_{N-1}^1], \ \ \mathbf{x}^1 \in \mathscr{C}_1,$$

are two codewords. By interleaving the two codewords $\mathbf{x}^0$ and $\mathbf{x}^1$, the following m-bit sequence is formed:

$$\mathbf{x}^1 * \mathbf{x}^0 = [x_0^1 x_0^0, x_1^1 x_1^0, \dots, x_{N-1}^1 x_{N-1}^0].$$

As can be seen, the coded bits in $\mathscr{C}_0$ and $\mathscr{C}_1$ correspond to the least significant bit (LSB) and the most significant bit (MSB) of the binary labels of the constellation points, respectively. We denote the label of every signal point $x_j^1 x_j^0$ for $j \in [0, N-1]$ by $v_1 v_0$ in the BPSK signal space S for simplicity. In general, we have $\mathscr{C} = [\mathscr{C}_0, \mathscr{C}_1, \dots, \mathscr{C}_{m-1}]$, where the position i in the constellation labels takes the value $v_i \in \mathscr{C}_i$. Finally, the mapping function $f(x_i^1 x_i^0)$, $i \in [0, N-1]$ gives

$$\mathbf{s} = [s_0, s_1, \dots, s_{N-1}].$$

Figure 12.3 illustrates the aforementioned encoding procedure called *multilevel coding* (MLC). The process of two-level coding above can be generalized to multilevel coding over any 2^m-ary PSK or QAM signal set. Suppose that we have m binary linear code $\mathscr{C}_i(N, K_i, d_i)$ for $i \in [0, m-1]$, and $\mathbf{x}^0, \mathbf{x}^1, \dots, \mathbf{x}^{m-1}$ are the codewords in $\mathscr{C}_0, \mathscr{C}_1, \dots, \mathscr{C}_{m-1}$. The interleaving results in

$$\mathbf{x}^{m-1} \cdots * \mathbf{x}^1 * \mathbf{x}^0 = [x_0^{m-1} \cdots x_0^1 x_0^0, \dots, x_j^{m-1} \cdots x_j^1 x_j^0, \dots, x_{N-1}^{m-1} \cdots x_{N-1}^1 x_{N-1}^0].$$

The signal mapping $f(x_j^{m-1} \cdots x_j^1 x_j^0)$ for $j \in [0, N-1]$ gives a sequence of N signals with labels $x_j^{m-1} \cdots x_j^1 x_j^0$. As a result, we have an m-level BCM code (see Figure 12.4) over S with dimension $K = \sum_{i=0}^{m-1} K_i$. Note that the component codes $\mathscr{C}_i, i \in [0, m-1]$ are linear, hence the BCM code

$$\mathscr{C} = f(\mathscr{C}_{m-1} * \cdots \mathscr{C}_1 * \mathscr{C}_0) = \{f(\mathbf{x}^{m-1} * \cdots * \mathbf{x}^1 * \mathbf{x}^0) : \mathbf{x}^i \in \mathscr{C}_i, i \in [0, m-1]\} \tag{12.8}$$

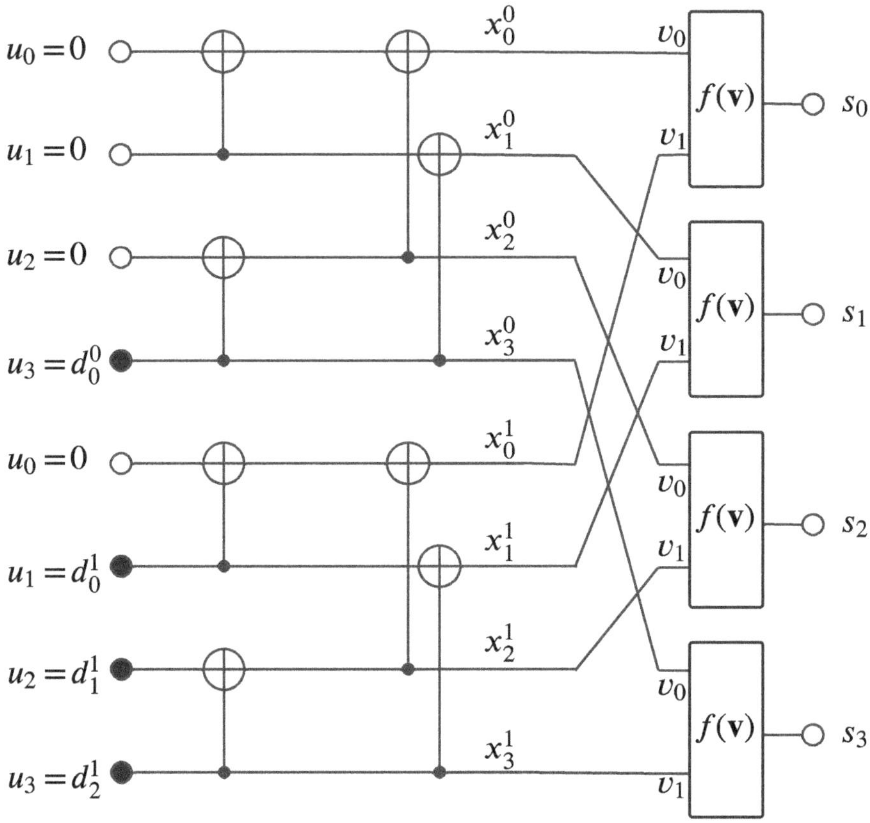

Figure 12.3 An example of polar multilevel coding (PMLC) for $K = 4$ and $m = 2$.

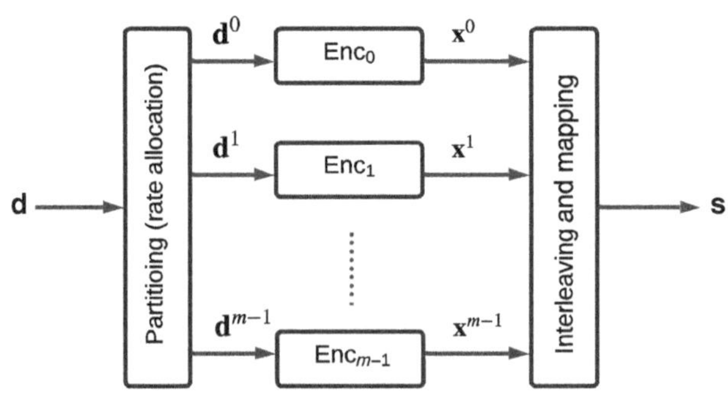

Figure 12.4 Multilevel coding (MLC) using a rate-flexible coding scheme.

is a linear code as well. Observe that the overall code rate is essentially the sum of the individual code rates, namely

$$R = \sum_{i=0}^{m-1} R_i = \frac{1}{N} \sum_{i=0}^{m-1} K_i = \frac{K}{N}. \tag{12.9}$$

Observe that in multilevel coding, we are virtually decomposing the overall transmission channel into m equivalent binary-input sub-channels, and each sub-channel transmits the coded bits in $\mathscr{C}_i$.

12.2.4 Multistage Decoding of BCM Codes

For decoding, the log-likelihood ratio (LLR) of each coded bit is computed from the received signal vector $y = [y_0, y_1, \dots, y_{N/m-1}]$. The LLR of every coded bit depends on the subchannel over which the coded bit is transmitted. That is, the performance of a code $\mathscr{C}$ depends on $\mathscr{C}_i$'s as $\mathscr{C}$ is decoded in a *multistage decoding* (MSD) fashion. As Figure 12.5 shows, in this decoding scheme, the LLRs corresponding to v_0 of the received signals are first decoded, then v_1 and so on. Hence, it is paramount to transmit a reliable $\mathscr{C}_0$ as the corresponding LLRs are computed without a priori knowledge about v_0. However, the LLRs of v_1 corresponding to $\mathscr{C}_1$ are calculated knowing the value of v_0. That is why the level of protection for each label position should be different.

The mapping function $\mathcal{M}$ is bijective independently of the labeling strategy, mutual information $I(Y; S)$ between the transmitted symbol $\mathbf{s} \in \mathcal{S}$ and the received symbol $\mathbf{y} \in \mathcal{Y}$ equals mutual information $I(Y; X_0, X_1, \dots, X_{m-1})$ between the bit levels $\mathbf{x} \in \{0, 1\}^m$ and the received symbol $\mathbf{y} \in \mathcal{Y}$. Therefore, by applying the chain rule of mutual information, we have

$$\begin{aligned} I(Y; S) = I(Y; X_0, X_1, \dots, X_{m-1}) &= I(Y; X_0) + I(Y; X_1 | X_0) \\ &+ \cdots + I(Y; X_{m-1} | X_0, X_1, \dots, X_{m-2}). \end{aligned} \tag{12.10}$$

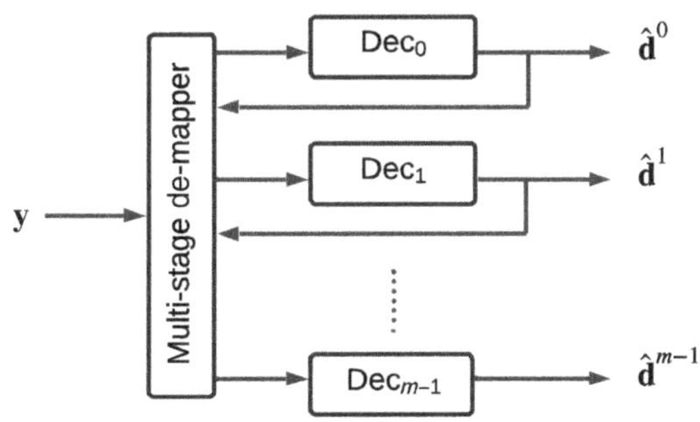

Figure 12.5 Multistage decoding (MSD).

From equation (12.10), the transmission of m-tuple of binary digits x_i, $i \in [0, m-1]$ can be separated into parallel transmission of binary digits over *m equivalent channels* given that $x_0, x_1, \dots, x_{i-1}$ are known.

The right-hand side of (12.10) suggests a successive decoding procedure called *multistage decoding* (MSD). Figure 12.5 illustrates the staged decoding of the component code $\mathscr{C}_i$ by the corresponding decoder Dec_i for $i = 0, 1, \dots, m-1$. At a decoding stage i, the decoder Dec_i requires not only the received signal points $\mathbf{y} = [y_0, y_1, \dots, y_{N-1}]$ but also the decisions made on the coded bits $\hat{\mathbf{x}}^j, j = 0, 1, \dots, i-1$ in the previous stages of decoding. Note that MSD requires the actual transmitted symbol $\mathbf{x}^j$, not the estimated $\hat{\mathbf{x}}^j$ according to the chain rule (12.10). That is, the decisions made in the previous stages must be error-free.

The structure of polar codes makes polar codes attractive for coded modulation schemes if the code is properly designed in conjunction with the underlying modulation scheme.

12.3 Bit-Interleaved Polar-Coded Modulation

The simplest and most popular coded modulation scheme is perhaps the bit-interleaving scheme, where the resulting coded modulation is called bit-interleaved polar-coded modulation (BI-PCM). Before discussing the bit-interleaving scheme, let us understand the concept of *coherence time*. The time duration over which the channel impulse response (CIR) is supposedly not varying is called the coherence time. During coherence time, the propagation wave through any medium, such as a wireless link or an optical fiber, remains coherent, that is the phase remains predictable. Channel variation is more significant in wireless communications, due to Doppler effects and propagation factors such as dispersion, scattering, and diffraction, resulting in frequency-selective fading.

Bit interleaving is based on the fact that symbols transmitted over different coherence times will experience different channel effects. Spreading the coded bits over time is a way to achieve diversity and avoid error bursts. One can determine a constraint length of the codewords such that the transmission time of the entire codewords would be longer than the coherence time; otherwise, the bit-interleaving will not be useful because all the bits will experience the same channel effect.

As shown in Figures 12.6 and 12.7, in bit interleaving, multiple codewords are concatenated, where the bits of each codeword are interleaved with other codewords. In contrast, bits in the same codeword can be bit interleaved as well, given that they fall apart exceeding the coherence time. Note that bit-interleaving decreases correlation among the bits of the same codewords and increases the correlation among different codewords. Clearly, maximum diversity is achieved

Figure 12.6 Bit-interleaved polar-coded modulation scheme.

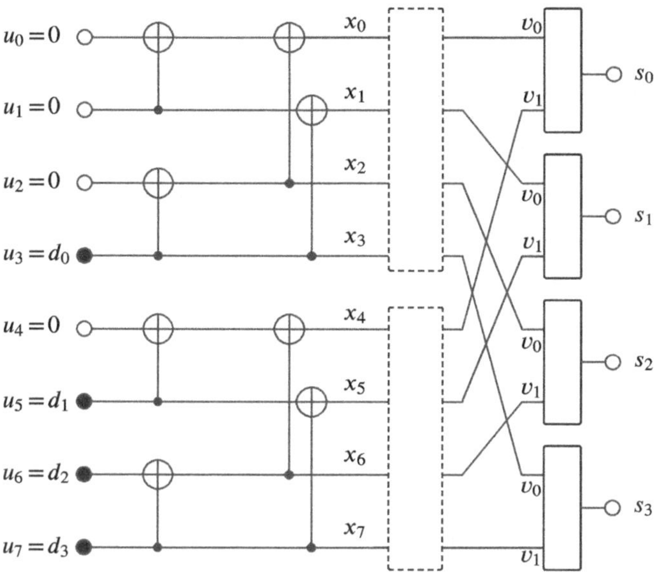

Figure 12.7 An example of bit-interleaved polar-coded modulation for $K = 8$ and $m = 2$.

in an ideal case, where every bit of the same codeword is transmitted at different coherence times.

In a BI-PCM scheme, a 2^m-ary additive white Gaussian noise (AWGN) channel W is decomposed into m binary sub-channels that are viewed as m parallel binary-input AWGN channels $W^{(1:m)}$, where $W^{(i)}, i \in [0, m-1]$ corresponds to the ith bit level in a modulated symbol. This channel transform is called *parallel binary partition* (PBP), denoted as

$$\varphi : W \to \left\{ B_\varphi^{(1)}, B_\varphi^{(2)}, \ldots, B_\varphi^{(m)} \right\},$$

where the channel W is mapped to a set of mutually independent binary sub-channels $B_\varphi^{(j)} : \{0,1\} \to \mathcal{Y}, j = 1, 2, \ldots, m$ for the jth bit level. For the sake of simplicity, we consider m a power of 2, for example, in QPSK, 16-QAM, and 256-QAM. In this case, a codeword of length N can be partitioned into m segments of length N/m. These sub-channels have symmetric capacities of

$$I\left(B_\varphi^{(i)}\right) := I\left(B_i; Y\right).$$

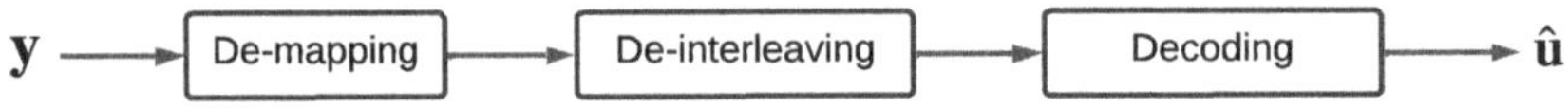

Figure 12.8 Decoding in bit-interleaved polar-coded modulation scheme.

Obviously, when compared to the corresponding sub-channel in SBP, for all pairs of sub-channels, the following holds:

$$\begin{aligned} I\left(B_{\varphi}^{(i)}\right) &= I\left(B_i; Y\right) \\ &\leq I\left(B_i; Y \mid B_0, \ldots, B_{i-1}\right) = I\left(B_{\varphi}^{(i)}\right). \end{aligned} \tag{12.11}$$

In PBP, each sub-channel $B_{\varphi}^{(j)}$ only uses the channel output $y \in \mathcal{Y}$, while SBP requires knowledge of previous bits $\{0,1\}^{j-1}$. The commonly used labeling rule for sub-channels in BI-PCM is gray labeling that generates bit levels that are as independent as possible, assuming the sub-channels correspond to bit levels with symmetric capacities that do not differ significantly.

After demodulation, as shown in Figure 12.8, the soft information of all mN bits is de-interleaved and fed to the de-rate-matcher. Note that to use a single polar decoder for BI-PCM, the order m of the constellation has to be a power of two.

12.4 Bit-Interleaving in 5G NR

Bit-interleaving, also known as (sub-)channel interleaving, is used to improve the error correction capability in uplink for physical uplink control channel (PUCCH), where the higher-order modulation QPSK is also employed for transmission of upload control information (UCI), unlike for physical downlink control channel (PDCCH) and physical broadcast channel (PBCH) in downlink where only the BPSK modulation is used.

Assume that we have a sequence of E rate-matched bits

$$\mathbf{e} = [e_0, e_1, \ldots, e_{E-1}].$$

In 5G NR (subclause 5.4.1.3 in 3GPP TS 38.212, Release 17), an isosceles right triangle-shaped array is adopted for bit-interleaving. The input of this interleaver is the sequence $\mathbf{e}$. The depth of the isosceles interleaver is denoted by T which is the side length of the right triangle as shown in Figure 12.9. Since the number of elements in an isosceles right triangle array with T elements on each side is $\frac{T(T+1)}{2}$, to have a triangle array that accommodates E bits, the side length T should be the smallest integer that satisfies

$$\frac{T(T+1)}{2} \geq E.$$

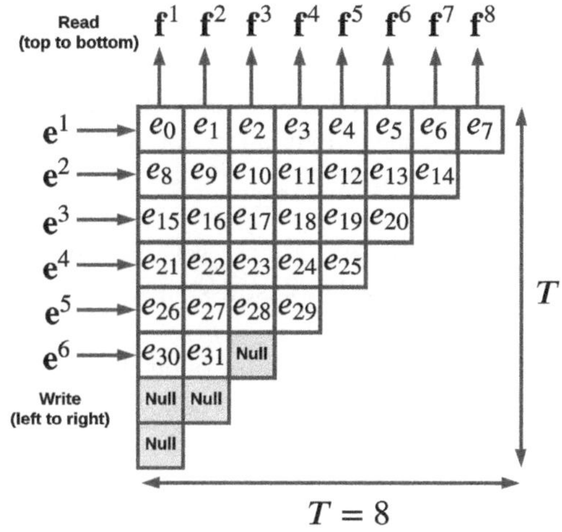

Figure 12.9 Write into and read from a right triangular interleaver with depth $E = 32$ and $T = 8$. The vector $\mathbf{f}$ is constructed by concatenation of the transpose of column vectors $\mathbf{f}^k$, $k \in [1, T]$.

As we have $T^2 + T - 2E \geq 0$, we can obtain the smallest integer T by

$$T = \left\lceil \frac{\sqrt{1+8E}-1}{2} \right\rceil,$$

The bit sequence $\mathbf{e}$ is written in the right triangle array row by row. Reading is performed column by column. When equality $\frac{T(T+1)}{2} = E$ holds, However, when inequality holds, there will be some unused space after all bits are loaded into the right triangle array. In this case, when writing row by row, these spaces are loaded by dummy elements (nulls), and when reading column by column, the dummy elements are discarded.

If $v_{i,j}$ represents the elements of the triangle, where rows and columns are indexed by $i,j \in [1, T]$ from top to bottom and left to right, respectively, then $v_{i,j} = e_k, k \in [0, E-1]$ starting from $i = 0$ and sweeping $j \in [0, T-1-i]$ and incrementing i and repeating until $i = T - 1$. Observe that after interleaving the right triangle, the spacing between each adjacent bit becomes at most T when we have $T(T+1)/2 = E$ (otherwise less), that is, between the adjacent elements in the first column and the second column, and at least 2, that is, between the adjacent element between the Tth column and the $T - 1$th column.

We denote the output of the interleaver by

$$\mathbf{f} = [f_0, f_1, \ldots, f_{E-1}].$$

The reverse operation is performed at the receiving end to deinterleave the LLR sequence corresponding to $\mathbf{e} = [e_0, e_1, \ldots, e_{E-1}]$.

Example 12.1 The interleaving operation shown in Figure 12.9 is for vector $\mathbf{e}$ with length $E = 32$ and hence $T = 8$.

Exercises

1 What are the trade-offs between using bit-interleaving and multilevel coding in polar-coded modulation in terms of error correction performance and decoding complexity?

2 In a multilevel coding scheme using set partitioning, how do polar codes ensure reliable transmission across different levels of modulation?

3 How does the reliability of bit channels influence the design of set partitioning and its integration with polar codes?

4 Given a signal space S of size 2^m points. Show that the minimum squared Euclidean distance $d_E^2(\mathscr{C})$ and minimum Hamming distance $d_H(\mathscr{C})$ of an m-level BCM code over S are

$$d_E^2(\mathscr{C}) \geq \min\{d_i\delta_i^2 : 0 \leq i < m\}. \tag{12.12}$$

$$d_H(\mathscr{C}) = \min\{d_i : 0 \leq i < m\}. \tag{12.13}$$

Bibliographical Notes

The multilevel coding (MLC) was initially proposed by Imai and Hirakawa [1977] and elaborated by Wachsmann et al. [1999]. This scheme was incorporated into the polar codes by Seidl et al. [2013] to construct polar-coded modulation schemes. Recall that Arıkan's polarized vector channel was also inspired by Imai and Hirakawa [1977]. On the other hand, the bit-interleaved coded modulation (BICM) technique was originally proposed by Zehavi [1992], where it was shown that the reliability of coded modulation over a Rayleigh fading channel can be improved by bitwise interleaving at the encoder output. This scheme was further modeled by Caire et al. [1998] as a set of parallel independent binary-input channels under the assumption of an infinite-length interleaver.

Seidl et al. [2013] provide a detailed study of the polar-coded modulation schemes. For further reading on different coded modulation techniques, the reader is referred to the following papers, out of many papers existing in the literature: Mahdavifar et al. [2015], Chen et al. [2013], and Bocherer et al. [2017]. The BI-PCM technique was standardized by the third-generation partnership project (3GPP) as described in the technical specification report by 3rd Generation Partnership Project (3GPP) [2022].

Bibliography

3rd Generation Partnership Project (3GPP). NR; multiplexing and channel coding. Technical Speification (TS) 38.212 V15.13.0 (Release 15), Sophia Antipolis, France, 2022.

Georg Bocherer, Tobias Prinz, Peihong Yuan, and Fabian Steiner. Efficient polar code construction for higher-order modulation. In *2017 IEEE Wireless Communications and Networking Conference Workshops (WCNCW)*, pages 1–6. IEEE, 2017.

Giuseppe Caire, Giorgio Taricco, and Ezio Biglieri. Bit-interleaved coded modulation. *IEEE Transactions on Information Theory*, 44(3):927–946, 1998.

Kai Chen, Kai Niu, and Jia-Ru Lin. An efficient design of bit-interleaved polar coded modulation. In *2013 IEEE 24th Annual International Symposium on Personal, Indoor, and Mobile Radio Communications (PIMRC)*, pages 693–697. IEEE, 2013.

Hideki Imai and Shuji Hirakawa. A new multilevel coding method using error-correcting codes. *IEEE Transactions on Information Theory*, 23(3):371–377, 1977.

Hessam Mahdavifar, Mostafa El-Khamy, Jungwon Lee, and Inyup Kang. Polar coding for bit-interleaved coded modulation. *IEEE Transactions on Vehicular Technology*, 65(5):3115–3127, 2015.

Mathis Seidl, Andreas Schenk, Clemens Stierstorfer, and Johannes B Huber. Polar-coded modulation. *IEEE Transactions on Communications*, 61(10):4108–4119, 2013.

Udo Wachsmann, Robert F H Fischer, and Johannes B Huber. Multilevel codes: Theoretical concepts and practical design rules. *IEEE Transactions on Information Theory*, 45(5):1361–1391, 1999.

Ephraim Zehavi. 8-PSK trellis codes for a Rayleigh channel. *IEEE Transactions on Communications*, 40(5):873–884, 1992.

13

Performance Comparisons

The purpose of computing is insight, not numbers.

— Richard Hamming

In this chapter, we numerically compare the error correction performance of polar codes and polar-like codes under various decoding algorithms. The results are plotted in figures that depict the frame (or block) error rate versus SNR per information bit, E_b/N_0. These results are obtained based on the 5G reliability sequence introduced in Chapter 6. We limited our results to short codes and medium-length codes under various code rates. Then, these comparisons are performed with Turbo codes that were relevant in the previous generations of cellular networks, 3G and 4G, and LDPC codes used in 5G networks, to demonstrate the advantage of polar codes for short block lengths.

13.1 Error Correction Performance of Polar Codes

In this section, we compare known variants of polar codes, namely CRC-polar codes and PAC codes (with and without CRC concatenation). These results are obtained for the AWGN channel and binary phase-shift keying (BPSK) modulation. For this purpose, we choose two codes: the short code (128, 64) and the medium-length code (1024, 512). They are within the length range used for 5G. The CRC and convolutional polynomials used for precoding in CRC-polar and PAC codes are $g_{\text{CRC11}}(D) = D^{11} + D^{10} + D^9 + D^5 + 1$ and $g_{\text{Conv6}}(D) = 1 + D^2 + D^3 + D^5 + D^6$, respectively. To achieve a thorough comparison, we consider three sets of setups where, in each setup, only one coding factor changes. The factors are as follows:

Polar Codes: From Theory to Practice, First Edition. Mohammad Rowshan and Emanuele Viterbo.

Table 13.1 Evaluation of the frame error rate performance versus E_b/N_0 for variants of polar codes with two different constructions.

Channel	AWGN
Modulation	BPSK
Coding variant	CRC-Polar | PAC | CRC-PAC
Constructions	5G, 5G-RM
Code rate, R	$1/2$
Info. length, K (bits w/o CRC)	64, 512
Decoding algorithm	SCL, $L = 32$

- Coding scheme and code construction: As Table 13.1 shows, we choose CRC-polar, PAC and CRC-PAC constructed using a 5G reliability sequence and its modified version called 5G-RM (see Chapter 6).
- Code rate: $1/4, 1/2, 3/4$, as Table 13.2 states.
- Decoding algorithms according to the setup in Table 13.3.

Note that there are numerous methods for constructing a code. However, they are either not explicit or have high complexity, which makes them impractical and unsuitable for standard adoption. Similarly, there are other decoding algorithms that we do not include in this comparison. In the following, we discuss each scenario in the same order as mentioned above.

Figure 13.1 illustrates the frame (or block) error rate (FER) for variants of the polar codes (128, 64) and (1024, 512), respectively. In this scenario, summarized in Table 13.1, we use the SCL decoder with the list size $L = 32$ for all codes. As can be seen in Figure 13.1a, for the relatively short code (128, 64), the PAC code constructed with the 5G-RM construction outperforms the PAC codes with the 5G construction and the CRC-polar codes. Note that the 5G construction gives the minimum distance of $d = 8$ for both PAC and CRC-polar codes, while the

Table 13.2 Evaluation of the frame error rate performance versus E_b/N_0 for variants of polar codes at three different rates.

Channel	AWGN
Modulation	BPSK
Coding variant	CRC-Polar | PAC | CRC-PAC
Constructions	5G, 5G-RM
Code rate, R	$1/4, 1/2, 3/4$
Info. length, K (bits w/o CRC)	64, 512
Decoding algorithm	SCL, $L = 8$

Table 13.3 Evaluation of the frame error rate performance versus E_b/N_0 for variants of polar codes under different decoding algorithms.

Channel	AWGN
Modulation	BPSK
Coding variant	CRC-Polar | PAC
Constructions	5G
Code rate, R	1/2
Info. length, K (bits w/o CRC)	64, 512
Decoding algorithm	SC | BP | SCL | SCLF | SCL

5G-RM's minimum distance is $d = 16$. Furthermore, CRC concatenation with short codes is detrimental as it may decrease the minimum distance and utilize more low-reliability sub-channels. When it comes to medium-length codes such as (1024, 512) in Figure 13.1b, the 5G-RM construction performs poorly. The reason lies in employing severely bad sub-channels though the corresponding rows of $\boldsymbol{G}_N$ have weight that is larger than minimum weight/distance. Therefore, it is recommended to use the 5G construction for PAC coding in medium block length. To obtain results comparable with CRC-polar codes in this block length range, we need to concatenate PAC codes with CRC.

Now, let us compare the codes considered earlier under different code rates. Figure 13.2 demonstrates the FER performance of the CRC-polar and PAC codes for the two codes of (128, 64) and (1024, 512), respectively. In this scenario, summarized in Table 13.2, we use the SCL decoder with list size 8 and 5G construction for all codes. As can be seen, PAC codes constructed with the 5G sequence behave differently for short codes and medium-length codes. A short PAC code can outperform the corresponding CRC-polar code at low SNR regimes. At high code rates, this is also the case in medium SNR regimes. Note that short PAC codes can outperform CRC-polar codes when constructed with 5G-RM, as we observed in Figure 13.1a; however, here we use 5G sequence for all PAC codes. In contrast, at medium block lengths, the PAC codes cannot compete with CRC-polar codes, and we need a CRC concatenation to improve them, as we observed in Figure 13.1b. However, there is one exception for high-rate PAC codes.

Finally, we compare the error correction performance of the CRC-polar codes and PAC codes under various decoding algorithms. Figure 13.3 compares the FER of the CRC-polar code (128, 64) under the successive cancellation decoding (SC) (see Chapter 4), belief propagation (BP) decoding (see Chapter 10), SC List (SCL) decoding (see Chapter 8), and SCL-flip decoding (also known as SCL decoding with shifted pruning, see Chapter 8) with different decoding

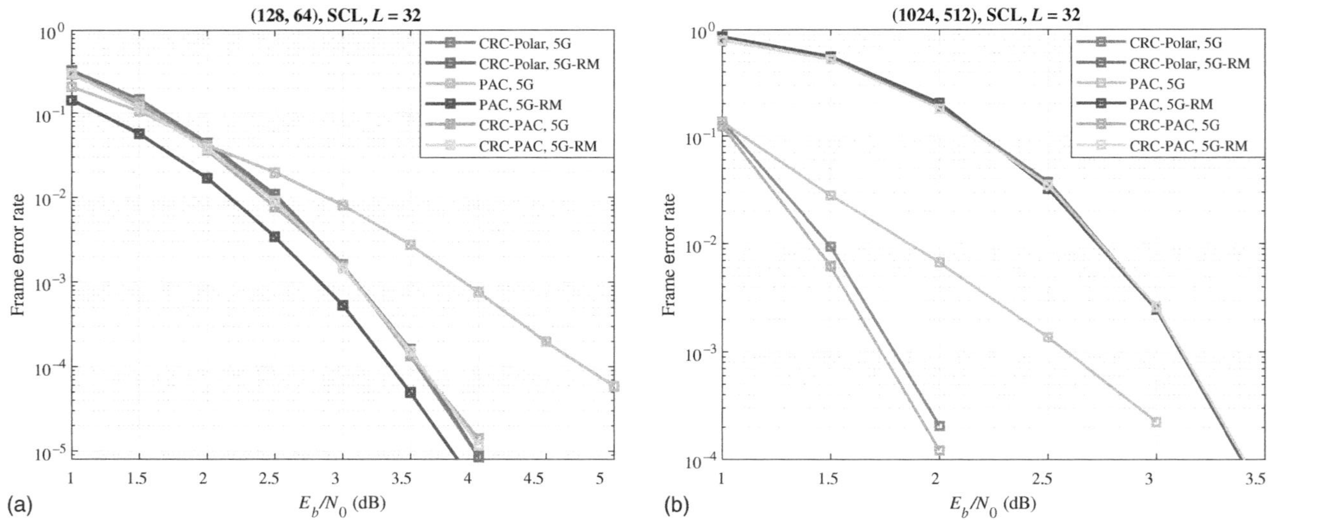

Figure 13.1 FER comparisons of different variants of polar codes constructed with two rate profiles under SCL decoding for two half-rate codes: (a) $N = 128, K = 64$. (b) $N = 1024, K = 512$.

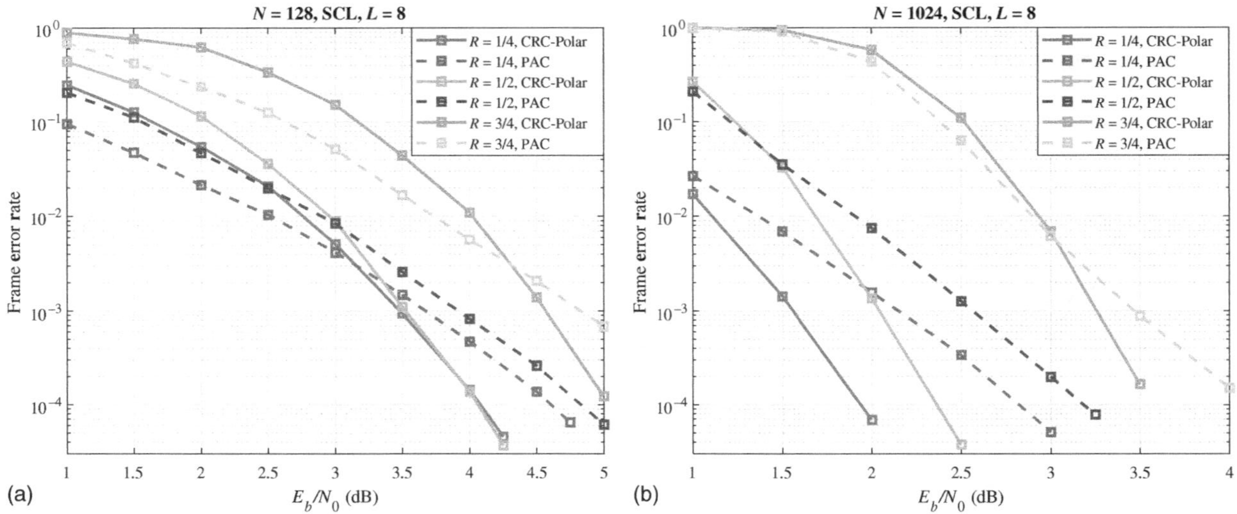

Figure 13.2 FER comparisons of polar codes with different rates constructed with 5G rate profile under SCL decoding: (a) $N = 128, K = 64$. (b) $N = 1024, K = 512$.

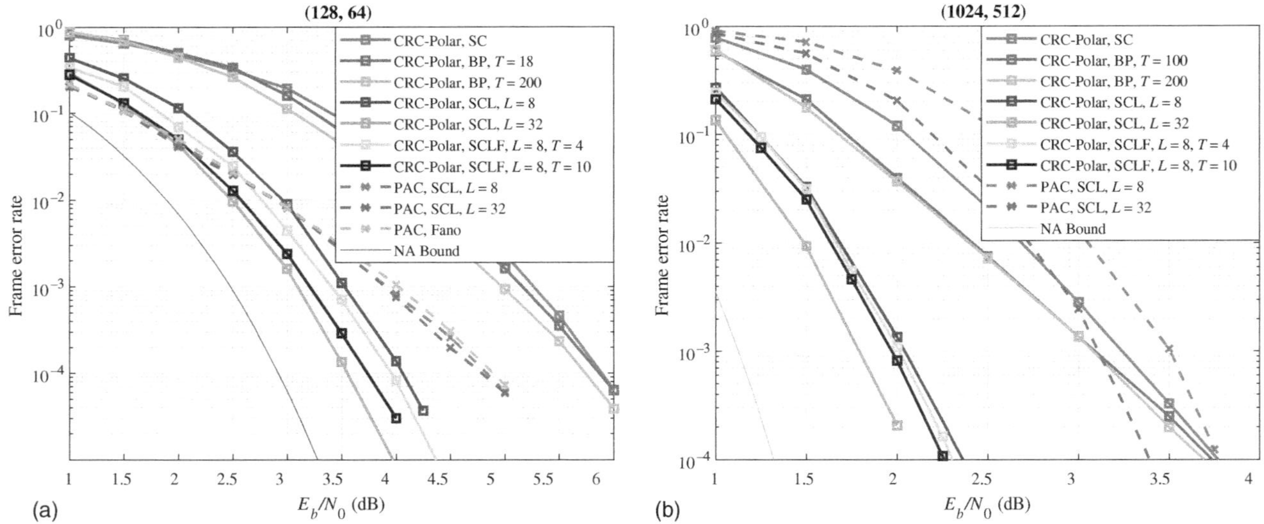

Figure 13.3 FER comparisons of different variants of half-rate polar codes constructed with 5G rate profile under different decoding algorithms: (a) $N = 128, K = 64$. (b) $N = 1024, K = 512$.

parameters, including list sizes ($L = 8, 32$), iterations (T), or both. In this scenario, summarized in Table 13.3, we use the 5G construction for all codes. As the figures demonstrate, the SCL decoding algorithm with a large list $L = 32$ appears to be the best option, as it outperforms the rest. Under resource constraints such as limited silicon area, SCLF with list size $L = 8$ could alternatively be used at the cost of longer latency due to multiple iterations ($T = 4, 10$). Note that the PAC codes demonstrated here are constructed with 5G reliability sequences, not 5G-RM. As a result, it does not outperform a short CRC-polar code.

13.2 Performance Comparison of Turbo, LDPC, and Polar Codes

In this section, we compare 5G polar codes (adopted for control channels in the 5G standard) with 5G LDPC codes (adopted for traffic/data channels in the 5G standard) and turbo codes (used in 3G and 4G standards). We consider the information lengths $K = 100, 400, 1000, 4000,$ and 8000 bits and the code rates $R = 1/5, 1/3, 1/2, 2/3,$ and $8/9$. Note that information lengths $K = 96, 992$ are employed for turbo codes, whereas $K = 100, 1000$ are used for LDPC and polar codes. The extrinsic information for the Max-log-MAP implementation of BCJR (Bahl–Cocke–Jelinek–Raviv) algorithm is scaled as $s^{(1)} = 0.6$, $s^{(I_{\max})} = 1$, and $s^{(I)} = 0.7, \forall I \in \{2, \ldots, I_{\max} - 1\}$, where $I_{\max}$ denotes the maximum number of iterations. The setups for the numerical evaluations are summarized in Table 13.4. Note that puncturing, shortening, and repetition techniques are used to have the same rate (or the same block lengths) for all three codes.

As can be seen in Figure 13.4, polar codes outperform LDPC and turbo codes when the length of the information block is $K \leq 400$ bits. For $K = 1000$ bits, the performance of polar codes is inferior compared to that of LDPC codes at rates

Table 13.4 Evaluation of the frame error rate performance versus SNR of coding schemes.

Channel		AWGN	
Modulation		QPSK	
Coding scheme	Turbo	LDPC	Polar
Code rate R		$1/5, 1/3, 1/2, 2/3, 8/9$	
Decoding algorithm	BCJR (max-log-MAP), $I_{\max} = 8$	BP (adjusted min-sum), $I_{\max} = 25$	SCL, $L = 32$
Info. length K (bits w/o CRC)		$100, 400, 1000, 4000, 8000$	

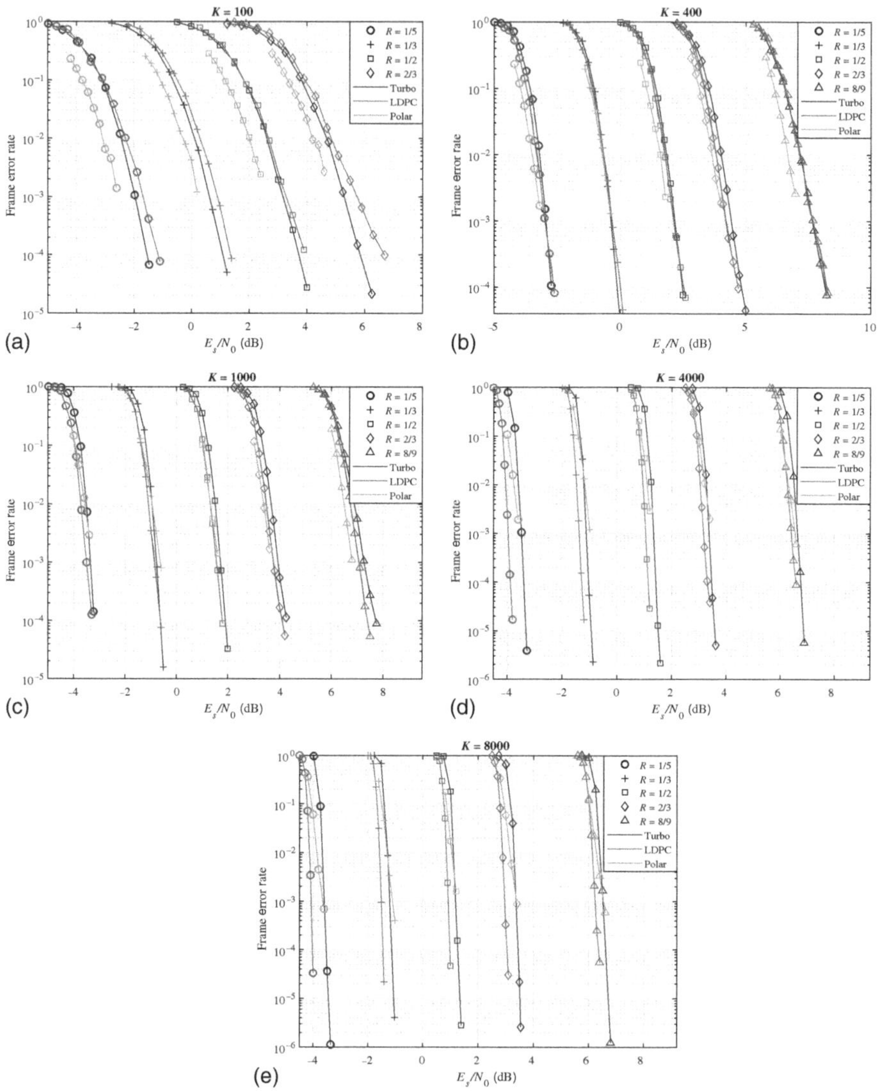

Figure 13.4 FER comparisons of polar, LDPC, and turbo codes with different information block sizes K. (a, b) $K = 100$ and $K = 400$. (c, d) $K = 1000$ and $K = 4000$. (e) $K = 8000$.

$R \leq 1/3$. As the length of the information block increases to $K = 4000$ and 8000 bits, LDPC codes outperform turbo and polar codes at all rates, and the power gain becomes more significant as K increases.

Note that we did not consider recent improvements of the coding schemes discussed earlier, in particular with regard to the polar coding scheme. Recent developments are still the subject of research and need to be further investigated and improved to the level needed for adoption into a standard. Nevertheless, we have discussed and evaluated them in their own section in this book.

Bibliographical Notes

The results presented in this chapter are obtained from Rowshan et al. [2024] and 3GPP technical reports R1-1611071 [2016] and R1-1613029 [2016]. For LDPC and polar codes, we use the 5G standard's constructions. Turbo codes in the comparisons are enhanced turbo codes with tail-biting termination from 3GPP technical report [R1-167413], [2016], where particular interleavers and puncturers are adopted. For further details about various turbo and LDPC coding schemes used for these numerical results, the readers may refer to Rowshan et al. [2024].

Bibliography

R1-1611071. NR channel coding BLER database. 3GPP TSG-RAN WG1 meeting 86, Orange and Institut Mines-Telecom, Qualcomm, Huawei, October 2016.

R1-1613029. Enhanced turbo codes for NR: Performance evaluation for eMBB and URLLC. 3GPP TSG-RAN WG1 meeting 87, Orange and Institut Mines-Telecom, October 2016.

R1-167413. Enhanced turbo codes for NR: Implementation details. 3GPP TSG-RAN WG1 meeting 86, Orange and Institut Mines-Telecom, August 2016.

Mohammad Rowshan, Min Qiu, Yixuan Xie, Xinyi Gu, and Jinhong Yuan. Channel coding toward 6G: Technical overview and outlook. *IEEE Open Journal of the Communications Society*, 5:2585–2685, 2024.

14

5G New Radio (NR) Polar Coding with MATLAB®

The real problem is what can man and machine do together and not in competition.

— Richard Hamming

This chapter is a tutorial on how to use MATLAB's *5G Toolbox* to obtain the block error rate (BLER) of 5G polar codes over the additive white Gaussian noise (AWGN) channel. We describe the components involved, such as code construction, encoding, rate-matching, and decoding. We use an example to show how to use these components in MATLAB. To follow this tutorial and run the example script, you need to add the 5G Toolbox to your MATLAB software using "Get Add-Ons" in the Home menu. You may also need some other toolboxes, such as *Communications Toolbox.*

14.1 Polar Encoding

Suppose that we want to transmit a random message *msg* with length A bits. This message is encoded using a systematic cyclic code with a generator polynomial $poly \in \{$'6','11','24C'$\}$, and *crcLen* redundancy bits called cyclic redundancy check (CRC) bits are added to the message and form *msgcrc* with length K bits, as discussed in Section 2.6.3. As discussed in Chapter 6, a polar code is constructed offline using a signal-to-noise ratio (SNR)-independent method. Then, based on the reliability-ordered sequence and utilizing the nested property of polar codes, we can construct any $N = 2^n$-length code sequence Q_0^{N-1} up to maximum block length $N_{\max} = 1024$ from the mother code sequence $Q_0^{N_{\max}-1}$, and information-CRC sequence length, K, by the function *nrPolarEncode*$(\cdot)$. This function also performs the non-systematic encoding of the information sequence including CRC bits.

Polar Codes: From Theory to Practice, First Edition. Mohammad Rowshan and Emanuele Viterbo.

```
msg = randi([0 1],K-crcLen,1);                  % Generate a random ...
    binary message
msgcrc = nrCRCEncode(msg,poly);                 % Attach CRC bits in ...
    distributed fashion
encOut = nrPolarEncode(msgcrc,E,nMax,iIL);      % Polar encoding
N = length(encOut);
```

There are differences between the uplink encoding and the downlink encoding as follows:

14.1.1 Uplink

For uplink in physical uplink control channel (PUCCH)/physical uplink shared channel (PUSCH), depending on $A \in [12, 1706]$, the *crcLen* $\in \{6, 11\}$ and the generator polynomial are chosen as

$$\begin{cases} g_{CRC6}(D) = \left[D^6 + D^5 + 1\right], & A \in [12, 19], \\ g_{CRC11}(D) = \left[D^{11} + D^{10} + D^9 + D^5 + 1\right], & A \in [20, 1706]. \end{cases} \quad (14.1)$$

When we have $A \geq 360$, the message is segmented into $C = 2$ equilength segments. The CRC bits are appended to each segment, and the rest of the processes including encoding and rate-matching are performed on each segment separately. Note that for uplink, as shown below, if $3 \leq A \leq 11$, then Reed–Muller (RM) coding is employed. If $A = 2$, simplex coding is used. Lastly, if $A = 1$, repetition coding is used.

$$\begin{cases} \text{Reed–Muller (RM) codes}, & A \in [3, 11], \\ \text{Simplex codes}, & A = 2, \\ \text{Repetition codes}, & A = 1. \end{cases}$$

Furthermore, when we have $A \in [12, 19]$, parity check bits of length $nPC = 3$ in addition to the CRC bits are used as well.

The maximum value of $N = 2^{nMax}$, the maximum length of the encoded message, is determined as a function of the message length A and the desired encoded block length G for both uplink and downlink, except for physical broadcast channel (PBCH), where we always use $N = 512$. For uplink configuration, we use $nMax = 10$.

14.1.2 Downlink

For all block lengths in physical downlink control channel (PDCCH) and PBCH, the *crcLen* is 24 bits where the generator polynomial is

$$\begin{aligned} g_{CRC24C}(D) = \big[&D^{24} + D^{23} + D^{21} + D^{20} + D^{17} + D^{15} \\ &+ D^{13} + D^{12} + D^8 + D^4 + D^2 + D + 1\big]. \end{aligned} \quad (14.2)$$

Table 14.1 The transmitter configurations used for uplink and downlink.

	A	C	*crcLen*	*poly*	nPC	iIL	iBL
UpLink (UL)	[12,19]	1	6	'6'	3	false	true
	≥ 20	1,2	11	'11'	0	false	true
DownLink (DL)	≥ 12	1	24	'24C'	0	true	false

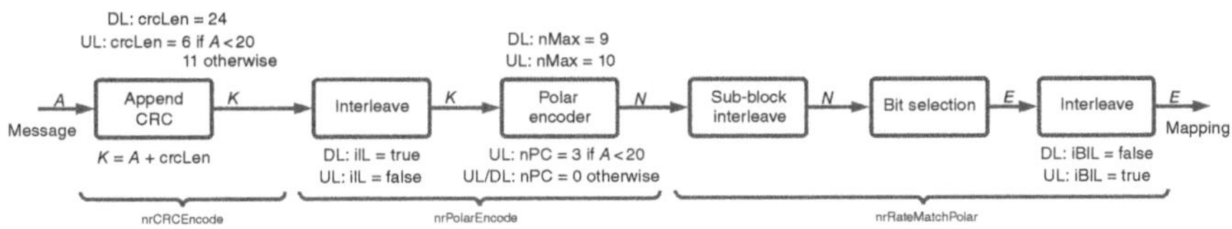

Figure 14.1 The operations at the transmitter end.

If the message length A is less than 12 bits, it should be zero-padded until it becomes 12 bits. The encoded message's maximum length for downlink configuration is set to $nMax = 9$.

Unlike the uplink, the input bits for the downlink are interleaved before polar encoding. Hence, we need to set the input interleaving to $iIL = true$. In summary, the CRC polynomial and other parameters are chosen from Table 14.1.

Independent of uplink or downlink configuration, the rate-matched output length E, which depends on K, is selected as follows:

$$\begin{cases} K+3 < E \leq 8192, & 18 \leq K \leq 25 \text{ or } 12 \leq A \leq 19, \\ K < E \leq 8192, & K \geq 31 \text{ or } A \geq 20. \end{cases} \tag{14.3}$$

The desired encoded block length is denoted by G. This length in the case of long PUCCH blocks is obtained after the concatenation of the $C = 2$ segments. Then, the rate-matched output length E is $E = \lfloor G/C \rfloor$. Note that if we do not provide the parameters $nMax$ and iIL to the encoding function as below:

```
encOut = nrPolarEncode(msgcrc,E);  % Polar encoding
```

then, by default, the input interleaving is considered enabled, $iIL = true$, and the maximum length of the encoded message 2^{nMax} is set to 512. Therefore, this syntax can be used for downlink configuration. Figure 14.1 illustrates the encoding and rate-matching operations of the transmitter end.

Let us summarize the key parameters used for the encoding process in this section in Table 14.2.

Table 14.2 The defnition of key parameters in the encoding and rate-matching processes.

Parameter	Description
A	Message length
$crcLen$	CRC length
K	$A + crcLen$
N_{max}	Maximum block length
$nMax$	$\log_2(N_{max})$
N	Length of codeword (or code sequence)
nPC	Number of parity check bits
C	Number of equilength segments
G	Desired block length
E	$\lfloor G/C \rfloor$, Length of rate-matching output

14.2 Rate-Matching and Rate-Recovery

Rate-matching is performed to map the N coded bits to the E bits available for transmission. The rate-matching schemes supported in 5G New Radio (NR) are puncturing, shortening, and repetition. For more details on the rate-matching techniques in 5G NR, see Chapter 11. As discussed in Chapter 11, these operations are performed on a circular buffer of length N. The N coded bits are subblock-wise interleaved before being inserted into the circular buffer.

The choice of rate-matching scheme depends on the payload size K, the mother code length N, and the rate-matching output length E as indicated in (14.4).

$$\begin{cases} \text{repetition}, & E > N, \\ \text{puncturing or shortening}, & E < N. \end{cases} \tag{14.4}$$

After rate-matching, the selected bits from the circular buffer are passed to a triangular bit-interleaver before the modulation mapper for uplink, as discussed in Chapter 12. However, interleaving is not applied for downlink where high-order modulation is not supported. The rate-matching process is performed by the function *nrRateMatchPolar*. As mentioned, we need to set the bit-interleaving for uplink only; i.e. *iBIL* = *true*.

```
modIn = nrRateMatchPolar(encOut,K,E,iBIL);   % Rate match
```

14.2.1 Modulation and Channel Effect

We need to map the output of rate-matching to modulated symbols, pass them through a noisy channel, and then demodulate them by obtaining the log-likelihood ratio (LLR) values. For this purpose, the channel should be configured as follows:

```
noiseVar = 1./(10.^(snrdB/10));
chan = ...
    comm.AWGNChannel('NoiseMethod','Variance','Variance',noiseVar);
```

Then, modulation, channel emulation, and demodulation can be performed as follows.

```
modOut = nrSymbolModulate(modIn,'QPSK');        % Modulate
rSig = chan(modOut);     % Add White Gaussian noise
rxLLR = nrSymbolDemodulate(rSig,'QPSK',noiseVar);     % Soft ...
    demodulate
```

At the receiver end, to prepare the N LLR values for decoding, the rate recovery depending on the rate-matching scheme is performed as follows.

- For repetition, the first N LLRs are selected.
- For puncturing, the $N - E$ missing LLRs are set to 0, as the coded bit could take either value of 0 or 1 on the transmitter side although the corresponding sub-channels are selected to be frozen.
- For shortening, the $N - E$ missing LLRs are set to a large value as the coded bit could be guaranteed to be 0 on the transmitter side while, similarly to puncturing, the corresponding sub-channels are selected to be frozen as well.

The rate recovery is performed by the function *nrRateRecoverPolar*, as seen in the following. Recall that the parameter *iBIL* for bit-interleaving is set for uplink only.

```
decIn = nrRateRecoverPolar(rxLLR,K,N,iBIL);
```

14.3 Polar Decoding

After rate recovery, the sequence "decIn" is decoded by function *nrPolarDecode*. As the underlying decoder is a successive cancellation (SC) list decoder, we also need to specify the list size L as a power of two. The rest of the decoding arguments should match the transmitter end. Recall that the bits are interleaved before

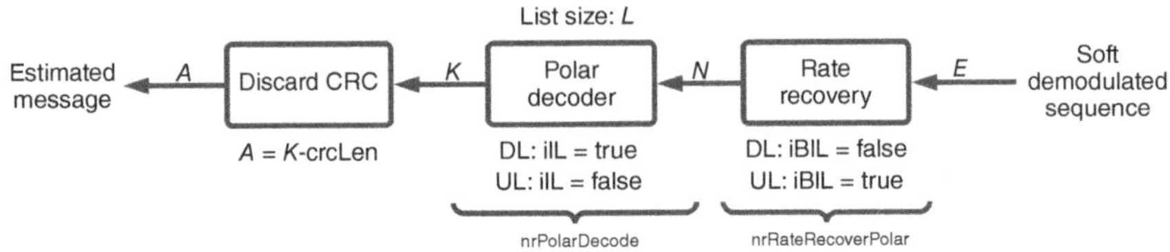

Figure 14.2 The operations at the receiver end.

polar encoding in downlink; hence, we need to set $iIL = true$ for downlink to apply deinterleaving after decoding.

```
decbits = nrPolarDecode(decIn,K,E,L,nMax,iIL,crcLen);
```

Note that the default values for the last three arguments are $nMax = 9$, $iIL = true$, and $crcLen = 24$, which are used for downlink. Hence, if you want to use the function for downlink, you can alternatively use the following:

```
decBits = nrPolarDecode(decIn,K,E,L);
```

This function has the arguments *padCRC*, which is used for messages prepadded with ones before CRC encoding, and *rnti*, which is a radio network temporary identifier (RNTI) used to mask the CRC bits after CRC encoding, but they are not within the scope of this book.

Eventually, we need to discard the CRC bits from K decoded bits in *decBits*. This is given by $decBits(1 : K - crcLen)$. Figure 14.2 illustrates the rate recovery and decoding operations at the receiver end.

If you are running a simulation to count the bit errors and block errors, you can compare the transmitted message *msg* and the decoded bits to count the errors as follows:

```
numBITerr = numBITerr + biterr(double(decBits(1:K-crcLen)), ...
    msg);
numBLerr = numBLerr + any(ne(decBits(1:K-crcLen) = msg));
```

14.4 BLER Versus SNR of a Code—Full Script

In this section, we provide a full MATLAB script to collect numerical results for the error correction performance of a specific 5G polar code in uplink or downlink scenarios for a range of E_b/N_0 in dB.

In the *snrdB* loop, for every SNR point, the random messages are processed up to *numErrors* frame errors, for example, *numErrors*=200, or a maximum of numFrames frames, for example, *numFrames*=1e6, whichever occurred first.

```
function FiveG_polar(linkDir, A, E, L, numFrames, ...
    numErrors, SNRdB_set, isEbN0, modul)
    % linkDir: 'UL' for uplink, 'DL' for downlink,
    % A: The number of information bits,
    % E: The length of rate-matcher output.
    % L: List size for SC lis tdecoding,
    % numFrames: Maximum number of frames to transmit for ...
        every SNR, ex:1e6
    % numErrors: The number of errors to count for every ...
        SNR, ex: 100
    % SNRdB_set: [starting pint: step: end point], ex: ...
        [1:0.5:4]
    % isEbN0: Is the SNRdB_set in Eb/N0 or in Es/N0? true ...
        or false
    % modul: BPSK or QPSK
    % Example:
    % FiveG_polar('UL', 16, 64, 32, 1e6, 100, [2:0.5:5], ...
        true, 'BPSK')


    s = rng(100);          % Seed the RNG for repeatability %100
    bps = 1;                    % bits per symbol, 1 for BPSK, ...
        2 for QPSK
    if modul == 'QPSK'
        bps = 2;

    end

    if strcmpi(linkDir,'DL') % Downlink scenario (K ≥ 36, ...
        including CRC bits)
        crcLen = 24; %24;          % Number of CRC bits for DL, ...
            Section 5.1, [6]
        poly = '24C'; %'24C';        % CRC polynomial
        K = A + crcLen;
        nPC = 0;               % Number of parity check bits, ...
            Section 5.3.1.2, [6]
        nMax = 9;              % Maximum value of n, for 2^n, ...
            Section 7.3.3, [6]
        iIL = true;            % Interleave input, Section ...
            5.3.1.1, [6]
        iBIL = false;          % Interleave coded bits, Section ...
            5.4.1.3, [6]
    else        % 'UL': Uplink scenario (K ≥ 21, including ...
        CRC/Parity bits)
        if A ≥ 20
            crcLen = 11;
            poly = '11';
            nPC = 0;
        elseif A ≥ 12
```

```
            crcLen = 6;
            poly = '6';
            nPC = 3;
        else
            disp('A is smaller than 12 bits!');
            exit;
        end
        numParities = crcLen + nPC;
        K = A + numParities;
        nMax = 10;
        iIL = false;
        iBIL = true;
    end


    for snrdB = SNRdB_set
        R = A/E;                    % Code rate, not R = K/E;
        if isEbN0 == true
            noiseVar = 1./(bps*R*10.^(snrdB/10)); % N0: ...
                noise variance
        else
            noiseVar = 1./(10.^(snrdB/10)); % N0: noise ...
                variance
        end
        chan = comm.AWGNChannel('NoiseMethod','Variance',
                                      'Variance',noiseVar);   ...
                                        % Channel


        numFerr = 0;
        numBITerr = 0;
        for blck = 1:numFrames
            msg = randi([0 1],K-crcLen,1);     % Generate a ...
                random message
            msgcrc = nrCRCEncode(msg,poly);    % Attaching CRC
            encOut = nrPolarEncode(msgcrc,E,nMax,iIL);    % ...
                Polar encoding
            N = length(encOut);
            modIn = nrRateMatchPolar(encOut,K,E,iBIL);     % ...
                Rate matching
            modOut = nrSymbolModulate(modIn,modul);     % ...
                Modulation
            rSig = chan(modOut);    % Add White Gaussian noise
            rxLLR = ...
                nrSymbolDemodulate(rSig,modul,noiseVar);     ...
                % Soft demodulate
            decIn = nrRateRecoverPolar(rxLLR,K,N,iBIL);     % ...
                Rate recover
            decBits = ...
                nrPolarDecode(decIn,K,E,L,nMax,iIL,crcLen);   ...
                  % Polar decode
            numBITerr = numBITerr + ...
                biterr(double(decBits(1:A)), msg);     % ...
                Compare msg and decoded bits
```

```
            numFerr = numFerr + any(decBits(1:A)≠msg);
            if numFerr == numErrors
                break;
            end
        end

        disp(['BLER: ' num2str(numFerr/blck) ...
              ', BER: ' num2str(numBITerr/(A*blck)) ...
              ', at Eb/N0 = ' num2str(snrdB) ' dB'])
    end
    rng(s);     % Restore RNG
end
```

Appendix A

Conceptual Channels in 5G New Radio

Polar codes and LDPC codes were selected as the two main channel coding schemes for the fifth-generation (5G) cellular network called new radio (NR) in the technical specification (TS) 38.212 of 3GPP Release 15 issued in 2018. To clarify where polar coding is exactly employed in the 5G radio access network (RAN), we briefly describe the architecture of the air interface to the extent that we can distinguish where polar coding is used. The air interface of a cellular network is at layers 1 and 2 of the open systems interconnection (OSI) model, a conceptual model that provides a basis for networks. Layer 1 is also called the physical (PHY) layer and is responsible for channel coding, hybrid automatic re-transmission request (HARQ) processing, modulation, multi-antenna processing, and allocation of signals to the appropriate time–frequency resources. Layer 2 is divided into multiple protocol layers. Among them are the radio link control (RLC) sublayer and the medium access control (MAC) sublayer. The base station, which is referred to as gNodeB (gNb), controls the air interface and schedules the downlink access and the uplink access of all devices.

To carry the data across the radio access network (RAN), the data is allocated to a number of conceptual channels, namely, logical, transport, and physical channels, between the layers, as shown in Figure A.1. This hierarchy helps to manage data transfers in an orderly manner.

Logical channels define what type of data is transferred. This data can be either control or user data. Depending on the type of data, the logical channels are divided into control or signaling channels and traffic channels. These channels logically exist in the interface between the MAC and the RLC protocol layers, that is in layer 2. A logical channel can be either "common" to all users in a cell (point-to-multipoint channel) or "dedicated" to a single user (point-to-point channel). After defining different types of logical channel, we briefly look at all logical channels in the 5G hierarchy. The logical channels used are the following.

Polar Codes: From Theory to Practice, First Edition. Mohammad Rowshan and Emanuele Viterbo.

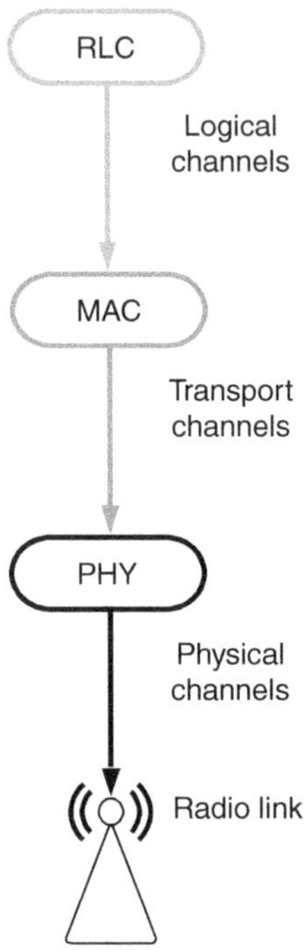

Figure A.1 The conceptual channels between layers.

- Broadcast control channel (BCCH), a downlink common channel used for sending broadcast-style information to the UEs within a cell (the identity of the network). The system information transmitted by this channel is divided into different blocks: (a) master information block (MIB) is mapped onto the transport-type broadcast channel (BCH) and then to the physical broadcast channel (PBCH) to allow a UE to access the network; (b) system information blocks (SIBs) are mapped onto the transport-type downlink shared channel (DL-SCH) and then onto the physical downlink shared channel (PDSCH).
- Paging control channel (PCCH), a downlink channel that transfers paging information (to find the cell of the user equipment (UE) such as mobile phones) and some other notifications.
- Common control channel (CCCH), used in both downlink and uplink for transmitting control information to and from the UE for initial access, that is those UEs that do not have a radio resource control (RRC) connection.
- Dedicated control channel (DCCH), used in both downlink and uplink to carry dedicated control information between the UE and the network after RRC connection has been established.

- Dedicated traffic channel (DTCH), dedicated to one UE and is used in both downlink and uplink for carrying user information to and from a specific UE and the network.

MIB is mandatory system information that is broadcasted by the gNB at a set periodicity. During the initial cell selection procedure, UE assumes that the synchronization signal block (SSB) is broadcasted every 20 ms by the gNB. UE gets MIB by decoding the PBCH from the SS-PBCH block beam.

Logical channels are mapped to transport channels in the MAC sublayer. The transport channels are used to multiplex the logical data to be transported by the physical channels over the radio interface. The transport channels defined in the 5G hierarchy are:

- The broadcast channel (BCH) is a downlink channel that broadcasts in the cell. It conveys the master information block (MIB) received from the BCCH logical channel. The MIB is a block of data that contains a limited amount of the total information that the user equipment (UE) requires to acquire the remaining system information (SIBs) broadcast by the network via the DL-SCH. The transport block that it delivers to the physical layer has a fixed format defined by the specifications.
- The downlink shared channel (DL-SCH): The DL channel used for the transmission of user data received from the DTCH above. It is also used for transmission of SIBs received from the BCCH above, transmission of UE-specific control information received from the DCCH above, and transmission of common control information received from the CCCH. It supports HARQ with soft combining and dynamic link adaptation via the variation of coding, modulation, and transmit power. It further supports both dynamic and semi-static resource allocation, discontinuous reception (DRX), and the possibility to use beamforming.
- The uplink shared channel (UL-SCH): Similar to the DL-SCH, the UL-SCH is the UL channel used for the transmission of user data. It supports HARQ with soft combining and dynamic link adaptation via the variation of coding, modulation, and transmit power. It further supports both dynamic and semi-static resource allocation and the possibility to use beamforming.

It is time to look at the physical channels. The physical channels are the closest to the actual transmission of the data over the radio access network by 5G radio frequency (RF) signals. The physical channels shown in Figure A.2 are as follows:

- The physical broadcast channel (PBCH): A DL channel fed by the BCH above, which carries some of the required system information to allow a UE to access the network.

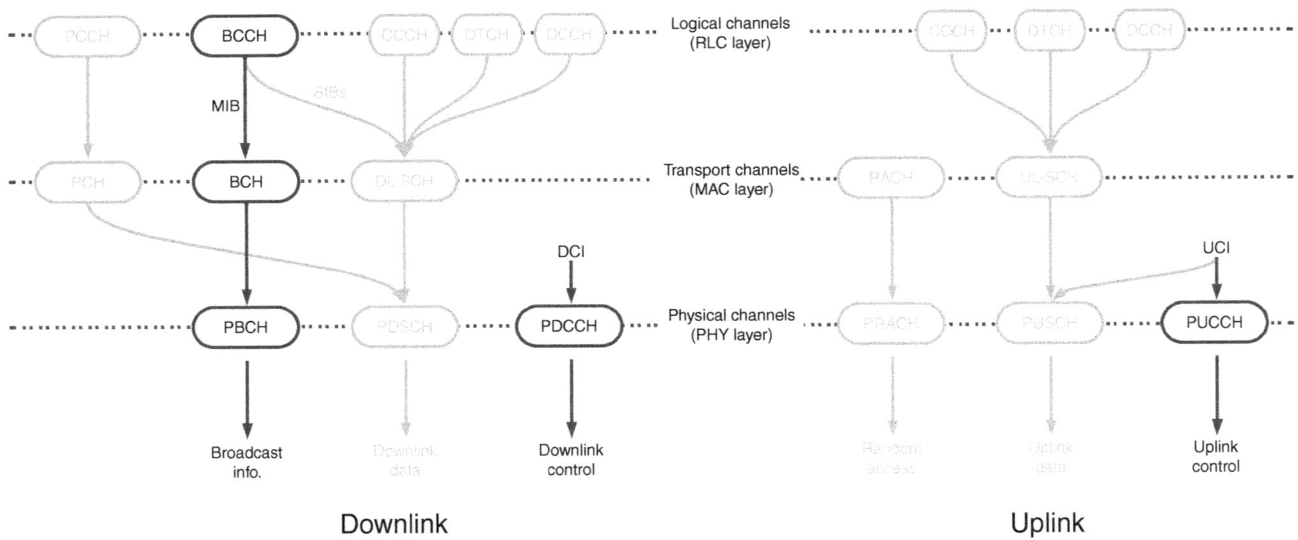

Figure A.2 Channel mapping in 5G NR hierarchy: logical, transport, and physical channels. Polar coding is used for the channels highlighted in black.

- The PDCCH transmits downlink control information (DCI) from the gNB to the UE. DCI includes necessary information for proper reception and decoding of downlink data transmission, such as time–frequency resource allocation and transport format used in both PDSCH and PUSCH transmission, hybrid automatic re-transmission request (HARQ) information for both uplink, downlink, and power control commands (used to the transmitter power to achieve better performance).
- On the other hand, the PUCCH transmits uplink control information (UCI) from the user equipment (UE) to the gNB. UCI information includes HARQ acknowledgments of received PDSCH transport blocks (TBs), channel state information (CSI) regarding the downlink channel status, and scheduling requests by the UE to be granted uplink resources for PUSCH transmission.

For the control channels (DCI, PBCH, and UCI) in NR, polar coding is adopted, which has a better performance compared to the convolution codes used in LTE-A at the cost of increased decoding complexity (see 3GPP TS 38.212). Polar codes are used for coding the downlink and uplink control information (DCI/UCI) as well as the broadcast channel's master information block (MIB). Depending on the payload size of the control channel, different types of polar codes can be used with a different number of cyclic redundancy check (CRC) bits.

Appendix B

Channel Coding from 2G to 5G

B.1 A Journey from 2G to 5G

As cellular networks have evolved from 2G to 5G, channel coding schemes have advanced significantly to meet the demands of increasingly complex channel conditions, higher data rates, and stringent reliability and latency requirements. Table B.1 outlines the coding schemes employed in various wireless communication standards from 2G to 5G. In the following, we concisely review the coding schemes used in various generations.

B.2 Channel Coding in 2G

Channel coding in 2G systems, as specified in the technical specification TS 45.003 by 3rd Generation Partnership Project (3GPP) [2005], involves a two-step process:

1) Error detection: A 40-bit Fire code or a 16-bit CRC with generator polynomial $g_{\mathrm{CRC16}}(D) = D^{16} + D^{12} + D^{5} + 1$ is added as a block check sequence (BCS).
2) Error correction: Half-rate convolutional coding is applied, followed by optional puncturing to achieve the desired code rate.

The specific coding scheme, denoted as CS-1 to CS-4, determines the BCS length and puncturing rate, as shown in Table B.2. CS-1, the most robust scheme, is used for distant UEs, while CS-4, with no convolutional coding, is used for faster but less reliable transmissions near BTSs. The bit rate per time slot increases from 8 kbps in CS-1 to 12, 14.4, and 20 kbps in CS-2, CS-3, and CS-4, respectively.

As GSM evolved and 8-PSK modulation was added to enhanced GPRS (EGPRS) in EDGE, link quality control (LQC) was adapted to adjust the modulation and coding scheme (MCS) to the most suitable one as per the channel condition. There

Polar Codes: From Theory to Practice, First Edition. Mohammad Rowshan and Emanuele Viterbo.

Table B.1 Coding schemes from 2G to 5G.

Gen.	Channel coding
2G	Cyclic codes (FIRE/CRC), (Punctured) convolutional codes
3G	Convolutional codes, turbo codes
4G	Tail biting convolutional codes, turbo codes
5G	Polar codes, LDPC codes

Table B.2 Convolutional coding parameters in GSM (2G).

Scheme	Code rate	Generator polynomial	BCS	Tail
CS-1	1/2	$1 + D^2 + D^3$	40	4
CS-2	~2/3	$1 + D^2 + D^3 + D^4 + D^5$	16	4
CS-3	~3/4	$1 + D^2 + D^3 + D^4 + D^5 + D^6$	16	4
CS-4	1		16	-

Table B.3 Convolutional coding parameters in EDGE (2.75G).

Scheme	Code rate	BCS	Tail bits	Modulation	Family
MCS-1	53/100	12	6	GMSK	C
MCS-2	33/50	12	6		B
MCS-3	17/20	12	6		A
MCS-4	1	12	6		C
MCS-5	37/100	12	6	8-PSK	B
MCS-6	49/100	12	6		A
MCS-7	19/25	2×12	2×6		B
MCS-8	23/25	2×12	2×6		A
MCS-9	1	2×12	2×6		A

are nine MCSs (MCS-1 to MCS-9) for link adaptation, as shown in Table B.3. The MCSs are divided into three families A, B, and C, as follows:

- Family A: Basic payload unit of 37 (or 34) octets, with 1,2, or 4 payload units transmitted per Radio Block.
- Family B: Basic payload unit of 28 octets, with 1, 2, or 4 payload units transmitted per Radio Block.

- Family C: Basic payload unit of 22 octets, with 1 or 2 payload units transmitted per Radio Block.

To ensure reliable transmission, a BCS is added to every two payload units for error detection. This BCS is then interleaved over two or four bursts, depending on the MCS. Subsequently, six or twelve tail bits are added, and rate-1/3 convolutional coding is applied, followed by puncturing to achieve the desired code rate.

This adaptive approach allows flexible data rates, ranging from 8.8 kbps in MCS-1 to 59.2 kbps in MCS-9, optimizing performance based on varying channel conditions.

B.3 Channel Coding in 3G

According to the technical specification TS 25.222 by 3rd Generation Partnership Project (3GPP) [2002],[1] 3G employs a two-step channel coding process:

- Error detection: A Cyclic Redundancy Check (CRC) of 24, 16, 12, 8, or 0 bits is added to each transport block, depending on its size.
- Error correction:
  - Convolutional coding: For shorter packets, convolutional coding with a constraint length of 9 is used. The generator polynomials of $(561, 753)$ for rate $1/2$ and $(557, 663, 711)$ for rate $1/3$ are used. Eight tail bits are added before encoding.
  - Turbo coding: For longer packets, parallel concatenated convolutional coding with a rate-1/3 code is used. This involves two constituent encoders with the transfer function $G(D) = \left[1, g_1(D)/g_0(D)\right]$, where $g_0(D) = 1 + D^2 + D^3$ and $g_1(D) = 1 + D + D^3$.

B.3.1 Block Segmentation and Rate Matching

To accommodate different data rates and channel conditions, the transport blocks are segmented and concatenated into larger blocks. If necessary, zero-padding is added to ensure that the block length is a multiple of the required size for the chosen coding scheme. The rate matching is then applied to adjust the code rate to the desired level.

Table B.4 provides a detailed overview of the coding schemes and parameters used in 3G, including the choice of convolutional or turbo coding, code rates, and modulation schemes. By combining these techniques, 3G systems achieve reliable and efficient data transmission over wireless channels.

1 3GPP stands for 3rd Generation Partnership Project. It is a collaboration of telecommunications standards development organizations responsible for developing protocols for mobile telecommunications.

Table B.4 Coding parameters in UMTS (3G).

Coding scheme	Code rate	Data block size (K)
Convolutional coding	1/2 & 1/3	≤ 504
Turbo coding	1/3	[40, 5114]
No channel coding	1	Unlimited

B.4 Channel Coding in 4G

The channel coding scheme used for the transport blocks in 4G, which is specified in the evolved universal terrestrial radio access (E-UTRA), is turbo coding with the code rate $R = 1/3$. This scheme incorporates two 8-state constituent encoders (identical to 3G) and a contention-free quadratic permutation polynomial (QPP) interleaver. Furthermore, trellis termination is used for turbo coding. Before turbo coding, transport blocks are segmented into byte-aligned segments with a maximum information block size of 6144 bits.

According to the technical specification TS 36.212 by 3rd Generation Partnership Project (3GPP) [2009], the transport blocks (TBs) are protected by appended CRC bits with a length of 24 (calculated by type-A generator polynomial, CRC24A). The detected error by CRCs is reported to higher layers. The transport block is segmented into code blocks (CBs) of no larger than 6144 bits, where each CB is protected by type-B 24-bit CRC (CRC24B). If the code block is smaller than 40 bits, filling bits are appended. The CRC bits for control channels are shorter as broadcast channel (BCH) and physical downlink control channel (PDCCH) use 16-bit CRC, while physical uplink control channel (PUCCH) employs 8-bit CRC. The channel coding scheme for the aforementioned control channels is tail-biting convolutional coding (TBCC) whereas the transport channels (TrCH) use turbo coding. Tail-biting convolutional coding with constraint length 7 is performed by generator polynomials (in octal) $G_0 = 133$, $G_1 = 171$, and $G_2 = 165$ for the code rate 1/3. The details of the coding schemes used in 4G are summarized in Table B.5.

Table B.5 Coding parameters in E-UTRA (4G).

Coding scheme	Code rate	Data block size (K)
Tail-biting convolutional coding	1/3	
Turbo coding	1/3	[40, 6144]
Block code	1/16, variable	
Repetition code	1/3	

B.5 Channel Coding in 5G

G supports new channel coding schemes. According to the technical specification TS 38.212 by 3rd Generation Partnership Project (3GPP) [2022], 5G incorporates Low-Density Parity-Check (LDPC) codes for data channels including physical downlink shared channel (PDSCH) and physical uplink shared channel (PUSCH), replacing turbo codes in 4G. Likewise, for control channels (to protect the downlink control information (DCI), the uplink control information (UCI), and the system information in the physical broadcast channel (PBCH)), polar codes are introduced in lieu of the TBCC used in 4G. The key advantages of LDPC codes (compared to turbo codes) are improved performance with very low error floors, reduced decoding complexity and latency, better power and area efficiency, and support of multi-gigabit data rates. Polar coding yields better performance at moderate payload sizes (in the order of $K \leq 250$ bits); however, it comes at the cost of higher complexity compared to TBCC. Note that the minimum supported payload size K in 5G polar codes is 12 bits. For the control information sequence $K < 12$ in the uplink, the short block codes used in 4G, namely, repetition codes, simplex codes, and Reed–Muller codes, are used.

Designed to target good performance and decoding latency for the full range of code rates and information block sizes, 5G supports two LDPC base matrices (which are, in turn, constructed using a photograph). Base matrix 1 was optimized for large information block sizes K and high code rates $R \geq 1/3$. On the other hand, base matrix 2 is suitable for small information block sizes and lower code rates than base matrix 1, down to $1/5$, which is lower than the code rate of 4G turbo codes (where for less than $1/3$, repetition is used).

5G polar codes employ 6, 11, and 24 CRC bits for error detection with the corresponding generator polynomials $g_{\text{CRC6}}(D) = D^6 + D^5 + 1$, $g_{\text{CRC11}}(D) = D^{11} + D^{10} + D^9 + D^5 + 1$, and $g_{\text{CRC24C}}(D) = D^{24} + D^{23} + D^{21} + D^{20} + D^{17} + D^{15} + D^{13} + D^{12} + D^8 + D^4 + D^2 + D + 1$. However, the 5G LDPC codes use 16 and 24 CRC

Table B.6 Channel coding schemes in 5G.

	Channel	**Coding scheme**	**Info. block, *A***	**CRC+PC**	**Encoded block (*G*)**
Downlink	PDSCH	LDPC	[80,8448]	16,24B	$[384, 6480]$
	PDCCH	Polar	$[12, 140]$	24C	$[A + 24, 8192]$
	PBCH	Polar	32	24C	864
Uplink	PUSCH	LDPC	$[48, 3840]$	24A,24B	$[192, 15360]$
	PUCCH	Polar	$[12, 19]$	6+3	$[A + 9, 8192]$
			$[20, 1706]$	11	$[A + 11, 16385]$

bits with generator polynomials $g_{\mathrm{CRC16}}(D) = D^{16} + D^{12} + D^5 + 1$, $g_{\mathrm{CRC24A}}(D) = D^{24} + D^{23} + D^{18} + D^{17} + D^{14} + D^{11} + D^{10} + D^7 + D^6 + D^5 + D^4 + D^3 + D + 1$, and $g_{\mathrm{CRC24B}}(D) = D^{24} + D^{23} + D^6 + D^5 + D + 1$.

Table B.6 lists the parameters of the coding schemes in the 5G standard.

Bibliography

3rd Generation Partnership Project (3GPP). Multiplexing and channel coding (TDD). Technical Speification (TS) 25.222 v3.10.0 (Release 1999), Sophia Antipolis, France, 2002.

3rd Generation Partnership Project (3GPP). GSM/EDGE channel coding. Technical Speification (TS) 45.003 v4.4.0 (Release 4), Sophia Antipolis, France, 2005.

3rd Generation Partnership Project (3GPP). Evolved universal terrestrial radio access (E-UTRA); multiplexing and channel coding. Technical Speification (TS) 36.212 v8.8.0 (Release 8), Sophia Antipolis, France, 2009.

3rd Generation Partnership Project (3GPP). NR; multiplexing and channel coding. Technical Speification (TS) 38.212 v15.13.0 (Release 15), Sophia Antipolis, France, 2022.

Appendix C

Scripts

C.1 Meta-Converse Bound with a Normal Approximation

The MATLAB script below plots the meta-converse bound with a normal approximation for a given range of $E_b/N0$ and code parameters of block length N and code rate R:

```
N=2^7; % block length
R=0.5; % code rate
EbN0dB=1:0.1:4; % the range of SNR per info bit in dB
EbN0=10.^(EbN0dB/10);
SNR=EbN0*2*R;
Zmin = -9; Zmax = 9;
epsilonNA = [];
for rho = SNR
    f = @(z) exp(-z.^2/2)/sqrt(2*pi).* (log2(2) - ...
        log2(1+exp(-2*rho-2*sqrt(rho)*z)));
    C = integral(f, Zmin, Zmax);    % channel capacity
    g = @(z) exp(-z.^2/2)/sqrt(2*pi).* (log2(2) - ...
        log2(1+exp(-2*rho-2*sqrt(rho)*z))-C).^2;
    V = integral(g, Zmin, Zmax);    % channel dispercion
    %V=(rho/2).*((rho+2)./(rho+1).^2)*(log2(exp(1))^2);    ...
        % alternative
    epsilonNA(end+1)=qfunc((C-R+0.5/N*log2(N))./(sqrt(V/N)));
end
figure
semilogy(EbN0dB,epsilonNA);
```

C.2 Shannon Limit

The MATLAB script below gives the Shannon limit in dB:

```
eta = 2;            % spectral efficiency R/B in bits/s/Hz
EbN0_limit_dB = 10*log10((2^eta-1)/eta);
```

C.3 Standard Array in Syndrome Decoding

The MATLAB script below generates the standard array of the (6, 3) code $\mathscr{C}$ generated by (2.14).

```
% generator matrix
G=[1 0 0 1 0 1;
   0 1 0 0 1 1;
   0 0 1 1 1 0];
% codebook
CodeBook=[]; Info=[];
for u1=0:1
  for u2=0:1
    for u3=0:1
      Info = [Info;[u1,u2,u3]];
      CodeBook=[CodeBook;rem([u1,u2,u3]*G,2)];
end; end; end;
% parity check matrix
H=[G(:,4:6)',eye(3)]
% compute the syndrome for all vectors in F_2^6
X=[];
for x1=0:1
  for x2=0:1
    for x3=0:1
      for x4=0:1
        for x5=0:1
          for x6=0:1
           s=rem([x1,x2,x3,x4,x5,x6]*H',2); % syndrome 3 bits
            X=[X;bin2dec(char(s+48)),[x1,x2,x3,x4,x5,x6]];
end; end; end; end; end; end;
% group vectors with the same syndrome
[a idx]=sort(X(:,1)); UU=X(idx,:)
% arrange the standard array table (latex tabular format)
S_arr= char(UU(:,2:7)+48);
sp=kron(ones(8,1),'&'); sp1 = kron(ones(8,1),'\\');
[S_arr(1:8,:),sp, S_arr(9:16,:),sp, ...
    S_arr(17:24,:),sp,S_arr(25:32,:),sp, ...
 S_arr(33:40,:),sp,S_arr(41:48,:),sp,S_arr(49:56,:),sp, ...
     S_arr(57:64,:),sp1]
```

C.4 Polar Sequence in 5G

The MATLAB script below sets the reliability sequence and shows how to use it in the rate-profiling stage:

```
%Reliability sequence
Q=[0 1 2 4 8 16 32 3 5 64 9 6 17 10 18 128 12 33 65 20 256 ...
    34 24 36 7 129 66 512 11 40 68 130 19 13 48 14 72 257 ...
    21 132 35 258 26 513 80 37 25 22 136 260 264 38 514 96 ...
    67 41 144 28 69 42 516 49 74 272 160 520 288 528 192 ...
    544 70 44 131 81 50 73 15 320 133 52 23 134 384 76 137 ...
    82 56 27 97 39 259 84 138 145 261 29 43 98 515 88 140 ...
    30 146 71 262 265 161 576 45 100 640 51 148 46 75 266 ...
    273 517 104 162 53 193 152 77 164 768 268 274 518 54 83 ...
    57 521 112 135 78 289 194 85 276 522 58 168 139 99 86 ...
    60 280 89 290 529 524 196 141 101 147 176 142 530 321 ...
    31 200 90 545 292 322 532 263 149 102 105 304 296 163 ...
    92 47 267 385 546 324 208 386 150 153 165 106 55 328 ...
    536 577 548 113 154 79 269 108 578 224 166 519 552 195 ...
    270 641 523 275 580 291 59 169 560 114 277 156 87 197 ...
    116 170 61 531 525 642 281 278 526 177 293 388 91 584 ...
    769 198 172 120 201 336 62 282 143 103 178 294 93 644 ...
    202 592 323 392 297 770 107 180 151 209 284 648 94 204 ...
    298 400 608 352 325 533 155 210 305 547 300 109 184 534 ...
    537 115 167 225 326 306 772 157 656 329 110 117 212 171 ...
    776 330 226 549 538 387 308 216 416 271 279 158 337 550 ...
    672 118 332 579 540 389 173 121 553 199 784 179 228 338 ...
    312 704 390 174 554 581 393 283 122 448 353 561 203 63 ...
    340 394 527 582 556 181 295 285 232 124 205 182 643 562 ...
    286 585 299 354 211 401 185 396 344 586 645 593 535 240 ...
    206 95 327 564 800 402 356 307 301 417 213 568 832 588 ...
    186 646 404 227 896 594 418 302 649 771 360 539 111 331 ...
    214 309 188 449 217 408 609 596 551 650 229 159 420 310 ...
    541 773 610 657 333 119 600 339 218 368 652 230 391 313 ...
    450 542 334 233 555 774 175 123 658 612 341 777 220 314 ...
    424 395 673 583 355 287 183 234 125 557 660 616 342 316 ...
    241 778 563 345 452 397 403 207 674 558 785 432 357 187 ...
    236 664 624 587 780 705 126 242 565 398 346 456 358 405 ...
    303 569 244 595 189 566 676 361 706 589 215 786 647 348 ...
    419 406 464 680 801 362 590 409 570 788 597 572 219 311 ...
    708 598 601 651 421 792 802 611 602 410 231 688 653 248 ...
    369 190 364 654 659 335 480 315 221 370 613 422 425 451 ...
    614 543 235 412 343 372 775 317 222 426 453 237 559 833 ...
    804 712 834 661 808 779 617 604 433 720 816 836 347 897 ...
    243 662 454 318 675 618 898 781 376 428 665 736 567 840 ...
    625 238 359 457 399 787 591 678 434 677 349 245 458 666 ...
    620 363 127 191 782 407 436 626 571 465 681 246 707 350 ...
    599 668 790 460 249 682 573 411 803 789 709 365 440 628 ...
    689 374 423 466 793 250 371 481 574 413 603 366 468 655 ...
    900 805 615 684 710 429 794 252 373 605 848 690 713 632 ...
    482 806 427 904 414 223 663 692 835 619 472 455 796 809
```

```
714 721 837 716 864 810 606 912 722 696 377 435 817 319 621 ...
    812 484 430 838 667 488 239 378 459 622 627 437 380 818 ...
    461 496 669 679 724 841 629 351 467 438 737 251 462 442 ...
    441 469 247 683 842 738 899 670 783 849 820 728 928 791 ...
    367 901 630 685 844 633 711 253 691 824 902 686 740 850 ...
    375 444 470 483 415 485 905 795 473 634 744 852 960 865 ...
    693 797 906 715 807 474 636 694 254 717 575 913 798 811 ...
    379 697 431 607 489 866 723 486 908 718 813 476 856 839 ...
    725 698 914 752 868 819 814 439 929 490 623 671 739 916 ...
    463 843 381 497 930 821 726 961 872 492 631 729 700 443 ...
    741 845 920 382 822 851 730 498 880 742 445 471 635 932 ...
    687 903 825 500 846 745 826 732 446 962 936 475 853 867 ...
    637 907 487 695 746 828 753 854 857 504 799 255 964 909 ...
    719 477 915 638 748 944 869 491 699 754 858 478 968 383 ...
    910 815 976 870 917 727 493 873 701 931 756 860 499 731 ...
    823 922 874 918 502 933 743 760 881 494 702 921 501 876 ...
    847 992 447 733 827 934 882 937 963 747 505 855 924 734 ...
    829 965 938 884 506 749 945 966 755 859 940 830 911 871 ...
    639 888 479 946 750 969 508 861 757 970 919 875 862 758 ...
    948 977 923 972 761 877 952 495 703 935 978 883 762 503 ...
    925 878 735 993 885 939 994 980 926 764 941 967 886 831 ...
    947 507 889 984 751 942 996 971 890 509 949 973 1000 892 ...
    950 863 759 1008 510 979 953 763 974 954 879 981 982 927 ...
    995 765 956 887 985 997 986 943 891 998 766 511 988 1001 ...
    951 1002 893 975 894 1009 955 1004 1010 957 983 958 987 ...
    1012 999 1016 767 989 1003 990 1005 959 1011 1013 895 1006 ...
    1014 1017 1018 991 1020 1007 1015 1019 1021 1022 1023]+1;

  % An example for A value and CRC polynomial
  N = 1024;
  A = 500;
  gCRC11 = fliplr([1 1 1 0 0 0 1 0 0 0 0 1]); crcLen=11

  K = A + crcLen;

  Qseq = Q(Q<=N); % Polar sequence for N <= Nmax=1024
  QF = Qseq(1:N-K); % Frozen positions: Qseq(1:N-K), the ...
      least reliable ones.

  msg = randi([0 1],1,A); % Generate a random K-bit message
  % Generate CRC bits and append it to the message
  [quot,rem] = gfdeconv([zeros(1,crcLen) fliplr(msg)],gCRC);
  msgCRC = [msg fliplr([rem zeros(1,crcLen-length(rem))])];

  u = zeros(1,N);
  u(Qseq(N-K+1:end)) = msgCRC; % Insert message bits in ...
      rate-profile
```

C.5 Calculating CRC Bits

The following is a Python script to calculate the CRC bits of a binary message vector **m** using a generator polynomial **g**. To learn more about cyclic codes and CRCs, see Section 2.6.

Listing C.1: CRC calculation algorithm in Python

```
def crc_remainder(m, g):
    # Append r zero bits to the message
    r = len(g) - 1
    M = m + [0] * r

    # Perform binary division
    for i in range(len(m)):
        if M[i] == 1:  # Only divide if the leading bit is 1
            for j in range(len(g)):
                M[i + j] ^= g[j]  # XOR with the generator ...
                    polynomial

    # CRC remainder is the last r bits
    crc = M[-r:]
    return crc

# Example usage:
m = [1, 0, 1, 1]              # Message vector
g = [1, 1, 0, 1]              # Generator polynomial vector ...
    (e.g., g(x) = x^3 + x^2 + 1)
crc_bits = crc_remainder(m, g)
print("CRC bits:", crc_bits)
```

C.6 Polar Transform and Its Permutation Matrix

The following are the MATLAB scripts that return the permutation matrix B_N and the permuted polar transform $B_N G_N$.

Listing C.2: Permutation matrix R_N

```
function R_mat = R(N)
    matL = zeros(N,N/2);
    matR = zeros(N,N/2);
    for i=1:N/2
        matL(2*i-1,i)=1;
        matR(2*i,i)=1;
    end
    R_mat = [matL matR];
end
```

Listing C.3: Permutation matrix B_N

```
B_mat = B(N)
    B_mat = eye(2);
    for i=2:log2(N)
        B_mat = R(2^i)*kron(eye(2), B_mat);
    end
end
```

Listing C.4: Permuted polar transform

```
function G_mat = G(N)
    F2 = [1 0; 1 1];
    F_mat = F2;
    for i=2:log2(N)
        F_mat = kron(F2, F_mat);
    end
    G_mat = B(N)*F_mat;
end
```

C.7 Polar Code Construction

Listing C.5 returns the indices of K-most reliable sub-channels, set $\mathscr{A}$. The measure for sub-channels' reliability is Mean-LLR obtained from the Density Evolution using Gaussian Approximation (DEGA) by Chung et al.

Listing C.5: DE/GA code construction (rate-profile) for the code (N, K) with design SNR design_snr_db

```
function A =  rate_profile(design_snr_db, N, K)
    mllr = zeros(N,1);
    sigma_sq = 1/(2*K/N*power(10,design_snr_db/10));
    mllr(1) = 2/sigma_sq;
    for level = 1: log2(N)
        B = 2^level;
        for j = 1:B / 2
            T = mllr(j);
            mllr(j) = calc_phi_inv(T);
            mllr(B / 2 + j) = 2 * T;
        end
    end

    mask = zeros(N,3);
    for i = 0:N-1
        nat(i+1) = bitreversed(i,uint8(log2(N)));
```

```
    end
    for i = 1:N
        mask(i,:) = [nat(i), mllr(i), 1];
    end
    % sort sub-channels by mllr
    mask = sortrows(mask,2); %direction: ascend (default)
    % set info bits to 1 for sub-channels with K largest mllr ...
        values
    for i = 1:N-K
        mask(i,3) = 0;
    end
    % sort channels with respect to index in bitreversal order
    mask = sortrows(mask,1); %direction: ascend (default)
    A = find(mask(:,3)==1)-1;
end

function dec = bitreversed(i,n) % Alternatively, use ...
    bitrevorder() in "Singal Processing" toolbox.
    dec = bin2dec(fliplr(dec2bin(i,n)));
end

% returns Phi inverse based on piece-wise linear approximation. ...
    by Trifonov
function phi_inv = calc_phi_inv(x)
    if (x>12)
        phi_inv = 0.9861 * x - 2.3152;
    elseif (x<=12 && x>3.5)
        phi_inv = x*(0.009005 * x + 0.7694) - 0.9507;
    elseif (x<=3.5 && x>1)
        phi_inv = x*(0.062883*x + 0.3678)- 0.1627;
    else
        phi_inv = x*(0.2202*x + 0.06448);
    end
end
```

Listing C.7 returns the RM-polar rate-profile of the code $(2^m, K)$.

Listing C.6: RM-polar-coded construction (rate-profile) for the code $(2^m, K)$

```
function [I] = rm_polar(m,K)
    N=[0:2^m-1];
    I = []; Ir = zeros();
    Kp = 0; j = 0;
    while Kp + nchoosek(m,j) < K
        Kp = Kp + nchoosek(m,j);
        r = j; j = j+1;
    end
    for order = 0: r
        I = [I N(find(sum(¬(dec2bin(N)-'0'),2)==order))]; % ...
            0's represents variables corresponding to monomial
    end
```

```
    Ir = N(find(sum(¬(dec2bin(N)-'0'),2)==r+1)); % '0's ...
        represents variables corresponding to monomial
    Ir_contrib = [];
    for i = Ir %[1:length(Ir)]
        f = find(¬(reverse(dec2bin(i,m))-'0'));
        Ir_contrib = [Ir_contrib , r+1+lambda(f,f)];
    end
    Ir = sortrows([Ir' Ir_contrib'],2);
    I = [I Ir(1:K-Kp,1)'];
    I = sort(I')';
end
```

Listing C.7 returns the Reed–Muller rate-profile of the code $\left(2^m, \sum_{j=0}^{r} \binom{m}{j}\right)$.

Listing C.7: RM rate-profile for the code (r, m)

```
function I = RM_profile(r,m)
    N=[0:2^m-1];
    I = [];
     for order = 0: r
        I = [I N(find(sum(¬(dec2bin(N)-'0'),2)==order))];
     end
     I = sort(I)';
end
```

Listing C.8 returns the polarization weight of a sub-channel with index $i \in [0, N-1]$ based on equation 6.28.

Listing C.8: Polarization weight of a subchannel with index i

```
function w = pw(i,beta,n)
    k = 0;
    w = 0;
    for j = reverse(dec2bin(i,n))-'0'
        w = w + j * beta^k;
        k = k + 1;
    end
end
```

Listing C.9: Polarization weight of a subchannel with index i

```
function w = pw(i,beta,n)
    k = 0;
    w = 0;
    for j = reverse(dec2bin(i,n))-'0'
        w = w + j * beta^k;
        k = k + 1;
    end
end
```

C.8 Row Structure of Minimum-Weight Codewords

Listing C.10: Finding set $\mathcal{K}_i$

```
function Ki =  K_i(i,I,n)
    f = find((reverse(dec2bin(i,n))-'0')); % 1's
    Ki = [];

    for j = I(find(I>i))
        g = find((reverse(dec2bin(j,n))-'0')); % 1's
        if length(setdiff(g,f)) == 1
             Ki = [Ki j];
        end
    end
end
```

Listing C.11: Finding set $\mathcal{M}_{\mathcal{J}}$

```
function M = find_set_M(i_lead,J,n) %Progressively
    M = [];
    for x=2:length(J)
        A = nchoosek(J,x);
        for y = 1:size(A,1)
            Jj = A(y,:);
            m  = find_set_M1(i_lead,Jj,n);
            if m>0
                M = [M m];
            end
        end
    end
end

function m = find_set_M1(index,J,n)
    m = 0;
    supp_index = find_supp(index,n);
    supp_zeros = find(dec2bin(bitxor(2^n-1,index),n)-'0'==1); ...
        %complement of supp

    Sm_Ti = supp_zeros;

    row_sets = J; %nchoosek(J, s);
    valid_set_for_M = true;
    Sm_S = supp_index;
    Sm_T = [];
    for r = row_sets %(row_idx,:)
        supp_r = find_supp(r,n);
        Sm_S = intersect(Sm_S, supp_r);
        if ¬isempty(intersect(Sm_T,intersect(Sm_Ti, supp_r))) ...
            %distinctness condition
            valid_set_for_M = false; % No need to set M.
            break;
```

```
        end
        Sm_T = union(Sm_T, intersect(Sm_Ti, supp_r));
    end
    if valid_set_for_M == true
        Sm = union(Sm_T,Sm_S);
        m = supp2dec(Sm,n);
    end
end
```

C.9 Closed-Form Weight Enumeration

Listing C.12: Closed-form enumeration of minimum-weight (w_{min}) codewords

```
function [dmin, A_dmin] = err_coeff(A,N)
  d = min(sum(dec2bin(I)-'0',2));
  dmin = 2^d; n = log2(N); A_dmin = 0;
  B = find(sum(dec2bin(I)-'0',2)==d);
  for i = B'
    Ki_size = n - d;
    for x = find(dec2bin(I(i),n)-'0'==1)
      ii = dec2bin(bitxor(N-1,I(i)),n)-'0';
      Ki_size = Ki_size + sum(ii(1:x-1));
    end
    A_dmin = A_dmin + 2^Ki_size;
  end
end
```

Listing C.13: Closed-form enumeration of codewords of weights w_{min} and 1.5 w_{min}

```
function [w,A_w] = weight_enum(A,n)
    r = max(sum(¬(dec2bin(A)-'0'),2));
    Amr = A(find(sum(¬(dec2bin(sort(A))-'0'),2)==r))
    Amr_sub = Amr; w=zeros(1,2); A_w=zeros(1,2);  w(1)=2^(n-r);
        w(2)=1.5*w(1);
    for i = Amr
        f = find(¬(reverse(dec2bin(i,n))-'0'));
        A_w(1) = A_w(1) + 2^(r+lambda(f,f));
        Amr_sub(Amr_sub==i) = [];
        for j = Amr_sub
            g = find(¬(reverse(dec2bin(j,n))-'0'));
            h = intersect(f,g);
            if length(h)==r-2
                foh = setdiff(f,h); goh = setdiff(g,h);
                alpha = 1*(foh(2)>goh(2) & goh(2)>foh(1)) +
                    1*(goh(1)>foh(1));
                A_w(2) = A_w(2) + 2^(r+2 + lambda(h,h) +
                    lambda(f,foh) + lambda(g,goh) - alpha);
            end
```

```
        end
    end
end
function orbit = lambda(f,g)
    orbit = 0;
    for i = g
        orbit = orbit + length(setdiff([1:i-1],f));
    end
end
```

Listing C.14: Finding $\lambda_f(g)$

```
function orbit = lambda(f,g)
    orbit = 0;
    for i = g
        orbit = orbit + length(setdiff([1:i-1],f));
    end
end
```

Listing C.15: Orbit of monomial *i* with at most *n* variables

```
function orbit = orbit_i(i,n)
    f = find(¬(reverse(dec2bin(i,n))-'0'));
    dec = i;
    orbit = [];
    for j = f
        orbit = [orbit dec + 2^(j-1)];
    end
    ff = f;
    for j = f
        %ff(ff==j) = [];
        for k = setdiff([1:j-1],f)
            orbit = [orbit dec + 2^(j-1) - 2^(k-1)];
        end
    end
end
```

Index

IEEE Press Series on Digital and Mobile Communication

John B. Anderson, *Series Editor*
University of Lund

1. *Future Talk: The Changing Wireless Game*
   Ron Schneiderman
2. *Digital Transmission Engineering*
   John B. Anderson
3. *Fundamentals of Convolutional Coding*
   Rolf Johannesson and Kamil Sh. Zigangirov
4. *Mobile and Personal Communication Services and Systems*
   Raj Pandya
5. *Wireless Video Communications: Second to Third Generation and Beyond*
   Lajos Hanzo, Peter J. Cherriman, and Jürgen Streit
6. *Wireless Communications in the 21st Century*
   Mansoor Shafi, Shigeaki Ogose, and Takeshi Hattori
7. *Introduction to WLLs: Application and Deployment for Fixed and Broadband Services*
   Raj Pandya
8. *Trellis and Turbo Coding*
   Christian B. Schlegel and Lance C. Perez
9. *Theory of Code Division Multiple Access Communication*
   Kamil Sh. Zigangirov
10. *Digital Transmission Engineering,* Second Edition
    John B. Anderson
11. *Wireless LAN Radios: System Definition to Transistor Design*
    Arya Behzad
12. *Wireless Broadband: Conflict and Convergence*
    Vern Fotheringham and Chetan Sharma
13. *MillimeterWave Communication Systems*
    Kao-Cheng Huang and Zhaocheng Wang
14. *Channel Equalization for Wireless Communications: From Concepts to Detailed Mathematics*
    Gregory E. Bottomley
15. *Handbook of Position Location: Theory, Practice and Advances*
    Seyed A. Reza Zekavat and R. Michael Buehrer

16. *Digital Filters: Principles and Applications with MATLAB*
Fred J. Taylor
17. *Resource Allocation in Uplink OFDMA Wireless Systems: Optimal Solutions and Practical Implementations*
Elias Yaacoub and Zaher Dawy
18. *Non-Gaussian Statistical Communication Theory*
David Middleton
19. *Frequency Stability: Introduction & Applications*
Venceslav F. Kroupa
20. *Mobile Ad Hoc Networking: Cutting Edge Directions*, Second Edition
Stefano Basagni, Marco Conti, Silvia Giordano, and Ivan Stojmenovic
21. *Surviving Mobile Data Explosion*
Dinesh C. Verma and Paridhi Verma
22. *Cellular Communications: A Comprehensive and Practical Guide*
Nishith Tripathi and Jeffrey H. Reed
23. *Fundamentals of Convolutional Coding*, Second Edition
Rolf Johannesson and Kamil Sh. Zigangirov
24. *Trellis and Turbo Coding*, Second Edition
Christian B. Schlegel and Lance C. Perez
25. *Problem-Based Learning in Communication Systems Using MATLAB and Simulink*
Kwonhue Choi and Huaping Liu
26. *Bandwidth Efficient Coding*
John B. Anderson
27. *Visible Light Communications: Modulation and Signal Processing*
Zhaocheng Wang, Qi Wang, Wei Huang, and Zhengyuan Xu
28. *Switch/Router Architectures: Shared-Bus and Shared-Memory Based Systems*
James Aweya
29. *Handbook of Position Location: Theory, Practice, and Advances*, Second Edition
Edited by S. A. (Reza) Zekavat and R. Michael Buehrer
30. *Information and Communication Theory*
Stefan Höst
31. *Blockchains: Empowering Technologies and Industrial Applications*
Edited by Anwer Al-Dulaimi, Octavia A. Dobre, and Chih-Lin I
32. *Polar Codes: From Theory to Practice*
Mohammad Rowshan and Emanuele Viterbo